PICTURESQUE AMERICA

OR, THE LAND WE LIVE IN

A DELINEATION BY PEN AND PENCIL OF THE MOUNTAINS, RIVERS, LAKES, FORESTS, WATER-FALLS, SHORES, CAÑONS, VALLEYS, CITIES, AND OTHER PICTURESQUE FEATURES OF OUR COUNTRY

With Illustrations on Steel and Wood by Eminent American Artists

UM

EDITED BY
WILLIAM CULLEN BRYANT

REVISED EDITION

PREFACE.

IT is the design of the publication entitled "PICTURESQUE AMERICA" to present full descriptions and elaborate pictorial delineations of the scenery characteristic of all the different parts of our country. The wealth of material for this purpose is almost boundless.

It will be admitted that our country abounds with scenery new to the artist's pencil, of a varied character, whether beautiful or grand, or formed of those sharper but no less striking combinations of outline which belong to neither of these classes. In the Old World every spot remarkable in these respects has been visited by the artist; studied and sketched again and again; observed in sunshine and in the shade of clouds, and regarded from every point of view that may give variety to the delineation. Both those who see in a landscape only what it shows to common eyes, and those whose imagination, like that of Turner, transfigures and glorifies whatever they look at, have made of these places, for the most part, all that could be made of them, until a desire is felt for the elements of natural beauty in new combinations, and for regions not yet rifled of all that they can yield to the pencil. Art sighs to carry her conquests into new realms. On our continent, and within the limits of our Republic, she finds them—primitive forests, in which the huge trunks of a past generation of trees lie mouldering in the shade of their aged descendants; mountains and valleys, gorges and rivers, and tracts of sea-coast, which the foot of the artist has never trod; and glens murmuring with water-falls which his ear has never heard. Thousands of charming nooks are waiting to yield their beauty to the pencil of the first comer. On the two great oceans which border our league of States, and in the vast space between them, we find a variety of scenery which no other single country can boast of. In other parts of the globe are a few mountains which attain a greater altitude than any within our limits, but the mere difference in height adds nothing to the impression made on the spectator. Among our White Mountains, our Catskills, our Alleghanies, our Rocky Mountains, and our Sierra Nevada, we have some of the wildest and most beautiful scenery in the world. On our majestic rivers—among the largest on either continent—and on our lakes—the largest and noblest in the world—the country often wears an aspect in which beauty is blended with majesty; and on our prairies and savannas the spectator, surprised at the vastness of their features, finds himself, notwithstanding the soft and gentle sweep of their outlines, overpowered with a sense of sublimity.

By means of the railway communications established between the Atlantic coast and that of the Pacific, we have now easy access to scenery of a most remarkable character. For those who would see Nature in her grandest forms of snow-clad mountain, deep valley, rocky pinnacle, precipice, and chasm, there is no longer any occasion to cross the ocean. A rapid journey by railway over the plains that stretch westward from the Mississippi brings the tourist into a region of the Rocky Mountains rivalling Switzerland in its scenery of rock piled on rock, up to the region of the clouds. But Switzerland has no such groves on its mountain-sides, nor has even Libanus, with its ancient cedars, as those which raise the astonishment of the visitor to that Pacific region—trees of such prodigious height and enormous dimensions that, to attain their present bulk, we might imagine them to have sprouted from the seed at the time of the Trojan War. Another feature of that region is so remarkable as to have enriched our language with a new word; and *cañon*, as the Spaniards write it, or *canyon*, as it is often spelled by our people,

splendid trout with which the lakes are said to abound, and to go far down the bay for catches of cod and haddock, which here are of large dimensions and in great abundance. The bays, inlets, and sounds of the coast of Maine afford superb resources for the yachtman. The coast seems to have crumbled off from the main-land in innumerable islands, large and small, so that there is a vast area of inland-sea navigation, which, with infinite variety of scene, gives ample space for boating. A yachting-party might spend a summer delightfully in threading the mazes of this "hundred-harbored Maine," as Whittier describes it. Abandoning the pleasant vision of such a summer, let us for the present remember that our special object is to visit and depict the scenery of Mount Desert.

The several points along the coast to which the visitor's attention is directed are the cliffs known as "The Ovens," which lie some six or seven miles up the bay; and "Schooner

View of Mount-Desert Mountains from Saulsbury-Cove Road.

Head," "Great Head," and "Otter-Creek Cliffs," lying on the seaward shores of the island. It will fall more duly in order to proceed first to "The Ovens," which may be reached by boat or by a pleasant drive of seven or eight miles.

With a one-armed veteran for an escort, Mr. Fenn and the writer set forth for a scene where we were promised many charming characteristics for pen and pencil. It was necessary to time our visit to "The Ovens"—the nomenclature of Mount Desert is painfully out of harmony with the scenes it verbally libels—so as to reach the beach at low tide. The cliffs can be approached only by boat at high tide, and the picture at this juncture loses some of its pleasing features.

The Mount-Desert roads for the most part are in good condition, and have many attractions. The forests are crowded with evergreens, and the firs and the spruce-trees marshal in such array on the hill-sides that, with their slender, spear-like tops, they look like armies of lancers. The landscape borrows from these evergreens an Alpine tone, which

THE CLIFFS NEAR "THE OVENS."

groups of pedestrians for the mountains, armed with alpenstocks, notably enhance. The fir, spruce, pine, and arbor-vitæ, attain splendid proportions; the slender larch is in places also abundant, and a few sturdy hemlocks now and then vary the picture. The forest-scenes are, many of them, of singular beauty, and in our long drives about the island we discovered many a strongly-marked forest-group.

At one point on our drive to "The Ovens," the road, as it ascends a hill near Saulsbury Cove, commands a fine, distant view of the mountains, which Mr. Fenn rapidly sketched. Clouds of fog were drifting along their tops, now obscuring and now revealing them, and adding often a vagueness and mystery to their forms which lent them an additional charm.

The cliffs at "The Ovens" contrast happily with the rocks on the sea-front of the island in possessing a delicious quiet and repose. The waters ripple calmly at their feet, and only when winds are high do the waves chafe and fret at the rocks. Here the perpendicular pile of rock is crowned by growths of trees that ascend in exact line with the wall, casting their shadows on the beach below. Grass and flowers overhang the edge; at points in the wall of rock, tufts of grass and nodding harebells grow, forming pleasant pictures in contrast with the many-tinted rocks, in the crevices of which their roots have found nourishment. The whole effect of the scene here is one of delicious charm. The wide and sunny bay, the boats that glide softly and swiftly upon its surface, the peaceful shores, the cliff crowned with its green forest, make up a picture of great sweetness and beauty. "The Ovens" are cavities worn by the tides in the rock. Some are only slight excavations, such as those shown in Mr. Fenn's drawing, but a little northward of the spot are caves of a magnitude sufficient to hold thirty or forty people. The rocks are mainly of pink feldspar, but within the caves the sea has painted them in various tints of rare beauty, such as would delight the eye and tax the skill and patience of a painter to reproduce. The shores here, indeed, supply almost exhaustless material for the sketch-book of the artist.

To this spot, at hours when the tide permits, pleasure-seekers come in great numbers. It is a favorite picnic-ground for the summer residents at Bar Harbor, whose graceful pleasure-boats give animation to the picture. The visitors picnic in the caves, pass through the archway of a projecting cliff, which some designate as "Via Mala," wander through the forests that crown the cliffs, pluck the wild-roses and harebells that overhang the precipice, and roam up and down the beach in search of the strange creatures of the sea that on these rocky shores abound. Star-fishes, anemones, sea-urchins, and other strange and beautiful forms of marine life, make grand aquaria of the caves all along the coast, and add a marked relish to the enjoyment of the explorer.

From the quiet beauty of "The Ovens" to the turbulence of the seaward shore there is a notable change. Our next point visited was "Schooner Head," which lies four or five miles southward from Bar Harbor, and looks out on the wide Atlantic. "Schooner Head" is so named from the fancy that a mass of white rock on its sea-face, viewed at a proper distance, has the appearance of a small schooner. There is a tradition that, in the War of 1812, a British frigate sailing by ran in and fired upon it, under the impression that it was an American vessel hugging the shore. "Schooner Head" derives its principal interest from the "Spouting Horn," a wide chasm in the cliff, which extends down to the water and opens to the sea through a small archway below high-water mark. At low water the arch may be

Great Head.

reached over the slippery, weed-covered rocks, and the chasm within ascended by means of uncertain footholds in the sides of the rocky wall. A few adventurous tourists have accomplished this feat, but it is a very dangerous one. If the foot should slip on the smooth, briny rock, and the adventurer glide into the water, escape would be almost impossible. The waves would suck him down into their depths—now toss him upon rocks, whose slippery surface would resist every attempt to grasp, then drag him back into their foaming embrace. When the tide comes in, the breakers dash with great violence through the archway described, and hurl themselves with resounding thunder against the wall beyond, sending their spray far up the sides of the chasm. But, when a storm prevails, then the scene is one of extreme grandeur. The breakers hurl themselves with such wild fury through the cavernous opening against the walls of rock, that their spray is thrown a hundred feet above the opening at the top of the cliff, as if a vast geyser were extemporized on the shore. The scene is inspiriting and terrible. Visitors to Mount Desert but half understand or appreciate its wonders if they do not visit the cliffs in a storm. On the softest summer day the angry but subdued roar with which the breakers ceaselessly assault the rocks gives a vague intimation of what their fury is when the gale lashes them into tumult. At such times they cast themselves against the cliffs with a violence that threatens to beat down the rocky barriers and submerge the land; their spray deluges the abutments to their very tops, and the thunder of their angry crash against the rock may be heard for miles. But at other times the ceaseless war they make upon the shore seems to be one of defeat. The waves come in full, sweeping charge upon the rocks, but hastily fall back, broken and discomfited, giving place to fresh and hopeful levies, who repeat the first assault, and, like their predecessors, are hurled back defeated. The war is endless, and yet by slow degrees the sea gains upon its grim and silent enemy. It undermines, it makes channels, it gnaws caverns, it eats out chasms, it wears away little by little the surface of the stone, it summons the aid of frost and of heat to dislodge and pull down great fragments of the masonry, it grinds into sand, it gashes with scars, and it will never rest until it has dragged down the opposing walls into its depths.

"Great Head," two miles southward of "Schooner Head," is considered the highest headland on the island. It is a bold, projecting mass, with at its base deep gashes worn by the waves. A view of its grim, massive front is obtained by descending a broken mass of cyclopean rocks a little below the cliff, where at low tide, on the sea-washed bowlders, the cliff towers above you in a majestic mass.

People in search of the picturesque should understand the importance of selecting suitable points of view. The beauty or impressiveness of a picture sometimes greatly depends on this. It is often a matter of search to discover the point from which an object has its best expression; and probably only those of intuitive artistic tastes are enabled to see all the beauties of a landscape, which others lose in ignorance of how to select the most advantageous standing-place. To the cold and indifferent, Nature has no charms; she reveals herself only to those who surrender their hearts to her influence, and who patiently study her aspects. The beauty of any object lies partly in the capacity of the spectator to see it, and partly in his ability to put himself where the form and color impress the senses most effectively. Not one man in ten discerns half the beauty of a tree or of a pile of rocks, and hence those who fail to discover in a landscape the charm others describe in it should question their own power of appreciation rather than the accuracy of the delineation. The shores of Mount Desert must be studied with this appreciation and taste, if their beauties

THE "SPOUTING HORN" IN A STORM.

are to be understood. No indifferent half glance will suffice. Go to the edge of the cliffs and look down ; go below, where they lift in tall escarpments above you ; sit in the shadows of their massive presence ; study the infinite variety of form, texture, and color, and learn to read all the different phases of sentiment their scarred fronts have to express. When all this is done, be assured you will discover that "sermons in stones" was not a mere fancy of the poet.

One of the characteristics of Mount Desert is the abundance of fog. In July and August especially it seriously interferes with the pleasure of the tourist. It often happens that, for several days in succession, mountain, headland, and sea, are wrapped in an inpenetrable mist, and all the charms of the landscape obscured. But the fog has frequently a grace and a charm of its own. There are days when it lies in impenetrable banks far out at sea, with occasional incursions upon the shore that are full of interest. At one hour the sun is shining, when all at once the mist may be discerned creeping in over the surface of the water, ascending in rapid drifts the sides of the mountains, enveloping one by one the islands of the bay, until the whole landscape is blotted from view. In another hour it is broken ; the mountains pierce the shadowy veil, the islands reappear in the bay, and the landscape glows once more in the sunshine. It is a rare pleasure to sit on the rocky headlands, on the seaward side of the island, on a day when the fog and sun contend for supremacy, and watch the pictures that the fog makes and unmakes. Sometimes the fog skirts along the base of the islands in the bay, leaving a long, slender line of tree-tops painted against the blue ether, looking like forests hung in the sky. Then a vessel may be seen sailing through a fog-bank, now looking like a shadowy ghost floating through the mist, when suddenly its topsails flash in the light, like the white wings of a huge bird. In another moment the fog shifts, and the under edge of the mainsail may be traced in a line of silver, while all the rest of the vessel is in the deepest shadow. Now one sail glitters a brilliant white, and the fog envelops all the rest of the vessel. The pictures thus formed vary like a succession of dissolving views, and often produce the most striking and unique effects. Sometimes there is the marvellous exhibition of a mirage, when fleets appear sailing through the air, and, as described by Whittier—

> "Sometimes, in calms of closing day,
> They watched the spectral mirage play;
> Saw low, far islands, looming tall and high,
> And ships, with upturned keels, sail like a sea the sky."

The fog-pictures at Mount Desert are by no means the least interesting feature of this strange shore.

Near a small stream, known as "Otter Creek," deriving its name from the otter which once abounded there, are a succession of cliffs, which possess characteristics quite distinct from those already described. They are more remote from the village than "Schooner

Head" or "Great Head," but the drive to them derives great interest from the wild and narrow notch between Green and Newport Mountains, through which the road lies for a mile or two. The sides of the mountains are high, precipitous, and savagely rugged. The lower base of each is covered with a thick and tangled forest-growth; half-way up, a few gnarled and fantastic growths struggle for place amid the scarred and frowning rocks,

Thunder Cave.

while the upper heights show only the bare, seamed, and riven escarpments. It is a wild picture, inferior, no doubt, to the famous Notch of the White Mountains, but possessing, notwithstanding, very strong and impressive features.

At "Otter-Creek Cliffs" we set out in search of what is known as "Thunder Cave." After leaving our vehicle, we had a long but superb forest-walk to reach it. There are numerous fine birches on Mount Desert, and more than once we saw groups of these trees that would have filled any artist with delight, and especially the painter Whittredge, whose birch-forests are so famous. Near Great Head are numerous splendid specimens of this tree, whose bark, of yellow, Indian red, and gray, afforded delicious contrasts of color. On the path to Thunder Cave we noted one forest-picture that comes vividly back to memory. The trees were mostly evergreen, and the surface of the ground covered with outcropping rocks and tangled roots, all richly covered with mosses. The broken light through the dark branches, the tint of the fallen pine-leaves, the many-colored mosses which painted every rock in infinite variety of hue, the low, green branches of the fir and the spruce, all made up a picture of ripe and singular beauty.

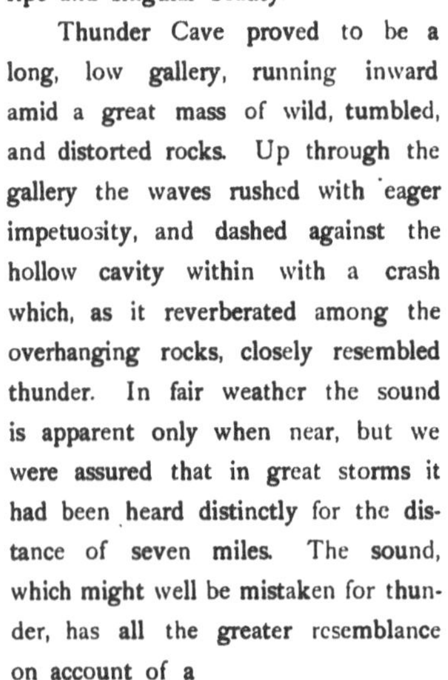

Thunder Cave proved to be a long, low gallery, running inward amid a great mass of wild, tumbled, and distorted rocks. Up through the gallery the waves rushed with eager impetuosity, and dashed against the hollow cavity within with a crash which, as it reverberated among the overhanging rocks, closely resembled thunder. In fair weather the sound is apparent only when near, but we were assured that in great storms it had been heard distinctly for the distance of seven miles. The sound, which might well be mistaken for thunder, has all the greater resemblance on account of a peculiarity which Mr. Fenn detected while making his sketch. Piled up within the cave at the end of the gallery are a great number of large stones, varying from one to probably three feet in length.

The Obelisk.

and of corresponding thickness. Every time the waves dash into the cave, they dislodge some of these stones, sometimes dragging them back, sometimes lifting them up and tossing them against the sides of the cavity, and, as these bowlders thus roll and grind together, they produce in the hollow of the cavern almost the exact mutterings and reverberations of thunder. The crash of the breakers against the wall is the clap of thunder; the rolling stones carry off the sound in its successive reverberations, making the resemblance complete.

Near Thunder Cave we discovered a natural obelisk. The woodland path at one point reaches the edge of a wide, precipitous break in the cliff. Forcing our way through tangled wood-growth to obtain a view of the cliff, we saw, situated directly under the bank, where the tourist ordinarily would not detect it, a tall, pointed column, with an apparently artificial base of steps, bearing a close semblance to a monument of stone. This singular freak of Nature the reader will find illustrated by Mr. Fenn's pencil.

Returning to our point of departure, we proceeded westward in search of other cliffs, where we made another discovery. The path lay along the top of the cliff, but, coming to a dislodgement of the perpendicular wall, where some convulsion had thrown down the cliff into a wild mass of rocks, we with no little difficulty clambered down the broken and jagged pile, with the purpose of getting from below a view of the cliffs. Fortunately, the tide was low; and this, the tourist should remember, is necessary, when he arranges his visits to the shores of Mount Desert. There is more animation when the tide is coming in, but high water cuts off access to many interesting points. Reaching a wet, barnacle-covered, projecting line of rocks, a picture presented itself that filled both artist and penman with surprise. "Why, this is an old Norman castle!" was our exclamation. The cliff, a little distant from our point of view, stood up in perpendicular lines of rock that assumed almost exactly the form of battlements. The upper line closely resembled the parapet of a castle-wall; there were in the sides deep embrasures; and the whole front had the aspect of a dark, broken, time-stained wall reared by the hand of man. It stood in grim and gloomy grandeur, fronting the sea in stern defiance of the world beyond. The waves chafed at its feet; wild sea-birds hovered about its crest; there was an air of neglect and desolation, as if it were an old ruin, and we found it impossible to dissociate the grim and frowning walls from the historic piles that look darkly down upon so many European landscapes. Finding afterward that the cliff was known by no name, we called it "Castle Head." The path followed by the customary visitor extends along the cliff above this strange pile, and hence its peculiarities escape the notice of all except those who boldly clamber down the broken wall just before it is reached, and survey it from the water's edge. The illustration of this striking scene is at the beginning of our article.

The interest of Mount Desert, as we have already said, is divided between its sea-cliffs and its mountain-views. It is customary for pedestrian parties to form at Bar

Harbor and walk to the mountain-top, and there remain overnight, in order to view the sunrise from this altitude. A convenient hotel, built a few years ago, stands on the extreme top of the mountain, and affords satisfactory accommodation for the tourists. A rude mountain-road, originally constructed by the United States Coast Survey, enables vehicles to ascend to the hotel; but pleasure-parties commonly prefer the ascent on foot. The distance from the village is four miles. The height of the mountain is fifteen hundred and twenty-seven feet.

The sunrise is a magnificent picture, but the prevalence of fogs is a continual cause of disappointment to people, who travel far and rise early often only to behold a sea of impenetrable mist. The prospect, however, whenever the fog permits it, is a splendid one at all hours, and possesses a variety and character quite distinct from the views usually obtained from mountain-heights. Here there is not only a superb panorama of

Eagle Lake.

hills and vales, but a grand stretch of sea, and intricate net-works of bay and islands which make up a picture marvellously varied both in form and color.

One of the most delightful features of the scene thus presented are the mountain-lakes that hang like superb mirrors midway in the scene. "Eagle Lake," so named by Church the artist, is visible at intervals during the entire ascent of the mountain, and at every point of view is beautiful. Half-way up, a short *détour* from the road will bring the tourist to its pebbly shore, where he may spend an hour or more watching its clear, mountain-encircled waters, or devote his entire day in pursuit of the trout with which it abounds. The largest lake in the island is on the western side of Somes's Sound, and is about four miles in length. There is a group of three lakes on each side of this sound, although to some of them the more prosaic designation of pond is applied.

Somes's Sound, which divides the lower portion of the island into two distinct portions, possesses many attractions for those who admire bold headlands. It bears a resemblance both to the shores of the Hudson and the Delaware Water-Gap. It is usual to ascend the sound in boats from Southwest Harbor; but explorers from Bar Harbor sometimes drive to Somesville, at the head of the sound, a distance of nine miles, and there take boat for a sail down the stream. The sound cuts through the centre of the mountain-range at right angles, between Dog Mountain and an elevation on the eastern side, to which the apellation of "Mount Mansell" has been given, in honor of Sir Robert

Eagle Cliff, Somes's Sound.

Mansell, after whom the island was at one time named by the English. Dog Mountain rises abruptly from the water's edge, and one of its cliffs, which is some eight hundred or a thousand feet in height, is called "Eagle Cliff." At the moment Mr. Fenn was sketching, a splendid bald-headed eagle was sailing in wide circles around the head of the cliff, thus giving, to the imagination of the artist, ample justification for the title.

We have now enumerated the principal features of this beautiful island. But there are hundreds of places that almost equally as well deserve the attention of pen and pencil. The shore varies in character and form at nearly every step, affording almost innumerable delightful pictures; while the lakes, the mountains, the forests, are endless in their long catalogue of rare and beautiful scenes. And in addition to scenes upon the island itself are the picturesque and rocky Porcupine Islands, the rugged shores of Iron-bound Island, on the eastern side of Frenchman's Bay, and Egg Rock, a few miles out at sea, upon whose narrow base stands a light-house. Artist and writer have been limited to giving mere indications of a locality that is almost exhaustless in its variety of scenery.

Mount Desert was discovered by the French, under Champlain, in the early part of the seventeenth century, who gave it the name by which it is now known. In 1619, the French formed a settlement, which was named "Saint-Sauveur," but in a few years it came to a cruel end. The Virginian settlers were accustomed to fish upon the New-England coast, and the captain of an armed vessel, hearing from the Indians of the settlement, sailed down upon it, and with a single broadside made himself its master. Some of the settlers were killed, and others carried away into captivity. The first permanent settlement was made by Abraham Somes, who in 1761 built a house at the head of the sound which now bears his name.

View from Via Mala, at The Ovens.

Mount Ranier, from the Columbia River.

UP AND DOWN THE COLUMBIA.

WITH ILLUSTRATIONS BY R. SWAIN GIFFORD.

MAPS are so unexpectedly made over nowadays, what with the Old World passion of conquest, and the New World instinct of truck and dicker, that even we young people, who are rather proud of not yet having forgotten our multiplication-table and syntax, are not a little put to it to bound American America, or United Germany, or dismembered France. There was a happy time when a "pent-up Utica" judiciously contracted our powers, and when we were limited toward the pole by undiscovered countries which we were taught to call respectively Russian Possessions and Upper and Lower Canada—which was which of the twins last mentioned the infant mind never clearly apprehending. In those days our national Northwestern estate was represented on the atlas by a green and a brown patch of uncertain outlines, severally labelled "Indian Territory" and "Oregon." Lewis and Clark were popularly believed to be the only civilized men who had ocular proof of their existence. In the common mind they stood only as irregular polygons on the map, and not as so many acres of soil, stones, forests, lakes, rivers, habitable places, over which familiar heavens arched, and where rains fell on just and unjust, the former class being represented by wild animals and the latter by wild men. Even the wise geographers skated nimbly over the thin ice of their ignorance, lingering only long enough for a single observation, to the effect that this vast, unexplored country was chiefly trackless desert and unexplored forest. And even Con-

MOUNT HOOD.

gressional orators, who spoke for "Buncombe," and went in, on all occasions, for river and harbor improvements, never could get beyond the third line of the sonorous—

> ". . . . the continuous woods
> Where rolls the Oregon, and hears no sound
> Save its own dashings."

Alas for the infant-schools! Out of that dull, green patch has broken a wealth of offshoots, as out of the scrimp and ugly cactus burst its superb blossoms. A list of States and Territories that dizzies the arithmetic of memory insists on place and nomenclature, and blessed be Providence which ordained that we should not be our own grandchildren, to encounter a tale of three hundred and sixty-five political divisions by them doubtless to be comprehended in the description of their dear, their native land! As the shoots increased the parent-stem dwindled, and now Oregon, pinched and shrivelled, is only a fourth larger than all New England, or rather less than twice New York in extent. And as for the vast Indian Territory, that would seem to exist variably wherever the Noble Savage is upon the war-path, and to comprise so much land as his blanket will cover.

In those better days we children used to have delightful thrills of horror at thought of the Great American Desert and far Pacific coast, peopled, as we believed, with lions, alligators, dragons, polar bears, anacondas, the "anthropophagi, and men whose heads do grow beneath their shoulders"—creatures all the more terrible by reason of the utter vagueness of their outlines and conditions. And we used to play at being Captain Cook, who, to our apprehension, was the very symbol and archetype of discoverers; and, as he and his heroic band, used to do much execution among the heaped-up sticks in the wood-shed, which alternately, or rather indiscriminately, represented the Rocky Mountains, hosts of savage foes, or such a menagerie of beasts as has not been seen since the creation. By-and-by one of us repeated the fable of "Rasselas," which is the apologue of Time, left behind him the Happy Valley of a delighted childhood, and went forth to explore the world. I do not remember that any wise Imlac began that long journey in his company, nor that he came to any Cairo where he spent two years in learning the Universal Language, and where every man was happy. On the contrary, I am afraid that Imlac, who stands for the lessons of experience, joined him only after long years and innumerable scrapes had cost him dear; and that the Cairo where all men are happy is not set down on any chart by which he took his way. At least it was not built between Boston and San Francisco, nor yet between that golden capital and Puget Sound, nor did any spire or minaret thereof glitter against the perfect skies of Oregon, whither the wanderings of the new Rasselas led him. But, to drop metaphor, which, like Malvolio's cross-garterings, "obstructs the blood," it was I who made the journey to Oregon, and I find that I cannot tell a comfortable story without saying so in the begin-

MULTANOMAH FALLS.

ning, with a heart-felt regret that I am not Wallace, or that most charming traveller, Mr. John Hay, instead of myself.

Perhaps oceans change their habits with time, like climates and individuals. It is easier to believe that in 1520 the Pacific lapsed on purple islands a summer sea, than to discredit the incorruptible Magalhaens, of Portuguese truth and directness, with wittingly bewraying the trust of unborn generations. In 1869, however, it had become the most deceitful of waters, with a horrible swell and pitch peculiar to itself, and caves full of head-winds, like Atlantic gales grown up, out on their travels, and equal to any mischief. Nor is the Pacific content to have its grim way with you only while you are its lawful prey. For it has set a bar at the mouth of the Columbia, which for nine days defied the best seamanship of Captain Robert Gray, of the good bark Columbia Rediviva, who named the beautiful river in 1792. And it is only by seizing the unwilling tide in the narrowest nick of time that the pilot compels it to float you beyond the dangerous shoal and into the safety of the stream. Once within the bar, the ship seems to relax every tense nerve and fibre, and to drift on the current like a spent deer which has escaped the hounds. And so, lazily, you come to Astoria.

Astoria, founded by the Northwestern Fur Company, was, I believe, our first white settlement in the Northwest, and it was named in honor of John Jacob Astor, who was the energetic spirit of the company. Astoria is a nice name enough, as names go, and certainly better than Astor's Corner, or Astorville, or New Astor. But to be a mighty trapper, or only to hire the skill of mighty trappers, hardly entitles a man to build himself a monument of imperishable earth, and wood, and water. The Astor Library commemorates in its name a noble benefaction. Astoria preserves the recollection of a sharp and lucky instinct of trade. However, for that matter, there are hardly ten men in a generation for whom a town should be named. Unless Astor, or Lansing, or Lawrence, be many-sided, hospitable, capable of large results and endless activities—unless there be broad avenues leading to temples in his soul, and straight ways to libraries, museums, gymnasiums, schools, in his brain—he has no business to impress his image and superscription on the possible germ of all this completeness. And, if he have this right and title, he will have modesty besides, and never claim it. Alexandria and Rome sound stately, and embalm the pagan virtues of hardiness, courage, force, invincibility. For the men who overran the younger world at least brought blood, brawn, and brains out of their tussle with Nature and man. But our century pretends to a different civilization, and condemns without hope the Anglo-Saxon idiots who, in this age and in this republic, have blasted nineteen post-towns with the name of Rome, and doomed sixteen to stagger under the weight of Alexandria, with the occasional suffix of *Centre*, *Four Corners*, and *Switch*. *Alexandria Switch!* Perhaps we all privately sympathize with the sentiment of Horace Walpole, who declared that he should be very fond of his country if it were not for his countrymen! In the name of grace and fitness, let us keep the sweet

Indian appellatives bequeathed our pleasant places by a vanishing race; and, for whatever other nomenclature we need, let us remember only the "high souls, like some far stars, that come in sight once in a century."

Well, we came to Astoria (which should have been Chetco) in the late afternoon of a perfect summer Sunday. The river, twelve miles wide, lay all aglow with color under the low sun, and out to the west the color deepened and deepened till it seemed to be no longer atmosphere, but substance, like some supernal gem. Astoria is such a tiny place to have set up in the world for itself, so far from civilization! The great river

Rooster Rock.

stretches like a sea to the north; the great ocean creeps close on the west; and on the south and east the forests crowd up to the very thresholds—such forests as only the cunning wolf and wild-cat can find their way in. Yet, as the twilight fell, the little church-bell rang with a sound of cheerful confidence in a responsive congregation, and men and women went churchward, and lights glanced in the windows, and a little, soft baby-cry trembled a moment in the air. So I suppose that the world goes on there just as it does in New York or Nova Zembla, with births and deaths and givings in marriage, and envies and heartaches and sweet charities. But to this hour I cannot think of that atom of civilization, made so pathetically small by the vastness of sea and river and woods without a little pang of pity for what seems its unutterable loneliness; and yet I dare say it sits by the fire in supreme satisfaction, finds the keenest zest in the excitement of the semi-weekly stopping of the steamer, and, if it condescended to make comparisons, would consider New York at a disadvantage as to situation. That beautiful and blessed quality of self-conceit, without whose protection the contusions of every day would keep us morally black and blue from head to foot, not only saves ourselves from the buffetings of the unworthy, but saves also our kin, our neighborhood, our township, even our select-man, unless he happen to belong to the opposite political party.

Very late the long twilight faded, and the darkness grew alive with sound. The soft slipping of the tide and the murmur of the great woods were the ever-recurring lovely air, as it were, with which unnumbered variations blended. The myriad creatures which, every summer-night, seem to be just born, and to try vainly to utter their joy in stridulous voices, piped the whole chromatic scale with infinite self-satisfaction. Innumerable crickets addressed us in cadence with cheery felicitations on our safe arrival among them; a colony of tree-toads interrupted everybody to ask, in the key of F sharp major, after their relatives in the East, and to make totally irrelevant observations, without ever waiting for a reply; and the swelling bass of the bull-frogs seemed to be thanking Heaven that they were not as these impertinents. This inarticulate welcome, this well-known iteration, made the Pacific seem no longer strange, but familiar as the shores of New-York Bay, and it would not greatly have surprised us to open our eyes, next morning, on the barrenness of Sandy Hook or the fair Heights of Brooklyn.

What they really saw, however, when daybreak found us far up the Columbia, was better than city or crowded anchorage. The great river, still lake-like in breadth and quietness, lay rosy in the dawn. The wonderful forests, whose magnificence our tame and civil imagination could not have conceived, came down from farthest distance to the very margin of the stream. Pines and firs two hundred feet high were the sombre background against which a tropical splendor of color flickered or flamed out, for, even in this early September, beeches and oaks and ash-trees were clothed with autumn pomp; and on the north, far above the silence of the river and the splendid shores, four snow-crowned, rose-flushed, stately mountains lifted themselves to heaven. For miles and miles and miles, Mount Adams, Mount Jefferson, Mount Rainier, and Mount St. Helens, make glad the way. Adams and Jefferson have an unvarying grandeur of form, a massive strength and nobility, as it becomes them to inherit with their names. Mount St. Helen's rises in lines so vague and soft as to seem like a cloud-mountain. Rainier, whose vastness you comprehend only when you see it from Puget Sound, looks, even from the river, immeasurable, lying snow-covered from base to peak.

Portland, one hundred and ten miles up the river, is the point of debarkation for the San-Francisco steamers, and there is much to be said about that busy and thrifty

CASTLE ROCK.

little clucking hen of a city. But as Portland is not on the Columbia at all, but on the Willamette, twelve miles from its mouth, it may not now be told what golden eggs she has laid. The little steamer which plies up and down the river leaves her dock at an uncomfortable early hour in the morning, and that must be very fine scenery, indeed, which reconciles one to being dragged out of bed at daybreak, and dumped, hungry, sleepy, and cross, in the chilly cabin of a day-boat; still, it is much more satisfactory than the trip by rail. The view which daylight brought us was a prospect of the boat's paddle-boxes. A gray mist swallowed up every thing beyond. But when it lifted, three hours later, it was worth while to have been chilled to the bone with its cold, and alarmed by its threat of showing us nothing, to see what it really had to show. For, as it slowly crept back to the shores and up the banks, and so away to the north-pole, which it must have come from, river and shores and mountains and sky, and the sun itself, came out upon us with such intensity of light and color that it seemed as if we or they were absolutely new that morning, and had never seen each other before.

Where the mists lifted, the stream flowed clear and smooth between mountain-shores a mile and a half apart, and rising sharp and bold thousands of feet in air. Forests covered their rocky sides, sometimes rising to the very top, sometimes dwindling into groups and thickets as they climbed. And on the very crest, standing alone and sucking their lusty life from the inhospitable stone, lone pines shot out of the crevices of rock, looking, so far above us, like the queer and graceless toy-trees in the shilling boxes of wooden soldiers, dear to the heart of boyhood. These mountains are a solid wall along the river for miles on miles. Sometimes there is neither rift, nor gorge, nor scar, in their huge sides. Then a canyon opens, and you see beyond and beyond other mountains coming down to link themselves in an unending chain, and glimpses of far-off levels or gray fields of rock bounding the vision. Sometimes a water-fall dazzles and dances out of the sky, a little, fluttering, quivering cobweb at first; then a floating ribbon; then a wind-blown veil of spray; then a cascade, leaping from rock to rock, forty, sixty, a hundred, three hundred feet; then a swift, resistless, triumphant rush of water, swirling and whirling toward the river of its love.*

Yet, if shores and water-falls were beautiful, the forests were the crowning glory of the place. First in rank, again, stood the pines and firs—if they *were* pines and firs. They looked to me like some celestial sort of grown-up, feathery ground-evergreen. And who could expect a pine to rise, straight and fair, three hundred feet, a glimpse of red-brown bolls warm through the foliage of the lesser trees, and a glory of spreading, plumy, dark-green boughs, so purely outlined that every little tuft of them looked as if it, and it alone, had been finished specially to show how perfect a thing a tree-branch may be made for the enjoyment of the woodpeckers and the slugs? Seeing these pines, one

* See Multanomah Falls.

understands the Northern myth of the tree Yggdrasill, at whose feet flowed sacred fountains and whose branches upheld the world.

Then came the cotton-woods, and the cotton-wood is to the Western settler the symbol of intermeddling and knavish incapacity. He considers it the "dead beat" of the vegetable kingdom, usurping ground that belongs to honester growths, making great pretensions to an early and useful maturity, and no better than a pipe-stem in value

The Cascades

when the axe claims it. Yet there crowded these plausible cotton-woods, standing so idly gracious and welcoming all along the shores in such gorgeousness of golden splendor, and in such royal ease and grace of attitude, that one forgets their good-for-nothingness and their general bad name among the virtuous and useful trees, and takes them to his heart at once. A tree whose polished, brilliant leaf looked like our maple, and whose scarlet, pendent swinging boughs looked like darting orioles, we were forbidden to

CAPE HORN.

consider a familiar friend, a very learned pundit assuring us that there were no maples on the river. That was the only vegetation with which we were bold enough to set up the plea of acquaintance, every thing else being quite too splendid to countenance any claim of kinship with the paler and punier growths of our ascetic climate.

Sometimes, so far above our heads that they looked like pigmies at play, we saw the lumbermen getting out logs which came tearing down the rugged sluice-ways to the river, More seldom, even, did a single logger's hut appear, like a hang-bird's nest, far up among the rocks, making the place look wilder than the wilderness, because this little struggle toward civilization and domesticity was so overborne by the savagery of Nature. These half-cleared places had a certain repulsiveness, too. Nature carefully hides unsightly spots with shrubs and bushes, and, when her dead trees fall, she tenderly adorns the wreck they make with vines and mosses. But, when man comes in with rude and indiscriminate rapine, she is profaned, and will long leave the place to his clumsy keeping, throwing neither vine nor moss, nor veiling shadow even, across the scars of his occupation. So that those half-clearings looked rough and coarse as the lives of the inhabitants.

Sometimes the river flowed straight and untroubled. Sometimes the mountains swept round into its path, and the stream bent and parted on rocky mounds or islands, and ran shallow, disturbed, and dangerous. Straightway it quieted into chains of narrow lakes without visible outlet, whereon we sailed close up to lofty, impassable shores, like the walls of Sinbad's valley, but, turning suddenly on our track, found unexpected deliverance. Then, looking back, the way was lost by which we had come. Here, in the solitary mountains, we were alone. No world behind—no world before. The sense of solitude was too vast to be painful. But we felt as escaping prisoners feel when we threaded our way through a narrow rift of stone into the wonderful stream that grew more wonderful as we sailed. For, just there, walls of basalt in vast ledges, rising sheer and straight from the shore, overtopped the farther mountains. Rifted bowlders, like Castle Rock, stood alone, their base washed by the river, their heads upholding the sky. Majestic ramparts, like Cape Horn, rose, a vast, columnar wall, sometimes seven hundred feet in height. Columns, and obelisks, and shafts, lifted themselves with a mightier strength and a more majestic grace than architecture has been able to achieve. And through these stately gate-ways we came to the Cascades.

The Cascades are the fierce and whirling rapids wherein the river falls forty feet, twenty feet of it being taken almost at a leap. But for six miles the river is a seething whirlpool, and a narrow-gauge railroad on the Washington side affords the portage. The track runs so near the water's edge that one has a view of these rapids for the whole way, from the Middle Block-house, relic of not unremote Indian wars, to the drowned forest above the upper landing. The whole river-bed is gigantic rocks, sometimes hidden by the water, sometimes tearing through the water to make sharp and naked islands, between which the current rushes down, white with foam and with a roar like the sea. Round the rocks and between the rocks and over the rocks, and almost burrowing under the rocks in its force, in those six miles the river takes on every possible form of cascade. Yet where we take steamer again, a moment before the river makes its first plunge, it is as quiet as the Connecticut, and washes along over submerged stumps like any slow bayou.

Being born under a lucky star, Imlac and I were invited to ride on the engine, nay,

Middle Block-house, Cascades.

on the very cow-catcher. It is impossible to imagine a madder excitement. With the whole tremendous motive-power behind you forgotten, you seem to be flying, without even the drawback of having to flap your wings. The wild river to the right of you, the wild mountains close on your left hand; your flight through rifts and chasms of stone which seem ever crowding forward with an evil-minded will to shut you in; just a glimpse of blue sky far above, such as miners see from the bottom of the black shaft; a fierce rush and roar of wind that strikes down your very eyelids—this is riding on the cow-catcher in the cañon of the Columbia. Half blinded as we were, we saw, as we passed it, the great sides of Rooster Rock and a little log-house beneath. This was the scene of General Philip H. Sheridan's first battle. Here, in 1856, a small party of white men was for two days besieged by a strong force of Indians; and here the irrepressible lieutenant, tired of his wise and masterly inactivity, determined to attack in his turn, and totally routed the enemy in a very whirlwind of a charge.

Now you are in the heart of the mountains. Soon the rock walls approach each other, and the stream flows narrower and fiercer. The wind roars through the gorges,

Peak of Red Rock

and in the spring, when the banks are full, beats up such waves that a boat cannot live in them, though these straits are two hundred miles from the sea. The walls are basaltic, columnar, rising in distinct, rudely-modelled pillars from four hundred feet to twelve hundred. Now and then, a bold rampart measures two thousand five hundred feet or even more. The receding or advancing cliffs break the river into a chain of tiny lakes. Wherever the mountains fall back on the south, Mount Hood fills the horizon, snow-covered, shining, vast. Mount Hood is eleven thousand feet in height, and it is mortifying to admit that Mont Blanc is almost sixteen thousand. But, with this foreground of river and forest, with all this blaze of color set against the cold splendor of the icy peak, and with the blue intensity of the warm sky above, Mount Hood is more magnificent than words can tell or brush can paint. And, if any "vagrom" man, having seen the two, pretends to think Mont Blanc the finer, let us, as Americans, laugh him to scorn.

Where the mountains recede before Mount Hood, the forest again encroaches, but it leaves bare a desolate peak called Coffin Rock, which was a place of Indian sepulture. Cairns of gray stones cover it, and rude monuments of rock. One is not near enough to see the vileness of the human taint upon it—for your true Indian in his death is little better than in his life, and bequeaths himself, a foul legacy, to the pure elements—and its gray melancholy is pathetic.

The Dalles is an important town in Oregon. The miners formerly made it their

Coffin Rock.

base of supplies. The gold was sent there for shipment, and this babe in the woods once dreamed of a mint. But its interest to the traveller is neither in grocery nor in ore, but in its wonderful outlook on river and mountain. For ten miles up the stream the *dalles*, or flag-stones, choke the way, and there you must take to the cars again. Here the strange, weather-beaten, weary-looking, old red rock reappeared, after a long absence, looking, amid the harder and bolder cliffs, like a poor relation, pathetic, but very seedy. The queer, battered, time-worn peak on the opposite page is of it.

The cliffs disappear above Dalles City, and lo, the sand-region! The endless wonder of the Pacific-coast journeys is the suddenness of their changes, as if supernatural scene-shifters were kept constantly busy in whipping off the old scenes, and setting new and unexpected ones for the next act. From forests of tropic splendor to mountains of northern bareness and grandeur, from still pond to roaring cataract, from verdure and cultivation into Sahara, you pass without the least hint from Nature of what she means to do five minutes hence. Possibly Science gets the better of her, and finds out her whimsical intentions; but to the unlearned she seems to have gotten a little tipsy on that wonderful air—which would intoxicate the soberest—and not to be quite sure of her own mind. Her desert on the Columbia is a lively suggestion of her greater works of the same order in Egypt or elsewhere. It looks a limitless plain of hot white sand.

Passage of the Dalles.

The wind is a hurricane, and seems to blow from every point at once, so that the heavens rain a sandy shower. The shifting, sifting sand covers the track. Men in sand-white garments, with sand-white beards and hair, blindly delve along the rails to clear them, and limp aside with sand-stiff joints that seem to creak, as we go by. The sky is a pale-blue vault, faded out by this torrid plain. The sun is veiled, intense, and colorless. The earth is like a place of graves, as if millions of men, whole peoples, whole races, had been buried there and forgotten. But, if Nature had ever set any race there, it must have been of the lowest—in mind vacant, in body vile, in worship regarding stones and wild animals, its only symbols of steadfastness and power. And when on the flat-shore rocks we saw the bark lodges of the Trascopin Indians, vile children and viler men and vilest women swarming within and without, we felt that they were accounted for—stupidity, dishonesty, beastliness, and all—and had no disposition to cast a stone at them.

The fifteen miles of portage show superb river scenery wherever the sand will let you see it. It is a succession of rapids, falls, and sucking currents, where the *dalles*, or *dales*—rough troughs or flag-stones, which have given their name to the place—make crooked and narrow channels for the stream. Every form which water may put on, every tint with which it can be beautiful, every caprice of motion with which it can move, finds illustration in this Columbia. Below the great fall, the whole volume of the stream —whose branches stretch north through British Columbia, east through Idaho and Montana, south and west into Nevada, and, reaching down, gather in the icy rivulets of the Rocky Mountains—pours through a gate-way not fifty yards in width, whose sides are perpendicular precipices, hewn as with implements. Smooth and green and glassy, it slides under brown shadows but to be torn again into a hundred ribbons by rocks below, as it has just been torn by rocks above. At the falls it is a mile wide, and plunges over a rocky wall twenty feet high and stretching from shore to shore.

Here are the famous Salmon Falls, up which the salmon go to the quiet reaches of the river to spawn, shooting the rapids with incredible agility. If you can keep your footing on the slippery ledges of rock, you watch them, fascinated. Up they come through the fierce and sucking rapids, gleaming white against the black stones that here and there tear the water; first come a few together; then a multitude swirls along; then the whole river from side to side is light with their innumerable host. And they mind that precipice and torrent no more than if it were a summer pool within its reedy margent. They swim swift and stately to the very foot, where you lose them in the seething, white whirlpool. Something flashes in the air, elastic, strong, light. Something glides up the stream above the fall. The daring, determined, wonderful thing has made that leap, defied rock and torrent, and found its safe shelter in the quiet pool beyond. Or, there is the flash, and then a struggle, and the poor bruised creature, wounded to death against the sharp-edged stones, drops back upon the current, and floats down a bloody track, dying after a little while. So they come, and come, and come—such myriads of them—and leap, and win, or lose, for all the hours of the day and for half the days of the year.

Above the Dalles the forests disappear; nay, every leaf vanishes, and for miles on miles the banks are covered with thick brown grass, wherein not even a mullein-stalk springs. The scenery is tame, and the most eager tourist seldom ventures above Wright's Harbor, two hundred and fifty miles from the sea. Steamers ply, however, for four hundred miles, and then, after a portage along impassable rapids, an odd little boat runs up the Snake River, in Idaho. Now that a great railroad connects the head-waters of the Missouri with the head-waters of the Columbia, the six hundred miles of track have opened an incalculable wealth to trade, and the most magnificent wilderness of the world to travel. But formerly, what with ubiquitous savages, and perils of cold and hunger, and lost trails, it was well to pause at Celilo, not far above the falls. There, after inspecting "the largest warehouse in the United States, being over eleven hundred feet in length, and built to receive the Idaho freights," as the station-master informs you in a solemn recitative, and there being nothing else in or of Celilo that unanointed eyes can behold, you are speedily ready for your train. And so back you go, leaving the

Salmon Falls.

falls and salmon and savage, leaving desert and whirlpool and whirlwind, at your back, and not reluctantly returning to the common-sense and conventions of decent and sober Dalles.

To come down the river in the early morning, with the clear eastern light behind you, is almost finer than to sail eastward, with the glow of the sunset over mountain and stream. Certainly Mount Hood lay more stately calm and fair, quite apart, rising lonely from a far, upward-going plain, white, glittering, perfect. Mount Adams and Mount Jefferson, also, seemed to win a charm from the presence of the pure morning; and we had not in the least understood Mount Rainier until this second coming before it. Under the blue heavens it rose in soft and tender liftings, till its triple crown overtopped Mount Hood itself. From Puget Sound the view of it is grander, but not so lovely; and, as we watched it, it seemed even more worthy to be remembered than sweet St. Helens.

As we reluctantly sailed away from the friendliness we had found, and from the majestic forests and the gracious mountains, we seemed to hear in the ripple of the waves these words of a most sweet philosopher: "Let us remember within what walls we lie, and understand that this level life, too, has its summit, and why from the mountain-top the deepest valley has its tinge of blue; that there is elevation in every hour, as no part of the earth is so low that the heavens may not be seen from it, and we have only to stand on the summit of our hour to command an uninterrupted horizon."

Indians on the Columbia.

THE NATURAL BRIDGE, VIRGINIA.

WITH ILLUSTRATIONS BY HARRY FENN.

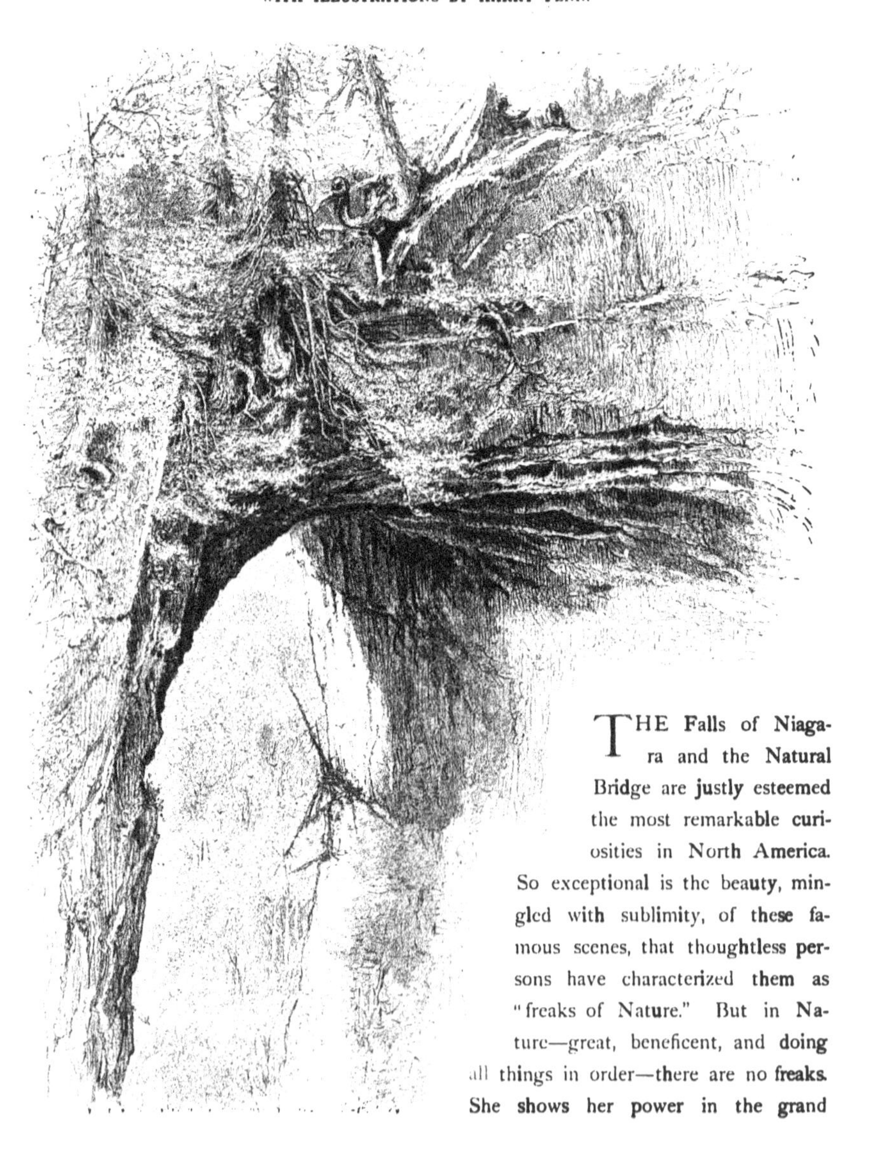

THE Falls of Niagara and the Natural Bridge are justly esteemed the most remarkable curiosities in North America. So exceptional is the beauty, mingled with sublimity, of these famous scenes, that thoughtless persons have characterized them as "freaks of Nature." But in Nature—great, beneficent, and doing all things in order—there are no freaks. She shows her power in the grand cataract, spanned with its rainbow, and in the dizzy arch of the Natural Bridge, as in the daisy and the violet she shows her grace and beauty.

The Natural Bridge, the character and formation of whose upper portion are displayed in the first of the accompanying sketches, has been, from about the middle of the eighteenth century, an object of curiosity and admiration in Europe as well as in America. Whatever traveller came to the Western World, to compare its natural

The Natural Bridge and its Surroundings.

grandeur with the grandeur of art and architecture in the countries he had left, went first, in the North, to the Falls of Niagara, and, in the South, to the world-famous bridge. Among these may be mentioned the distinguished Marquis de Chastellux, who in 1781 visited the place, and from whose rare volumes we present a few paragraphs:

"Having thus travelled for two hours," writes the marquis, "we at last descended a steep declivity, and then mounted another. . . . At last my guide said to me: 'You desire to see the Natural Bridge—don't you, sir? You are now upon it; alight and go twenty steps either to the right or left, and you will see this prodigy.' I had perceived that there was on each side a considerable deep hollow, but the trees had prevented me from forming any judgment or paying much attention to it. Approaching the precipice, I saw, at first, two great masses or chains of rocks, which formed the bottom of a ravine, or, rather, of an immense abyss. But, placing myself, not without precaution, upon the brink of the precipice, I saw that these two buttresses were joined under my feet, forming a vault of which I could yet form no idea but of its height. After enjoying this magnificently-tremendous spectacle, which many persons could not bear to look at, I went to the western side, the aspect of which was not less imposing, but more picturesque.

"But it is at the foot of these rocks, on the edge of a little stream which flows under this immense arch, that we must judge of its astonishing structure. There we discover its immense spurs, its back-bendings, and those profiles which architecture might have given it. The arch is not complete; the eastern part of it not being so large as the western, because the mountain is more elevated on this than on the opposite side. It is very extraordinary that at the bottom of the stream there appear no considerable ruins, no trace of any violent laceration which could have destroyed the kernel of the rock and have left the upper part alone subsisting; for that is the only hypothesis that can account for such a prodigy. We can have no possible recourse either to a volcano or a deluge, no trace of a sudden conflagration or of a slow and tedious undermining by the water."

The point here touched upon is one of the most interesting, in a scientific view, connected with this famous curiosity. The marquis, it will be seen, declares his conviction that the "prodigy" was neither caused by a volcanic upheaval, a conflagration burning in the heart of the rock-ribbed mountain, nor by the attrition of water slowly wearing away the stubborn limestone. These views are supported by men of science, as the following paragraphs will show:

"The mass of rock and stone which loads this arch," says De Turpin, "is forty-nine feet solid on the key of the great centre, and thirty-seven on that of the small one; and, as we find about the same difference in taking the level of the hill, it may be supposed that the roof is on a level the whole length of the key. It is proper to observe that the live rock continues also the whole thickness of the arch, and that on the opposite side it is only twenty-five feet wide in its greatest breadth, and becomes gradually narrower. The whole arch seems to be formed of one and the same stone; for the joints which one remarks are the effect of lightning, which struck this part in 1779. The other head has not the smallest vein, and the intrados is so smooth that the martins, which fly around it in great numbers, cannot fasten on it. The abutments, which have a gentle slope, are entire, and, without being absolute planes, have all the polish which a current of water would give to unhewn stone in a certain time. The four

UNDER THE NATURAL BRIDGE.

rocks adjacent to the abutments seem to be perfectly homogeneous, and to have a very trifling slope. The two rocks on the right bank of the rivulet are two hundred feet high above the surface of the water, the intrados of the arch a hundred and fifty, and the two rocks on the left bank a hundred and eighty."

The Baron de Turpin then proceeds to burst forth with:

"If we consider this bridge simply as a picturesque object, we are struck with the majesty with which it towers in the valley. The white-oaks which grow upon it seem to rear their lofty summits to the clouds, while the same trees which border on the rivulet appear like shrubs."

This exhibition of sentiment, however, appears to exhaust the baron's stock, and he returns to his better-loved science, adding:

"We see that these rocks, being of a calcareous nature, exclude every idea of a volcano, which, besides, cannot be reconciled with the form of the bridge and its adjacent parts. If it be supposed that this astonishing arch is the effect of a current of water, we must suppose, likewise, that this current has had the force to break down and carry to a great distance a mass of five thousand cubic fathoms, for there remains not the slightest trace of such an operation."

Mr. Fenn's second drawing furnishes a distant view of the bridge, the surrounding country, and objects in its vicinity. It will recall, doubtless, to many persons, agreeable recollections of the landscape which saluted their eyes as they first drew near the place —and the names of such are legion, for the spot has been, for more than half a century, the resort of parties led by a desire to explore the beauties of the romantic scene. Of the daring of some of these visitors, in climbing, or venturing to the brink of the precipice, we shall give one or two instances, kept alive by tradition. Among these traditions, the most thrilling is that of the unshrinking nerve displayed by Miss Randolph, a young *Virginienne*, a great belle of her time, which was the early portion of the present century. The young lady had ridden, with a gay party of youthful maidens and gallant cavaliers, to the bridge, and reached it on a beautiful evening of summer. Miss Randolph is said, by those who knew and remember her, to have been a young lady of surpassing loveliness—tall, slender, with sparkling eyes, cheeks all roses, and noted for her gayety and mirthful *abandon*. Reaching the summit of the bridge, the party dismounted, cautiously approached the brink, fringed with trees growing among the rocks, and gazed into the gulf beneath. Of the terrifying character of the spectacle, President Jefferson's words will give some idea:

"Though the sides of the bridge are provided, in some parts, with a parapet of rocks," he says, "yet few men have resolution to walk to them and look over into the abyss. You involuntarily fall on your hands and feet, creep to the parapet, and look over it. Looking down from this height about a minute gave me a violent headache; the view is painful and intolerable."

ABOVE THE NATURAL BRIDGE.

Reaching this dizzy brink, the party of young ladies and gentlemen gazed below when one of the gallants, pointing to the broken stump of a huge cedar which had once towered aloft upon a jagged abutment, separated by an intervening cleft from the main structure, expressed his conviction that no human being lived sufficiently daring to stand erect upon it. A gay laugh echoed the words, a silken scarf brushed by him, and the whole party uttered a cry of terror—Miss Randolph, at one bound, had reached and now stood erect upon the dizzy pinnacle. Tradition relates that her companions looked at her, white and speechless, as so many corpses. Her death seemed certain. For an instant, the daring young lady stood erect, riding-whip in hand, her scarf floating, her eyes sparkling with triumph; then, at a single bound, she regained her former position.

The most striking view of the Natural Bridge is that from below, and no better hour could be selected than that fixed upon by Mr. Fenn. As the sun rises and flashes its splendors through the gigantic arch, the scene becomes one of extraordinary beauty and sublimity—beauty from the exquisite flush which spreads itself over rocky mass and stately fir, over pendent shrub, and the fringe of evergreen; and sublimity from the well-nigh overpowering sentiment which impresses the mind in presence of the mighty arch of rock, towering far above, and thrown as by the hand of some Titan of old days across the blue sky, appearing both above and beneath. It has been well said that no one who has witnessed this extraordinary spectacle has ever forgotten it.

With the brilliant drawing of Mr. Fenn before his eyes, the reader would only be wearied by any description of the exquisite scene which it represents. The grandeur and serene loveliness of the spectacle are sufficiently indicated—the gentle stream which passes with a murmur from its hiding-place in the bosom of the hills—the lengthening vistas, cool and soft, and bathed in dawn—the silent mountains—and, in the midst of all this exquisite beauty, the great soaring arch, with its jutting buttresses and fringes of the evergreen pine, the shaggy eyebrows of the giant. They dwindle these heavy-headed evergreens into little fringes only—even that picturesque monarch, represented in the second drawing of Mr. Fenn, on the summit of the bridge, shows scarce so large as the spray of ferns and cedar held in the hand of a girl! From the summit to the surface of the stream below is two hundred and fifteen feet; and thus the Natural Bridge is fifty-five feet higher than Niagara.

It remains only, before terminating our brief sketch of this celebrated curiosity, to speak of the hazardous attempts, made by more than one person, to climb the rocky sides of the great arch and reach the summit. This has never yet been done, but a considerable distance has been attained by venturesome climbers, who have recorded their prowess by cutting their names on the surface, at the highest point reached by them. High up among these, it is reported, may be found the name of George Washington, who, strong, adventurous, and fond of manly sports, was seized with the ambition to ascend the precipice and inscribe his name upon the face of the rock.

The highest point ever reached by any one of these adventurous explorers is said to have been attained by Mr. James Piper, at the time a student of Washington College, and subsequently a State senator. It was about the year 1818, when, with some of his fellow-students, Mr. Piper visited the bridge, descended to the foot of the precipice, and determined to ascertain to what height it was possible for a human being to ascend by means of inequalities on the surface, the assistance of shrubs, or otherwise. He accordingly commenced climbing the precipice, and, taking advantage of every ledge, cleft, and protuberance, finally reached a point which, to his companions far beneath, seemed directly under the great arch. He was far above the names cut on the stone—fully fifty feet above that of Washington—and, standing upon a ledge, which appeared to his terrified fellow-students but a few inches in width, shouted aloud, waving one hand in triumph, while with the other he clung to the face of the precipice. They shouted back to him, begging him for God's sake to descend, but he only replied by laughter. He accordingly continued his way, working his toilsome and dangerous passage through clefts in the huge mass of rock. These were just sufficient, in many places, to permit his body to pass; and huge roots from the trees above, protruding through splits in the mass, curled to and fro, and half obstructed the openings. With unfaltering resolution, and not daring to look into the hideous gulf beneath him, the young man fought his way on, piercing by main force the dark clefts, crawling along narrow ledges, springing from abutment to abutment, until finally he stopped at an elevation of *one hundred and seventy feet* from the earth below. Here he was seen to look upward, but he did not move.

He slowly and cautiously divested himself first of one of his shoes, and then the other, next drew off his coat, and these articles he threw from him into the gulf beneath, without daring to look in the direction in which they fell. Then, clinging close to the face of the precipice, and balancing his body carefully as he placed each foot down, and raised each one up, he tottered along inch by inch, hanging between life and death until he reached a friendly cleft. Here pausing for a moment to brace his nerves, he continued his way in the same cautious manner, followed by the eyes of his pale and terrified friends; when, disappearing in a cleft, he reappeared no more. His friends had given him up, and agony had succeeded the long suspense, when suddenly, from behind a clump of evergreens, appeared the student—safe, sound, and smiling, after his perilous feat, during which he had stood face to face with the most terrible of deaths.

The Natural Bridge is in the southeastern corner of Rockbridge County, in the midst of the wild scenery of the Blue-Ridge region, and almost under its shadow upon its western side. It is a little over two miles from the railway station of Natural Bridge, on the Norfolk and Western Railway.

OUR GREAT NATIONAL PARK.

THE VALLEY OF THE YELLOWSTONE.

The Yellowstone.

THE Yellowstone River, one of the tributaries of the Missouri, has a long, devious flow of thirteen hundred miles ere it loses its waters in those of the larger stream. Its source is a noble lake in the State of Wyoming, and nestling amid the snow-peaks of the highest mountain-range in the country. The upper course of the river is through immense cañons and gorges, and its flow is often marked by splendid water-falls and rapids, presenting at various points some of the most remarkable scenery in the country. The entire region about its source is volcanic, and abounds in boiling springs, mud-volcanoes, soda-springs, sulphur-mountains, and geysers the marvels of which outdo those of Iceland.

This remarkable area has recently been set apart by Congress for a great national

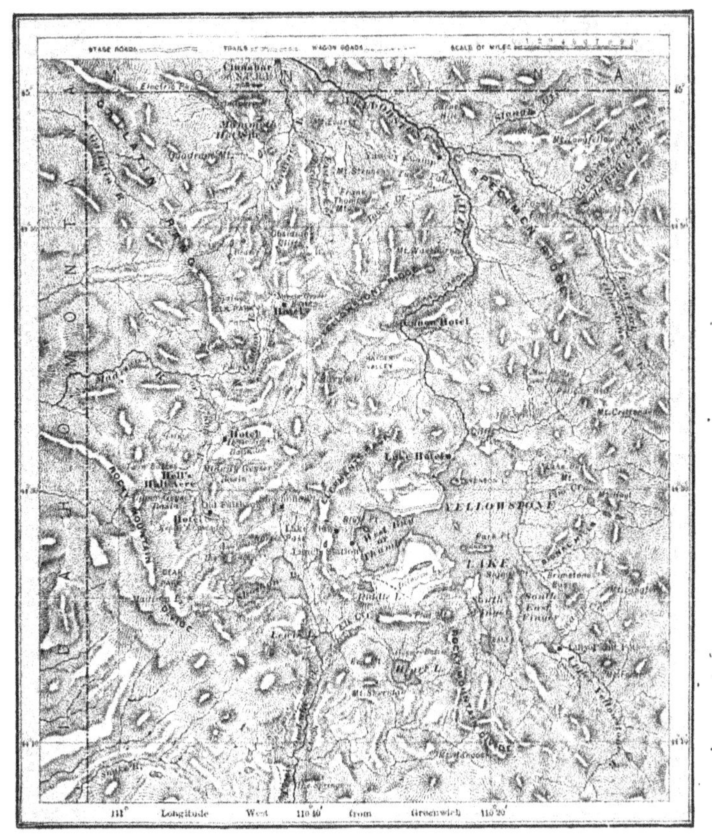

Map of the Yellowstone National Park.

park. It certainly possesses striking characteristics for the purpose to which it has been devoted, exhibiting the grand and magnificent in its snow-capped mountains and dark cañons, the picturesque in its splendid water-falls and strangely-formed rocks, the beautiful in the sylvan shores of its noble lake, and the phenomenal in its geysers, hot springs, and mountains of sulphur. It may be claimed that in no other portion of the globe are there united so many surprising features—none where the conditions of beauty and contrast are so calculated to delight the artist, or where the phenomena are so abundant for the entertainment and instruction of the student.

It is a magnificent domain in its proportions, extending nearly sixty-five miles from north to south, and fifty-five miles from east to west. The Yellowstone Lake lies near the southeasterly corner of the park, the Yellowstone River flowing from its upper boundary, and running almost due north. The lake is about twenty miles in length, and its average width from ten to fifteen miles. Its height above the level of the sea is seven thousand feet, while its basin is surrounded by mountains reaching an altitude of over ten thousand feet, the peaks of which are covered with perpetual snow. Numerous hot springs are found on the shores of the lake, and also along the banks of the river. About fifteen miles from its source, the river takes two distinct, precipitous leaps, known as the Upper and the Lower Falls, and beyond the falls cuts its way through an immense cañon, the vertical walls of which reach, at places, the height of fifteen hundred feet. Near the western boundary of the park, the Madison, an important tributary of the Columbia, takes its rise; and along one of the branches of this river, known as Fire-Hole River, are found numerous extraordinary geysers, some of which throw volumes of boiling water to a height exceeding two hundred feet. In the northwest corner of the park, the Gallatin, another tributary of the Columbia, takes its rise.

This wonder-land has only recently been explored. For years, marvellous stories have been rife among the hunters of the far West of a mysterious country in the heart of the Rocky Mountains, which the Indians avoided as the abode of the evil spirits, where the rumble of the earthquake is frequently heard, where great jets of steam burst through the earth, where volcanoes throw up mud instead of fire, and where a river flows through gorges of savage grandeur; but beyond these rumors, often apparently absurd exaggerations, nothing was known of that region. An exploring party, under Captain Raynolds, of the United States Engineer Corps, endeavored to enter the Yellowstone Basin in 1859, by way of the Wind-River Mountains, at the south, but failed on account of the ruggedness of the mountains and the depth of the snow. In 1870, an exploring party under General Washburn, escorted by Lieutenant Doane, of the United States Army, succeeded in entering the valley; and from this source the public obtained the first trustworthy accounts of the strange land. Immediately thereafter, an expedition, under sanction of Congress, was organized by the Secretary of the Interior, and placed in the charge of Professor F. V. Hayden, United States geologist; while, at the same time, a party under the command of Lieutenant Barlow, of the United States Engineer Corps, ascended the Yellowstone, and traversed the greater part of the area now included in the park. Professor Hayden's expedition made a thorough exploration of the whole region, and it is to his full and exhaustive report to Congress that we are indebted for an accurate detailed knowledge of the strange features of this remarkable land. It is to this scientist, probably more than to

CLIFFS ON THE YELLOWSTONE.

any other person, that we are indebted for the idea of converting the valley into a national park. The expedition, however, was organized by the Hon. Columbus Delano, Secretary of the Interior; and hence we may attribute the successful issue of the noble conception to the coöperation of the secretary with the purposes of the scientific explorers appointed by him. From the interesting pages of Professor Hayden's report we

Cañon of the Yellowstone.

mainly draw the subjoined particulars of the romantic wonders of our imperial pleasure-ground:

THE YELLOWSTONE BASIN.

"The Yellowstone Basin proper, in which the greater portion of the interesting scenery and wonders is located, comprises only that portion enclosed within the remark-

Gorge of the Yellowstone.

able ranges of mountains which give origin to the waters of the Yellowstone south of Mount Washburn and the Grand Cañon. The range of which Mount Washburn is a conspicuous peak seems to form the north wall, or rim, extending nearly east and west across the Yellowstone, and it is through this portion of the rim that the river has cut its channel, forming the remarkable falls and the still more wonderful cañon. The area of this basin is about forty miles in length. From the summit of Mount Washburn a bird's-eye view of the entire basin may be obtained, with the mountains surrounding it on every side, without any apparent break in the rim. This basin has been called, by some travellers, the vast crater of an ancient volcano. It is probable that during the Pliocene period the entire country drained by the sources of the Yellowstone and the Columbia was the scene of as great volcanic activity as that of any portion of the globe. It might be called one vast crater, made up of thousands of smaller volcanic vents and fissures, out of which the fluid interior of the earth, fragments of rock, and volcanic dust, were poured in unlimited quantities. Hundreds of the nuclei or cores of these volcanic vents are now remaining, some of them rising to a height of ten thousand to eleven thousand feet above the sea. Mounts Doane, Langford, Stevenson, and more than a hundred other peaks, may be seen from any high point on either side of the basin, each of which formed a centre of effusion. Indeed, the hot springs and geysers of this region, at the present time, are nothing more than the closing stages of that wonderful period of volcanic action that began in Tertiary times. In other words, they are the escape-

Column-Rocks.

pipes or vents for those internal forces which once were so active, but are now continually dying out. The evidence is clear that, ever since the cessation of the more powerful volcanic forces, these springs have acted as the escape-pipes, but have continued to decline down to the present time, and will do so in the future, until they cease entirely."

THE FALLS AND THE GRAND CAÑON.

"But the objects of the deepest interest in this region are the falls and the Grand Cañon. I will attempt to convey some idea by a description, but it is only through the eye that the mind can gather any thing like an adequate conception of them. As we approached the margin of the cañon, we could hear the suppressed roar of the falls, resembling distant thunder. The two falls are not more than one-fourth of a mile apart. Above the Upper Falls the Yellowstone flows through a grassy, meadow-like valley, with a calm, steady current, giving no warning, until very near the falls, that it is about to rush over a precipice one hundred and forty feet, and then, within a quarter of a mile, again to leap down a distance of three hundred and fifty feet.

"But no language can do justice to the wonderful grandeur and beauty of the cañon below the Lower Falls; the very nearly vertical walls, slightly sloping down to the water's edge on either side, so that from the summit the river appears like a thread of silver foaming over its rocky bottom; the variegated colors of the sides, yellow, red, brown, white, all intermixed and shading into each other; the Gothic columns of every form, standing out from the sides of the walls with greater variety and more striking colors than ever adorned a work of human art. The margins of the cañon on either side are beautifully fringed with pines. In some places the walls of the cañon are composed of massive basalt, so separated by the jointage as to look like irregular mason-work going to decay. Here and there, a depression in the surface of the basalt has been subsequently filled up by the more modern deposit, and the horizontal strata of sandstone can be seen. The decomposition and the colors of the rocks must have been due largely to hot water from the springs, which has percolated all through, giving to them the present variegated and unique appearance.

"Standing near the margin of the Lower Falls, and looking down the cañon, which looks like an immense chasm or cleft in the basalt, with its sides twelve hundred to fifteen hundred feet high, and decorated with the most brilliant colors that the human eye ever saw, with the rocks weathered into an almost unlimited variety of forms, with here and there a pine sending its roots into the clefts on the sides as if struggling with a sort of uncertain success to maintain an existence—the whole presents a picture that it would be difficult to surpass in Nature. Mr. Thomas Moran, a celebrated artist, and noted for his skill as a colorist, exclaimed, with a kind of regretful enthusiasm, that these beautiful tints were beyond the reach of human art. It is not the depth alone that gives such an impression of grandeur to the mind, but it is also the picturesque forms and coloring. After the waters of the Yellowstone roll over the upper descent, they flow with great rapidity over the apparently flat, rocky bottom, which spreads out to nearly double its width above the falls, and continues thus until near the Lower Falls, when the channel again contracts, and the waters seem, as it were, to gather themselves into one compact mass, and plunge over the descent of three hundred and fifty feet in detached drops of foam as white as snow; some of the large globules of water shoot down like the contents of an exploded rocket. It is a sight far more beautiful than, though not so grand or impressive as, that of Niagara Falls. A heavy mist

THE LOWER FALLS

always rises from the water at the foot of the falls, so dense that one cannot approach within two hundred or three hundred feet, and even then the clothes will be drenched in a few moments. Upon the yellow, nearly vertical wall of the west side, the mist mostly falls; and for three hundred feet from the bottom the wall is covered with a thick matting of mosses, sedges, grasses, and other vegetation of the most vivid green, which have sent their small roots into the softened rocks, and are nourished by the ever-ascending spray. At the base and quite high up on the sides of the cañon are great quantities of talus, and through the fragments of rocks and decomposed spring deposits may be seen the horizontal strata of breccia."

Tower Creek.

TOWER CREEK.

"Tower Creek rises in the high divide between the valleys of the Missouri and Yellowstone, and flows about ten miles through a cañon so deep and gloomy that it has very properly earned the appellation of the Devil's Den. As we gaze from the margin down into the depths

TOWER FALLS, YELLOWSTONE VALLEY.

below, the little stream, as it rushes foaming over the rocks, seems like a white thread, while on the sides of the gorge the sombre pinnacles rise up like Gothic spires. About two hundred yards above its entrance into the Yellowstone, the stream pours over an abrupt descent of one hundred and fifty-six feet, forming one of the most beautiful and picturesque falls to be found in any country. The Tower Falls are about two hundred and sixty feet above the level of the Yellowstone at the junction, and they are surrounded with pinnacle-like columns, composed of the volcanic breccia, rising fifty feet above the falls, and extending down to the foot, standing like gloomy sentinels or like

The First Boat on the Yellowstone.

the gigantic pillars at the entrance of some grand temple. One could almost imagine that the idea of the Gothic style of architecture had been caught from such carvings of Nature. Immense bowlders of basalt and granite here obstruct the flow of the stream above and below the falls; and, although, so far as we can see, the gorge seems to be made up of the volcanic cement, yet we know that, in the loftier mountains, near the source of the stream, true granitic as well as igneous rocks prevail." The pointed rocks and their resemblance to Gothic architecture are strikingly shown in the beautiful illustration by Mr. Fenn on the opposite page.

Yellowstone Lake.

YELLOWSTONE LAKE.

"On the 28th of July (1871)," says Professor Hayden, "we arrived at the lake, and pitched our camp on the northwest shore, in a beautiful grassy meadow or opening among the dense pines. The lake lay before us, a vast sheet of quiet water, of a most delicate ultramarine hue, one of the most beautiful scenes I have ever beheld.

"Usually in the morning the surface of the lake is calm, but, toward noon and after, the waves commence to roll, and the white caps rise high, sometimes four or five feet. This lake is about twenty-two miles in length from north to south, and an average of ten to fifteen miles in width from east to west. It has been aptly compared to the human hand; the northern portion would constitute the palm, while the southern prolongations or arms might represent the fingers. There are some of the most beautiful shore-lines along this lake that I ever saw. Some of the curves are as perfect as if drawn by the hand of art. The water of the lake has at all seasons nearly the temperature of cold spring-water. The most accomplished swimmer could live but a short time in it; the dangers attending the navigation of such a lake in a small boat are thereby greatly increased. The lake abounds in salmon-trout, and is visited by great numbers of wild-fowl.

"We adopted the plan of making permanent camps at different points around the lake while explorations of the country in the vicinity were made. Our second camp was pitched at the hot springs on the southwest arm. This position commanded one of the finest views of the lake and its surroundings. While the air was still, scarcely a ripple could be seen on the surface, and the varied hues, from the most vivid green shading to ultramarine, presented a picture that would have stirred the enthusiasm of the most fastidious artist. Sometimes, in the latter portion of the day, a strong wind would arise, arousing this calm surface into waves like the sea. Near our camp there is a thick deposit of the silica, which has been worn by the waves into a bluff wall, twenty-five feet high above the water. It must have originally extended far out into the lake. The belt of springs at this place is about three miles long and half a mile wide. The deposit now can be seen far out in the deeper portions of the lake, and the bubbles that arise to the surface in various places indicate the presence, at the orifice, of a hot spring beneath. Some of the funnel-shaped craters extend out so far into the lake, that the members of our party stood upon the silicious mound, extended the rod into the deeper waters, and caught the trout, and cooked them in the boiling spring, without removing them from the hook. These orifices, or chimneys, have no connection with the waters of the lake. The hot fumes coming up through fissures, extending down toward the interior of the earth, are confined within the walls of the orifice, which are mostly circular, and beautifully lined with delicate porcelain."

THE HOT SPRINGS.

"Upon the west side of Gardiner's River, on the slope of the mountain, is one of the most remarkable groups of hot springs in the world. The springs in action at the present time are not very numerous, or even so wonderful as some of those higher up in the Yellowstone Valley or in the Fire-Hole Basin, but it is in the remains that we find so instructive records of their past history. The calcareous deposits from these springs cover an area of about two miles square. The active springs extend from the margin of the river five thousand five hundred and forty-five feet, to an elevation nearly one thousand above, or six thousand five hundred and twenty-two feet above the sea by barometrical measurement. Our path led up the hill by the side of a wall of lower cretaceous rocks, and we soon came to the most abundant remains of old springs, which, in past times, must have been very active. The steep hill, for nearly a mile, is covered with a thick crust, and, though much decomposed and covered with a moderately thick growth of pines and cedars, still bore traces of the same wonderful architectural beauty displayed

Hot Springs.

in the vicinity of the active springs farther up the hill. After ascending the side of the mountain, about a mile above the channel of Gardiner's River, we suddenly came in full view of one of the finest displays of Nature's architectural skill the world can produce. The snowy whiteness of the deposit at once suggested the name of White-Mountain Hot Spring. It had the appearance of a frozen cascade. If a group of springs near the summit of a mountain were to distribute their waters down the irregular declivities, and they were slowly congealed, the picture would bear some resemblance in form. We pitched our camp at the foot of the principal mountain, by the side of the stream that contained the aggregated waters of the hot springs above, which, by the time they had reached our camp, were sufficiently cooled for our use. Before us was a hill two hundred feet high, composed of the calcareous deposit of the hot springs, with a system of step-like terraces, which would defy any description by words. The eye alone could convey any adequate conception to the mind. The steep sides of the hill were ornamented with a series of semicircular basins, with margins varying in height from a few inches to six or eight feet, and so beautifully scalloped and adorned with a kind of bead-work, that the beholder stands amazed at this marvel of Nature's handiwork. Add to this a snow-white ground, with every variety of shade, of scarlet, green, and yellow, as brilliant as the brightest of our aniline dyes. The pools or basins are of all sizes, from a few inches to six or eight feet in diameter, and from two inches to two feet deep. As the water flows from the spring over the mountain-side from one basin to another, it loses continually a portion of its heat, and the bather can find any desirable temperature. At the top of the hill there is a broad, flat terrace, covered more or less with these basins, one hundred and fifty to two hundred yards in diameter, and many of them going to decay. Here we find the largest, finest, and most active spring of the group at the present time. The largest spring is very near the outer margin of the terrace, and is twenty-five by forty feet in diameter, the water so perfectly transparent that one can look down into the beautiful ultramarine depth to the bottom of the basin. The sides of the basin are ornamented with coral-like forms, with a great variety of shades, from pure white to a bright cream-yellow, and the blue sky, reflected in the transparent waters, gives an azure tint to the whole, which surpasses all art. Underneath the sides of many of these pools are rows of stalactites, of all sizes, many of them exquisitely ornamented, formed by the dripping of the water over the margins of the basins.

" On the west side of this deposit, about one-third of the way up the White Mountain from the river and terrace, which was once the theatre of many active springs, old chimneys, or craters, are scattered thickly over the surface, and there are several large holes and fissures leading to vast caverns beneath the crust. The crust gives off a dull, hollow sound beneath the tread, and the surface gives indistinct evidence of having been adorned with the beautiful pools or basins just described. As we pass up to the base of the principal terrace, we find a large area covered with shallow pools, some of them containing water, with all the ornamentations perfect, while others are fast going to decay, and the decomposed sediment is as white as snow. Upon this kind of sub-terrace is a remarkable cone,

Liberty Cap.

about fifty feet in height, and twenty feet in diameter at the base. From its form we gave it the name of the Liberty Cap. It is undoubtedly the remains of an extinct geyser. The water was forced up with considerable power, and probably without intermission, building up its own crater until the pressure beneath was exhausted, and then it gradually closed itself over at the summit and perished. No water flows from it at the present time. The layers of lime were deposited around it like the layers of straw on a thatched roof, or hay on a conical stack.

"The entire Yellowstone Basin is covered more or less with dead and dying springs, but there are centres or groups where the activity is greatest at the present time. Below the falls there is an extensive area covered with the deposits which extend from the south side of Mount Washburn across the Yellowstone rim, covering an area of ten or fifteen square miles. On the south side of Mount Washburn there is quite a remarkable group of active springs. They are evidently diminishing in power, but the rims all around reveal the most powerful manifestations far back in the past. Sulphur, copper, alum, and soda, cover the surface. There is also precipitated around the borders of some of the mud-springs a white efflorescence, probably nitrate of potash. These springs are located on the side of the mountain nearly one thousand feet above the margin of the cañon, but extend along into the level portions below. In the immediate channel of the river, at the present time, there are very few springs, and these not important. A few small steam-vents can be observed only from the issue of small quantities of steam. One of these springs was bubbling quite briskly, but had a temperature of only one hundred degrees. Extending across the cañon on the opposite side of the Yellowstone, interrupted here and there, this group of springs extends for several miles, forming one of the largest deposits of silica, but only here and there are there signs of life. Many of the dead springs are mere basins, with a thick deposit of iron on the sides, lining the channel of the water that flows from them. These vary in temperature from ninety-eight to one hundred and twenty degrees. The highest temperature was one hundred and ninety-two degrees. The steam-vents are very numerous, and the chimneys are lined with sulphur. Where the crust can be removed, we find the under-side lined with the most delicate crystals of sulphur, which disappear like frost-work at the touch. Still there is a considerable amount of solid amorphous sulphur. The sulphur and the iron, with the vegetable matter, which is always very abundant about the springs, give, through the almost infinite variety of shades, a most pleasing and striking picture."

MUD-SPRINGS.

"We pitched our camp on the shore of the river, near the Mud-Springs, thirteen and a half miles above our camp on Cascade Creek. The springs are scattered along on both sides of the river, sometimes extending upon the hill-sides fifty to two hundred feet above the level of the river. Beginning with the lower or southern side of the group, I will attempt to describe a few of them. The first one is a remarkable mud-spring, with a well-defined circular rim composed of fine clay, and raised about four feet above the surface around, and about six feet above the mud in the basin. The diameter of the basin is about eight feet. The mud is so fine as to be impalpable, and the whole

Mud-Springs.

may be most aptly compared to a caldron of boiling mush. The gas is constantly escaping, throwing up the mud from a few inches to six feet in height; and there is no doubt that there are times when it is hurled out ten to twenty feet, accumulating around the rim of the basin. About twenty yards distant from the mud-spring just described is a second one, with a basin nearly circular, forty feet in diameter, the water six or eight feet below the margin of the rim. The water is quite turbid, and is boiling moderately.

Small springs are flowing into it from the south side, so that the basin forms a sort of reservoir. The temperature, in some portions of the basin, is thus lowered to ninety-eight degrees. Several small hot springs pour their surplus water into it, the temperatures of which are one hundred and eighty, one hundred and seventy, one hundred and eighty-four, and one hundred and fifty-five degrees. In the reservoirs, where the water boils up with considerable force, the temperature is only ninety-six degrees, showing that the bubbling was due to the escape of gas. The bubbles stand all over the surface. About twenty feet from the last is a small mud-spring, with an orifice ten inches in diameter, with whitish-brown mud, one hundred and eighty-two degrees. Another basin

Hot-Spring Cone.

near the last has two orifices, the one throwing out the mud with a dull thud about once in three seconds, spurting the mud out three or four feet; the other is content to boil up quite violently, occasionally throwing the mud ten to twelve inches. This mud, which has been wrought in these caldrons for perhaps hundreds of years, is so fine and pure that the manufacturer of porcelain-ware would go into ecstasy at the sight. The contents of many of the springs are of such a snowy whiteness that, when dried in cakes in the sun or by a fire, they resemble the finest meerschaum. The color of the mud depends upon the superficial deposits which cover the ground through which the waters of the springs reach the surface. They were all clear hot springs originally, perhaps geysers even; but the continual caving-in of the sides has produced a sort of mudpot, exactly the same as the process of preparing a kettle of mush. The water is at first clear and hot; then it becomes turbid from the mingling of the loose earth around the sides of the orifice, until, by continued accessions of earth, the contents of the basin become of the consistency of thick mush, and, as the gas bursts up through it, the dull, thud-like noise is produced. Every possible variation of condition of the contents is found, from simple milky turbidness to a stiff mortar. On the east side of the Yellowstone, close to the margin of the river, are a few turbid and mud springs, strongly impregnated with alum. The mud is quite yellow, and contains much sulphur. This we called a mud-sulphur spring. The basin is fifteen by thirty feet, and has three centres of ebullition, showing that, deep down underneath the superficial earth, there are three separate orifices, not connected with each other, for the emission of heated waters."

From Lieutenant Barlow's report we take the following description of the Giant Geyser: "Proceeding one hundred and fifty yards farther, and passing two hot springs, a remarkable group of geysers is discovered. One of these has a huge crater five feet in diameter, shaped something like the base of a horn—one side broken down—the highest point being fifteen feet above the mound on which it stands. This proved to be a tremendous geyser, which has been called the Giant. It throws a column of water the size of the opening to the measured altitude of one hundred and thirty feet, and continues the display for an hour and a half. The amount of water discharged was immense, about equal in quantity to that in the river, the volume of which, during the eruption, was doubled."

In the report to Congress by the Committee on Public Lands we learn that "the entire area comprised within the limits of the reservation is not susceptible of cultivation with any degree of certainty, and the winters would be too severe for stock-raising. Whenever the altitude of the mountain-districts exceeds six thousand feet above tide-water, their settlement becomes problematical, unless there are valuable mines to attract people. The ranges of mountains that hem the valleys in on every side rise to the height of ten thousand and twelve thousand feet, and are covered with snow all the year. These mountains are all of volcanic origin, and it is not probable that any mines or minerals of value will ever be found there. During the months of June, July, and August, the climate is pure and most invigorating, with scarcely any rain or storms of any kind; but the thermometer frequently sinks as low as twenty-six degrees. There is frost every month of the year." These statements make it evident that, in setting apart this area "as a great national park and pleasure-ground for the benefit and enjoyment of the people," no injury has been done to other interests. The land did not need to be purchased, but simply withdrawn from "settlement, occupancy, or sale"; and hence, by timely action, a great public benefit was secured, which in a few years would have been impracticable, or at least attainable only with great difficulty. The time is not dis-

THE GIANT GEYSER

tant, in the opinion of the Congressional committee, when this region will be a place of "resort for all classes of people from all portions of the globe." By means of the Northern Pacific Railroad the park is easily accessible; its north entrance being only seven miles from Cinnabar station, and in consequence the marvels of the strange domain tempt the curious in numbers to visit it. The Congressional enactment which creates the park amply provides for its control and management. "It shall," says the act, "be under the exclusive control of the Secretary of the Interior, whose duty it shall be, as soon as practicable, to make and publish such rules and regulations as he may deem necessary or proper for the care and management of the same. Such regulations shall provide for the preservation, from injury or spoliation, of all timber, mineral deposits, natural curiosities, or wonders, within said park. The secretary may, in his discretion, grant leases for building-purposes, for terms not exceeding ten years, of small parcels of ground, at such places in said park as shall require the erection of buildings for the accommodation of visitors; all of the proceeds of said leases to be expended under his direction in the management of the same, and the construction of roads and bridle-paths therein."

Soda-Springs.

THE WHITE MOUNTAINS.

WITH ILLUSTRATIONS BY HARRY FENN.

The White Mountains, from the Conway Meadows.

WE suppose that all our readers know that the White Mountains are in New Hampshire, and that they are the highest elevations in New England, and, with the exception of the Black Mountains of North Carolina, the highest in the United States, east of the Mississippi.

The mountains rise from a plateau about forty-five miles in length by thirty in breadth, and about sixteen hundred feet above the sea. This plateau, from which rise nearly twenty peaks of various elevations, and which is traversed by several deep, narrow valleys, forms the region known to tourists as the White Mountains. The peaks cluster in two groups, the eastern of which is known locally as the White Mountains, and the western as the Franconia Group. They are separated by a table-land varying from ten to twenty miles in breadth.

The principal summits of the eastern group are Mounts Washington, Adams, Jefferson, Madison, Monroe, Webster, Clinton, Pleasant, Franklin, and Clay. Of these, Mount Washington is the highest, being 6,293 feet above the level of the sea. The height of some of the other peaks is as follows: Adams, 5,819 feet; Jefferson, 5,736; Monroe, 5,396; Madison, 5,381; Franklin, 4,923; Pleasant, 4,781. The principal summits of the Franconia Group are Mounts Lafayette (5,269 feet), Moosilaukee (4,810 feet), Liberty (4,472 feet), Cherry Mountain (2,956 feet), and Pleasant (2,018 feet). Near the southern border of the plateau rise Whiteface Mountain (4,057 feet), Mount Ossipee (3,600 feet), Chocorua Peak (3,508 feet), and Red Hill (2,038 feet); and, in the southeast, Mount Kiarsarge (3,270 feet).

The rivers in the four great valleys that lead to the White Mountains—in the branches of the Connecticut Valley; in the Androscoggin Valley, that passes beyond these hills, commencing at a lake in Canada; in the Saco Valley, which begins here; and the Pemigewasset Valley, an off-shoot of the valley of the Merrimac—are fed by multitudes of little streams that force their way down steep glens from springs in the mountain-side, and flow through narrow valleys among the hills.

The course of these little rivulets, that break in water-falls, or whose amber flood runs over mossy beds among the forests, furnishes irregular but certain pathways for the rough roads that have been cut beside them, and by which the traveller gains access to these wild mountain-retreats.

In these modern times it is common for the hurried tourist, ambitious to see all he can in the briefest time possible, to pass through the White Mountains by railway; but it is far pleasanter to follow the old-fashioned way, and start from Center Harbor, a pleasant town on the northern end of Lake Winnepesaukee, a body of water surrounded by the Sandwich and Ossipee Hills, of which Whiteface and Chocorua are the loftiest peaks. Following the old coach-road toward Conway and the mountains, the tourist is soon among high hills, the ruggedness of which begins at once to develop itself. Winding in and out among them, he passes now under the dark, frowning brow of a cliff, and afterward by some deep ravine, and then comes upon a lofty plateau which overlooks the amphitheatre of hills; often the summit of Mount Washington may be seen, while its base is hidden by objects nearer. The most interesting feature of the ride, however, is Chocorua, and, to those unacquainted with mountain-scenery, the first impression of this peak is very striking. Driving over the mountain-road, one watches

MOUNT WASHINGTON, FROM THE CONWAY ROAD.

the great hill-tops come up, like billows, one after another, from the sea of mountains round about, as the vehicle winds and twists among them. The soft afternoon light and atmosphere rest over the land, which, as the sun sinks lower, becomes streaked with pale bars of light when the sides and shoulders of the hills are developed by the failing day. All at once, over their sides, bands of a still softer blue appear, which, after interlacing the mountains for a while, are succeeded by a cool purple that steals up these hill-sides, and chases in its path the sunny haze; and this in its turn gives place to a

Elephant's Head, Gate of Crawford Notch.

pinkish gray of almost rosy hue, each tint changing from minute to minute, till they are all finally merged in a dark-purple tone, over which rests a tint as soft as the bloom on a plum, enwrapping each mountain-peak clear cut against the evening sky.

No one who has been much in a forest-region can have failed to perceive and enjoy the delicious fragrance that emanates from the resinous woods when the cool air of evening develops the exhalations from their still and warm foliage. Descending into the damp, fresh valley, and making your way through the woods, the aromatic odor of a hundred different growing things greets your nostrils. A turn in the road, and a bit of open meadow, and a gust of air as warm as mid-day envelops you. So the ride goes on till the great stars quiver in the dark vault of the heavens, that seem the deeper and more mysterious from their framework of mountain-peaks. The hill-sides, fringed with trees that border the road, rise black and ghostly in the gloom, and only the tramp of the horses' hoofs on the hard ground break the intense stillness of the hour. One may not know the names of many of the mountains, but the peak of Chocorua, sharp and proud, crowns the view whenever a bluff of sufficient height is reached to overlook the landscape; and, after passing through a wood, it is always that lonely summit that rises first to the view.

It is after such a ride that the tourist strikes the valley of the Saco at Conway, where he may take the train for North Conway and the mountains. Leaving the peaks of Chocorua and Whiteface behind him over his left shoulder, Moat Mountain, with its long sweep, and the more broken outline of the Rattlesnake range, take the principal positions in the panorama, while the Ossipee Hills recede and retire toward the southern horizon. Soon the scene shows the level bank that rises about thirty feet above the intervales of the Saco, and, extending some three or four miles in length to the foot of Bartlett Mountain, reaches back two or three miles to the base of the Rattlesnake range and to Mount Kiarsarge, and forms the little plain where the township of North Conway nestles against the mountain-side. No one who has ever visited this valley can fail to remember the exquisite view when it first opened before him, and, varied slightly along the whole length of the ridge till arriving at the farther end of the village, the low hills at Bartlett shut off the chief features of the scene.

At the foot of the bank, and bathed in the morning sunshine, extends, far up the valley, a flat, velvety meadow of the freshest green, and dotted over it, in lines or little groups, rises the very ideal of elm-trees, as pure in form as a fountain or a vase. The Saco glimmers here and there in the morning light, its course nearly hidden by bands of dark-hued maples. Above these bands of trees are the purple slopes of Moat Mountain, which descends abruptly to the plain, when the steep face of the Conway ledges makes a sheer descent of from six to eight hundred feet to the valley of the Saco.

At the northern end of the valley, Moat Mountain bends down till it becomes a low ridge in what is called the "Devil's Arm-chair," and Bartlett slopes gently away to give place to a broad opening, across which, extending its entire length, lies Mount Washington and the other peaks of the White-Mountain range, each one being well separated from the other, and the outline of Mount Washington itself one of the best afforded from any position. The lower flanks of these mountains reach to the plain of the Saco, and, if one has watched this scene when the purple shades of evening gather on the mountain-sides long after the valley and the lower hills are wrapped in gloom, he may have seen the pink hues of the evening sky still lingering on those mountain-peaks till they melt from the dome-shaped summit of Washington, and, with a little quiver of the light, its huge side joins the purple mass of the valley and the hills that lie beneath it.

Every view of the mountains has its own peculiar type of expression; and each aspect on the north side is more or less bold and abrupt, and the lines of the hills, though they are fine, grand, and impressive, are not graceful. But the character of the scenery at Conway is peculiar for its loveliness. You see the curves of the hills on their long swell, rising slowly from valley to summit; and, on the northern slope, the mountain-wave appears to have broken and rushed abruptly to the plain. Such is the

The Willey Slide.

general aspect of the landscape, and one can easily picture to himself a beauty of the scenery that is almost feminine, as it appears at Conway. Not only the hills, but the village itself, and the gentle meadows of the Saco, add to the soft charm of this very Arcadia of the White Hills.

Mount Kiàrsarge, at the northern end of the Rattlesnake range, is the highest peak this side of the White Mountains, and rises in an almost perfect cone from the ridge on which Conway stands. The mountain is so near the town that the trees on its sides are distinctly seen, and partake of the greenish-purple hue of all near mountains.

GATE OF THE CRAWFORD NOTCH.

A very pleasant day may be spent at the Conway ledges, which are perhaps the finest cliffs in the whole White-Mountain region. A broken rock, six hundred feet high, is colored with the most delicate shades of buff, purple, and gray, with small birches growing out from the clefts in the fractured surface of the stone here and there, where a little earth and moisture have collected. To the rear of the lower ledge, Thompson's Falls break over a spur of Moat Mountain, where the broken rock is thrown about in the wildest confusion. The highest of the ledges rises more than nine hundred feet above the bed of the Saco. A little scramble of a hundred feet or so through herbage and over rocks brings you into a shallow cave below the cliff, whence the rocks have been split away for nearly a hundred feet high, and the wide front of the recess is almost choked with trees. This spot, a favorite resort for picnickers, is named the Cathedral, and shares with Diana's Bath the interest of the visitor as a place of rest. Diana's Bath, a little farther up the valley, is formed from a succession of water-falls that striking upon several tiers of rock, have worn wells into its substance, with perfectly smooth walls. The largest of these wells is about ten feet across, and as many deep. Looking into the clear depths of the water, one sees at the bottom small, round rocks, the cause of the excavation, which the water has used as pestles with which to scrape, and grind, and polish out these natural basins. Echo Lake, directly at the foot of Moat Mountain, has the character of numberless of these still mountain-ponds, hidden among the forests, deep and quiet.

Continuing on the mountain-road after leaving North Conway, you wind along the ridge of land that forms the town, till the valley becomes narrow and broken, and the hills abrupt. Brooks cross the road at several points, and the way winds round the lone flank of Bartlett Mountain, wooded from base to summit, passing the beautiful falls at Jackson, and Goodrich's Falls, near where the Ellis River joins the Saco. By-and-by the abrupt sides of Mount Crawford bound the road on one side, and when you have reached the little house that stands under Willey Mountain the sunbeams have already stolen far up the mountain. A bugle blown at this spot starts the echoes, repeating them back and forth heavier and louder than the first blast; one almost fancies it the music of a band of giants hidden among the trees on the mountain-slope. From the Willey House to the gate of the Notch the path becomes constantly narrower and sterner, though the common idea of the awfulness and almost horror of the passage of this portion of the journey is a somewhat erroneous one. The slope of the mountain-sides, here two thousand feet high, is very abrupt, and the narrow ravine is nearly unbroken for three or four miles, till one has passed the gate of the Notch; but, comparing this point with many others, its picturesque and romantic charm is the predominant impression. The river boils and plunges over broken rocks, and the narrow passage twists and winds, crossing the torrent at intervals over slender bridges, till, at the gate of the Notch, an opening, hardly wide enough to allow the passage of a team of horses, and the raging river, is bounded on each side by a sheer wall of rock, on the projections of which harebells and maiden's-hair are waving, and down whose steep sides leap the tiny waters of the silver cascade, whose course can be detected several hundred feet up the side of Mount Webster, sparkling in the sunlight.

Passing the gate of the Notch, you come out upon a little plateau of a few hun-

The Descent from Mount Washington.

dred acres, surrounded by hills, except at its upper and lower ends, which form the pass of the mountains, in the midst of which stands the Crawford House.

The ascent of Mount Washington—the great point of interest, of course—was formerly made by a bridle-path from this point. Most tourists, however, prefer the less difficult route afforded by the railway, which makes the ascent in three hours, or that by the carriage-road from the Glen House. To one unacquainted with mountain-scenery, the ascent by the bridle-path from the Crawford Notch affords more new sensations than can, perhaps, be gained anywhere else in this region in so few hours.

In the ascent the kind of trees changes constantly, turning from the yellow-birches, the beeches, with mossy trunks, and sugar-maples, in the valley, where are also mountain-ash trees, aspen-poplars, and striped maples, to white-pine and hemlock, white-birch and spruce, and balsam-fir, hung with a fine gray moss, much like that which drapes the trees of the Southern forests, till you reach the upland with an arctic vegetation and a sort of dwarf-fir, so intertangled with moss that you can often walk over the tops

Tuckerman's Ravine, from Hermit's Lake.

of these trees as if over thick moss. On the ground is an undergrowth of ferns, brakes, and mountain-vines, and near the summit of Mount Clinton you come upon a region of dead trees, their branches and trunks bleached and white as ghosts, until you emerge on the barren summit of the mountain.

The path is rather to the north of the top of Mount Clinton, and we wind around it over bare rocks, when the first noble mountain-prospect opens before us. In front is the conical peak of Kiarsarge, and seemingly quite near it are some small, shining lakes amid their hazy setting of mountains; behind rises Mount Willard and the group that surrounds the Notch, the clouds chasing wild shadows over their deep-blue sides. As we begin to descend to the narrow ridge which unites this mountain to the one next it, we catch a glimpse of a valley two thousand feet below, through which flows the Mount-Washington River at the base of a vast forest. On the left, at an equal depth, runs the Ammonoosuc, and you gain your first experience of mountain peril when the horses, planting their four feet close together on some rock in the narrow pathway, jump from this rough elevation three or four feet to the rocks beneath, where a slip or false leap would precipitate horse and rider down many hundreds of feet over the side of the mountain to sure destruction. The mountain on its almost perpendicular eastern slope is deeply seamed by a slide which happened during a severe storm in 1857. Passing around the side of Mount Monroe, which is little inferior to Mount Washington, one gazes into a frightful abyss, known as Bates's Gulf.

Crystal Cascade.

From Monroe is the first near view of Mount Washington, which rises in a vast cone, and shines with bare, gray stones fifteen hundred feet above, and across a wide plateau strewed with great numbers of bowlders. This elevated plain is about a mile above the sea. Patches of grass ,and hardy wild-flowers appear in the crevices of the rocks, and one comes upon small "tarns," or mountain-ponds, formed from springs or by the frequent storms that pass over these high regions. The "Lake of the Clouds," the head-waters of the Ammonoosuc, is the most beautiful of them. Having crossed the plateau, the last four or five hundred feet are best climbed on foot, for the stones are so loose, and the ascent so steep, that it is best not to trust to horse-flesh. The rocks are clean cut and glistening, as if fresh from the quarry, among which scarcely a living thing can be discovered;

Mount Washington, from top of Thompson's Falls, Pinkham Pass.

but, by-and-by, as one emerges upon the summit, the delicate Alpine plant and little white flowers appear among the rocks. From the summit you can have a view more extended and exciting than any this side of the Rocky Mountains. A sea of mountains stretches on every hand; the near peaks, bald and scarred, are clothed with forests black and purple, and sloping to valleys so remote as to be very insignificant. Beyond the near peaks, grand and solemn, the more distant mountains fall away rapidly into every tint of blue and purple, glittering with lakes, till the eye reaches the sea-line ninety miles away.

The summit of Mount Washington, from the plateau at the Crawford House, is five thousand feet high, and this plateau in its turn is fourteen or fifteen hundred feet above the sea.

Tuckerman's Ravine lies a few hundred feet down the side of the mountain, and the ridges in its rough, craggy wall form the faint, pink-gray lines that scar the summit of Mount Washington as seen at North Conway. If there is time, one can visit this ravine from the top of Mount Washington, and by a steep climb reach the summit again before night from the Snow Arch.

The ravine is an immense gully in the side of Mount Washington, the steep sides of which storms and frost are constantly changing, so that no vegetation has a chance to take root, except the little yearly plant whose seeds may be scattered here, for the next winter's storms are sure to wash away the scanty growth. Against the head of the ravine, where it abuts against the summit of Mount Washington, the lofty wall sparkles with a thousand streams that filter through its crevices or run over its summit. The Snow Arch is formed at first from the immense snow-drifts blown over the top of the mountain, which settle against this wall of the ravine in piles sometimes a hundred feet deep, and in the short summer of this great altitude scarcely have time to melt from year to year.

The tourist to the summit of Mount Washington may descend, if he chooses, by the carriage-road to the Glen House, which is approached from Glen through the Pinkham Notch, that runs nearly parallel with the Willey Notch, north and south, and is separated from it on the west by two ranges of mountains, Mount Crawford being one of the peaks; and, on the other side, it is bounded by Carter Mountain and the range of Mount Moriah. The stage from Glen follows the course of the Ellis River, which connects this narrow valley with the broad intervals where the Ellis joins the Saco, till a little plateau is reached, from which rise the whole group of the White Mountains, without any intervening peak to conceal any portion of them, from their base to the summit—a sheer ascent from the valley of more than five thousand feet.

Here, by the road-side, not very remotely set in the forest, is the Crystal Cascade, whose waters fall in an unbroken sheet from the summit to the base of the rock.

It is a wonderful view which opens before the tourist when he enters the glen, either from Gorham, by the course of the Peabody River, or, coming from Glen Station and through the wild Pinkham Notch, by the rushing Ellis, with its Glen-Ellis Falls, one of the famous cascades of the mountains. The five highest mountains of New England lie before him, dense forest clothing their lower flanks, the ravines, landslides, and windfalls clearly defined; and above all tower their desolate peaks. These little plateaus, scattered here and there—at the Crawford House, at Franconia, and at the Glen House—seem to be darker than ordinary places, for the sky is cut off many angles above the horizon on every hand, and the sun has a shorter transit across the diminished heavens, leaving a long period of twilight both at morning and evening, even during fair weather; but, when the heavy fog-banks collect on these lonely mountain-sides, and the storm-clouds muster over every peak, the sensation of solitary gloom is most impressive. In this valley lies the Emerald Pool, a sunny basin, bright and still.

Cliffs above Dismal Pool.

Leaving Gorham, and following the stage-road to the west, you soon emerge on a hill-side, leaving the Androscoggin Valley behind; and, when about a mile up this little valley, at a turn in the road, you suddenly find yourself gazing up at the steep side of Mount Madison, which rises with a clear sweep from its base, washed by the rocky Moose River, and its flanks clothed with huge forest-trees to its gray and rocky summit. Now we see one slope of the mountain, and now another, as the road winds along, till at length the twin peak of

Mount Adams, very like in form to Madison, peeps over one of the immense shoulders of Adams, and soon its sides rise to view. Mount Jefferson, in its turn, comes in sight, and the deep gullies in its sides and its rocky flanks present the same unbroken and satisfactory slopes which had made Madison at first seem quite the ideal mountain

Emerald Pool, Peabody-River Glen.

of one's imagination. From the moment this journey is begun at the hill-top in Gorham, it is most interesting, but, to be fully enjoyed, it should be taken with the afternoon light purpling the mountain-sides, and when the large, picturesque trees, twisted and bent, stand, like sentinels, profiled against the broad, soft light of the hills. Driving along, one flank after another comes into view, shutting off the previous one, filling one with an ever-new surprise at the number and variety of these mountains, which yet are always immense in their sweep and grand in curve. The mountains from this side are much more abrupt than when seen on their western declivity, and the rocky structure of their formation is more conspicuous. At the Glen, flanks and ravines cut up the sweep of the hills, but here they rise in an unbroken view to a height greater than the walls of the mountains at the Willey Notch, and far more impressive.

The cliffs above Dismal Pool, near the Crawford House and the Willey Notch, are among the loftiest and steepest to be found in the mountains. Our illustration gives a very good impression of these stupendous precipices.

From the Crawford House, on its little plateau, turning northward, the road, passing through dense woods, after a short space, enters the little valley, through which the infant stream of the Ammonoosuc issues from near the base of Mount Monroe. Nothing can be more charming than the trickle of waters by the side of these mountain-roads—"noises as of hidden brooks in the leafy month of June"—when the stage toils and creaks slowly over the rocky hills. We do not know the origin of the valleys, though they are probably volcanic, and the roads are apparently much more important than the little streams that rush along beside them, seeming like mere ornaments to the landscape; but, whatever their apparent uselessness, these mountain-torrents have carved out the natural roads through the hills, and it is by the ridges that bound them that nearly every person is made familiar with the glories and beauties of this region.

Following along the Ammonoosuc, the forest opens here and there, disclosing the White Mountains in all their beauty, until at the White-Mountain House, beyond the Ammonoosuc, the range of hills that connects the White Mountains with the Franconia range, rises before you. This stream, which is often named the wildest in New Hampshire, on account of the rapid flow of its waters, that descend more than a mile between its source and where it joins the Connecticut, is broken by many water-falls, that gleam among the trees along the road. Along the valley toward the eastward rise the White Mountains and their attendant ranges; on the south, the range of the Franconia Mountains and Mount Lafayette, towering majestically above the rest, shut in the plain; while to the north appear the mountains of Vermont. At one's feet on every side lie the valleys, and above this plain rise the mountain-peaks. Removed from the solemn gloom of the ravines, and from the exciting impressiveness of the mountain-tops, it would seem that dwellers in these elevated homes among the hills might have a healthier and serener life than anybody else.

The Franconia range, though of the same group of hills as the rest, has a character as distinct from the austere forms of the White-Mountain range as from the soft swells of the Green Mountains of Vermont, and is eminently charming and picturesque.

A little way from the Profile House the traveller finds himself beside the Echo

Profile Mountain.

Lake, surrounded by hills, with Mount Lafayette, the highest peak of any in that region, overlooking it, and the scene resembles that described by Sir Walter Scott, in his "Lady of the Lake," where he says:

> "Mountains that like giants stand,
> To sentinel enchanted land."

In a fresh, cool morning, after a good night's rest under the comfortable roof of the Profile House, you wander down to the little pebbly beach that edges the lake-shore. Green woods tangled over your head protect you from the heat of the summer sun, and before you lies this little lake, each mountain clearly reflected in its pure depths as if in a mirror. While you sit enjoying the quiet beauty of the scene, and watching one or two eagles circling about the near hills, a note from a bugle sounds from the little boat that takes passengers to the middle of the lake. Immediately the echo repeats itself against the mountain-side, and, as it jumps from point to point, almost instantly the woods seem filled with a band of musicians till the echoes fade off and off, unconsciously almost suggesting the words of Tennyson's "Bugle Song":

> "Oh, hark! Oh, hear! How thin and clear,
> And thinner, clearer, farther going;
> Oh, sweet and far from cliff and scaur
> The horns of elf-land faintly blowing!
> Blow, let us hear the purple glens replying.
> Blow, bugle; answer, echoes dying, dying, dying!"

Leaving the lake, and following the path that leads back to the Profile House, you come to the broken, scarred wall of Eagle Cliff, that rises directly in front of the hotel. Eagles build their nests here, whence its name, and there are various traditions of children and lambs being snatched away and borne up to their lofty eyries.

Nearly opposite Eagle Cliff, Profile Mountain rises abruptly from the margin of a little lake familiarly known as the "Old Man's Wash-basin," covered with forest-trees far up its side, over which, looking down the valley from its lofty position, nearly two thousand feet up the mountain, appears the wonder of this region, the "Old Stone Face," as firmly defined as if chiselled by a sculptor. The rocks of which it is formed are three blocks of granite so set together as to make an overhanging brow, a powerful, clearly-defined nose, and a chin sharp and decisive. Many of the pictures made on rocks by fissures and discolorations require an effort of the imagination to make out any meaning from the tangle of involved lines. Such are the figures on the ledge at Conway, and the Indian Chief on one of the mountains in the Notch. "Arm-chairs," "Graves," and "Seats," are always being pointed out, and give little satisfaction to eye or mind; but this view of the old man's profile is startling, and requires no description or suggestions to make it real.

Following the course of the Pemigewassett, whose source is in the "Old Man's Wash-basin," as that of its sister-stream the Ammonoosuc is in Echo Lake, with only the rise of a little mound between them to turn the waters north or south, one comes upon beautiful cascades, where the little stream rushes over its rocky bed, fashioning itself as it moves along through green moss, wet at noonday with the spray from the falling water, till you come to the Flume House, where the narrow gorge of the Pemigewassett River widens out to the long, flowing sweep of the open valley that closes no more, but sweeps down amid constantly lower hills till it reaches the sea, and the wild woods with their beauty are left behind in the mountains. Leaving the main road at the Flume House, you strike into a rough path, following it where the sound of falling water attracts you not in vain.

Here you come upon smooth, flat rocks, over which flows the pure, colorless sheet of the mountain-water. Above this rocky stairway the water dashes over a green, mossy bed, the rich hues of which are seen in the sparkling sunshine that penetrates below the flood, revealing the golden and amber tints on sand and pebbly floor. Above this mossy bed we reach a fissure in the hill, with steep, rocky sides fifty feet or more in elevation and hundreds of feet long, narrowing at its upper end till it is only ten or twelve feet wide. Stepping from one stone to another, and then threading the narrow footpath, crossing and recrossing the ravine, alternately climbing rocks and traversing rude tree-trunks thrown across for bridges, at length a little point is gained in the narrowest part of the ravine. The rocky walls are dark, and the little stream bounds along between them. Emerald mosses hang from the sharp angles of the ledge or from the tree-trunks on its side. Just above the place where you are standing, a huge bowlder is wedged, seemingly just ready to slip from its uncertain resting-place, and this is the famous Flume.

The Flume.

The Dixville Notch is a remarkable pass, in a group of hills some sixty miles to the north of the White Mountains, and, though but recently explored, the region abounds in scenery of the finest kind. Even the White Mountains, it is said, do not surpass it in sublimity and desolate and wild grandeur.

Following the track of the Grand Trunk Railroad by its course along the Androscoggin, at length the train turns into the more cheerful valley of the Connecticut River till you come to North Stratford, where you take another train to Colebrook, a flourishing village on the New-Hampshire side of the Connecticut, from which you can easily reach the Dixville range of hills, some ten miles from the village. The route is through the best farming region of New Hampshire, and a person would never imagine there could be mountain-scenery of any degree of impressiveness near at hand. Suddenly the heavy walls of the Dixville Mountains show themselves, rising like thunder-clouds above the tree-tops of the forest. While you are admiring the gloomy sides of these hills, covered by dark woods, a turn in the road brings you in front of the savage opening of the Notch at its west end—a region of vast and mysterious desolation. The pass is narrower than either one of the great Notches of the White Hills, and the scenery is much bolder and sublimer.

Nothing can give an adequate impression of these bare and decaying cliffs, which shoot out into fantastic and angular projections on every side. The side-walls of this narrow ravine—for it can scarcely be called a pass—are strewed with *débris*. The only plant that appears to have maintained itself is the raspberry-vine. The great distinctive feature of this Notch is barrenness; and very great, therefore, is the transition of feeling from desolation and gloom, when you ride out from its slaty teeth into a most lovely plain called the Clear-Spring Meadows, embosomed in mountains, wooded luxuriantly from base to crown. It is in this Notch that you come upon one of the most characteristic formations of this region—Column Rock.

Column Rock, Dixville Notch.

HARPER'S FERRY.

WITH ILLUSTRATIONS BY GRANVILLE PERKINS.

AFTER a short but heavy rain the air was fresh and bracing on the October day when we started for Harper's Ferry. There is no season so glorious in any country as an American autumn, and it is, above all, the time to see the mountains to the best advantage. The atmosphere, bright, clear, and bracing, acts upon the frame like champagne; the forests put on their livery of splendid dyes, and gold and crimson and sober brown are massed on all the hills, or set in a dark background of pine and hemlock. For this reason, seated in the cars of the Baltimore and Ohio Railroad, and with the arriving and departing trains making discordant noises in our ears, we congratulated ourselves on the beauty of the day. Patiently waiting, we watch the passengers upon the platform, uniting and dispersing, aggregating in little groups, only to dissolve and form again. We soon, however, leave these scenes behind us, and are skirting the brick-fields lying on the western suburbs of Baltimore, and by hillocks covered with low and stunted shrubbery of cedar and oak, past the Relay and on westward.

The road winds along the Patapsco, and only leaves that picturesque, artist-haunted river for short distances until it strikes the Potomac at Point of Rocks, and follows that river to its junction with the Shenandoah at Harper's Ferry.

Stretching far away to the right, dimly outlined in their characteristic smoky blue, appears the range of mountains, nestled in a gorge of which the gate-way to the wild and magnificent scenes beyond lies our objective point, Harper's Ferry. As we approach, the smoky whiteness of the enveloping haze is dissipated, and gives place to a more pronounced blue; the billowy hills roll more sharply clear to the eye; the irregular lines of the foliage stand out distinct, and here and there shaggy and wind-dishevelled pines cut the sky-line upon the summit-ridge.

While we have been watching the scenery we have passed over the costly and graceful bridge built by the Baltimore and Ohio Railroad Company, which spans the Potomac, on five substantial stone piers, just at its junction with the Shenandoah. When fairly on the platform and the train has left the view unobstructed, we see rising, sheer and inaccessible, directly before us the rocky sides of the Maryland Heights. Upon their laminated surface the curious eye ranges among the overhanging masses of projecting stone, to this point and to that, in search of the well-known Profile Rock. It catches, jutting out from a crevice in the wall-like side, a mass of shrubbery—the hair; a little lower down a patch of stunted bushes—the epaulets; glancing then a little to the left, the imagination quickly forms the features. Curiosity being thus satisfied, we turn to the foot of these frowning cliffs, where are seen the slow, languid canal-boats, almost imperceptibly in motion, crawling into the lock at the bridge, and vanishing by inches around the mountain. On the left is Bolivar Heights, and below them is the ruined armory at the ferry—a long row of walls, windowless, roofless, blackened and desolate. Far up, the mountains recede and become low hills, and close in upon the rocky-islanded Potomac, which comes boldly sweeping down past the town, meets the Shenandoah, and unites with it on its way to the ocean.

The town of Harper's Ferry is built at the foot of the narrow tongue of land that thrusts itself out like a cutwater, separating the Potomac and the Shenandoah, known as Bolivar Heights. It lies in Jefferson County, West Virginia. Just across the Potomac are the Maryland Heights, in Washington County, Maryland, and over the Shenandoah, beyond Loudon Heights, lies Virginia proper. The principal street runs parallel with the Shenandoah, with a side-street ascending the hill to the right, perpendicular to which numerous stairs, cut in the solid rock, lead up still steeper ascents.

HARPER'S FERRY

Quaint and old-fashioned in its best estate, two causes have contributed to make the town still more sleepy and dilapidated than is its normal condition. The recent war stunned it. Then came the disastrous flood of October, 1870, in the Shenandoah. Pass where you will, there are evidences of the desolation left behind by these two occurrences. And the people of the Ferry have very naturally lost heart. They talk about the old days when the Shenandoah ran the mills and the government rifle-works on its banks; when the armory was in busy activity, and a regiment of lusty workmen hammered and rolled and moulded the arms which it was then thought would never be used

View from Jefferson Rock.

except against foreign foes; when many millions of solid dollars—a golden Pactolus—poured into the arms of the thriving little village from the national treasury. The inhabitants now talk of these days of prosperity with regret, with even a mild kind of hope for better things in the future, but with no buoyancy of spirit.

The place takes its name from Robert Harper, a native of Oxford, England, who emigrated to this country in 1723. Harper settled at Philadelphia, and seems to have been a man of much ingenuity. At this time the infant colonies were offering high prices to skilled workmen, and Harper, being by profession an architect, was frequently required to travel to distant parts of the country. It was when on his way to erect a

Loudon Mountain and the Shenandoah.

meeting-house for the members of the Society of Friends, whose settlement, near where Winchester now stands, in the rich Valley of Virginia, was even then, in 1747, flourishing and increasing, that Harper, as a short route to his destination, first saw this pass. He was so attracted by it that he bought a tract of land here, which was subsequently confirmed to him by Lord Fairfax. In time, as the country became more settled, and the passage through the barrier of the Blue Ridge better known, he established a ferry here. The house erected by Harper, on what is now High Street, is still to be seen. In outward appearance it is one of the newest in the town; and, if it were not for the semicircular, latticed window in the side-wall, which betrays its antiquity, it might, like a well-preserved old beau, hold its own with its younger contemporaries, and deny its years. In 1794 the prosperity of Harper's Ferry, for half a century and more, became assured. It was at that date, and during the administration of General Washington, that the town, on account of the many advantages it offered, but more especially for its unrivalled water-power, was chosen as the site of the national armory. Land was purchased along the Potomac and Shenandoah. Subsequently Bolivar and Loudon Heights were acquired, and the buildings of the armory and the dwellings of the operatives gradually formed in themselves a small but thriving settlement. So the Ferry prospered until the night of Sunday, October 16, 1859. From that night the town was doomed. Stealthily, at ten o'clock, a band of twenty men crossed the railroad-bridge—then a clumsy, covered structure—over the Potomac. They came from the Maryland side. Quietly the watchmen were captured and the armory seized. They at once proceeded to establish and fortify themselves against an attack. They then threw out pickets, and arrested all persons who ventured abroad. A colored man, who incautiously approached too near the guarded railroad-bridge, was shot down and died soon after. At the dawn of the next day, as the sun struggled through the rising

Pleasant Valley.

THE POTOMAC FROM MARYLAND HEIGHTS

mists of the river, the little town was all excitement. All through the long morning a scattered fusillade was kept up between the armory and the neighboring houses; and the sharp crack of the rifle from the overshadowing hills was quickly returned from some one of the barred windows below. Gradually the toils tightened around the party desperately at bay in the armory. Fighting now with the energy of despair, Brown—for, of course, it is of the celebrated "Brown raid" that we are speaking—retreated to the engine-house, since removed to Chicago. The conflict was stayed, but the hours wore away in unceasing watchfulness. Colonel Robert E. Lee arrived early on Tuesday morning with a force of marines from Washington. The strong doors of the engine-house were soon battered in, and, with the loss of one man killed, the invaders were captured. Brown was executed at Charlestown soon after. So ended the raid. During the civil war, Harper's Ferry was alternately in the possession of the Northern and Southern forces, and suffered from both. When the ordinance of secession was passed, the Ferry was the station of a company of United States regulars. Soon came rumors that the Virginia militia were marching to capture the Ferry. So threatening was the aspect of affairs that a retreat was ordered, and the armory fired. The smoke of the burning buildings curled up black and ominous through the still air, and loud detonations shook the ground as the explosive material stored within took fire. Much of the armory was saved, but the arsenal was completely destroyed, with about fifteen thousand stand of arms. Then the Southern forces came in, but soon they abandoned the Ferry as of no strategic importance. They completed the destruction already begun.

The last important scene that closes this eventful history was in September, 1862, when Colonel Miles was caught in an untenable position on Bolivar. Here the record of the civil war, as regards the Ferry, rightly stops. After Antietam, McClellan concentrated his army here. The whole peninsula formed by the Potomac and the Shenandoah, from Smallwood's Ridge to the junction of the rivers, as well as the surrounding heights, was dotted with tents, and at night was aglow with thousands of watch-fires. A hum of voices, like that of an immense city or the hoarse murmur of the ocean, rose from the valleys on either side, and filled the air with a confusion of sounds.

We are now on our way to Jefferson's Rock. Perched high up to the right are the bare walls of the Episcopal and Methodist churches, whose joyous bells, in other times, aroused the echoes of the mountains on the calm Sabbath, while the worshippers wound their slow way up the steep hill, and perhaps paused at the church-door to take a last look at the glorious scene below, the wooded heights, the shining river, the sleeping town, and to thank God that their little home, secure among its sheltering peaks, was so peaceful and unthreatened. We pass by the side of the Episcopal church, which, in its time, must have been an imposing structure. We scramble over the rubbish and look in, and find all foulness and pollution. The four bare walls are open to the sky; the windows are seamed and broken; the place where the altar stood is vacant, and the marks of the gallery-stairs still wind their way upward into vacancy. Every trace of wood-work has vanished. It was not burned, but torn away gradually in the mere wanton riot of desecration and destruction.

A few steps farther bring us to Jefferson Rock, a remarkable stratified formation that

Maryland Heights.

rises abruptly from the street below. It is the pride of the town, and, among the townspeople, is almost a name "to conjure by." Upon it, according to one account, Jefferson inscribed his name; other authorities say that it was here that he wrote his "Notes on Virginia," "in answer to a foreigner of distinction." The first is, of course, the fact, and the other the accretion that time has added. Here we have the best view attainable of the mountains from their base, and of the meeting of the waters in this Vale of Avoca. Beyond the town loom up the Maryland Heights; to the left, Loudon frowns, crowned with its wealth of shaggy foliage, its sides seamed with innumerable fissures and dry ravines made in the crumbling rock by the winter-torrents. At the foot of these ravines the loose earth and gravel washed down has been piled in high and conical heaps. In the gap between these two mountains the Shenandoah, which comes down with many a curve and deflection, skirting the Blue Ridge from Bath County, and the Potomac, which flows south from the table-lands of the Alleghanies and divides the water-shed of the Ohio River and the Chesapeake Bay, unite. The spectator will have no wish to indulge in dry speculations in the presence of the scene that meets his eyes as he turns at the Rock and follows the broad river through the rugged gap, while on either side stand, in silent guard, the Sentinel Peaks. There is no grandeur in the scene—none claimed for it. Life, brightness, and quiet beauty, distinguish it. Over the Shenandoah the ferry-boat turns and twists among the bowlders, and seeks the deeper pools and the slow eddies that give it a passage. The fair river, viewed so near, is spread out between wide, enclosing banks, and catches the silvery glitter of the sunlight and the dark shadows of the hills on its ample bosom, dotted with the black, obtruding forms of rocks, around which the slow current chafes gently in swirls and circling ripples. Around the Maryland Heights run both the railroad and the canal, and the long trains and the unwieldy, cabined, awning-sheltered boats turn the foot of the ridge at intervals, and follow the sinuous river, ever trending southward.

Before visiting Maryland Heights and the superb panoramic view that there sweeps around almost from horizon to horizon, a few moments will be well spent in seeing the less striking scenery of the Heights of Bolivar. The road around Bolivar is the segment of a circle, the first part of which lies along the Shenandoah and the unused Slackwater Canal, bordered by majestic cotton-woods, their wide, gaunt, flecked branches spreading weirdly over the dismantled Government Rifle-Works, the empty, crumbling canal, and the havoc that war and flood have made on every side. Midway of the ascent of the hill, the scenery, looking back toward the Ferry, is soft and beautiful, water and mountain toned by distance, and, in the foreground, the long, straggling street of the ancient town. As we reach the top, we pass the remains of the Federal fortifications and the deep, bush-covered valley where the balloon was kept secure from stray shells. Nearly three hundred houses stood upon the western slope of these heights, and now hardly a trace of them remains. From here we get a nearer and less elevated view of Loudon and North Mountains over a rich and well-tilled farming country. We return through the neat village of Bolivar, created by the Armory. Its inhabitants, since its abandonment by the Government, are loath to leave their homes, and find livelihoods on canal or railroad wandering.

With a sharp deflection to the left, we pass through Harper's Ferry and over the

The Potomac above Harper's Ferry.

sounding plank-roadway of the railroad-bridge, creaking metallically with all its interwoven iron net-work. Our road is a narrow one, leading along the canal and past the old ferry-house, brooding under the beetling cliffs that overshadow it. As we looked at the placid, sluggish waters by whose walled margin we rode, there was in them but little suggestive of danger or of the tragic. But, as we heard afterward, at a spot that was then pointed out to us, was drowned the young son of the good old lady at whose house we were to stop.

Turning to the right, the ponies tug and strain up the steep roadway that ascends the mountain. Under the overhanging boughs of the chestnut and the oak we go; over tiny rivulets, and with a final pull, heavier than any yet, the panting horses come to a willing halt. From this point the remainder of the ascent must be made on foot.

Maryland Road.

The landscape below, seen from the north side of the Heights, tempts us to linger a moment, and then, plunging into the woods, we begin the ascent up a dry ravine that leads directly to the summit. We find out before long why so many persons are content with the tame views from the Ferry itself. We have been over other mountains, but the steady, knee-breaking climb up the nearly perpendicular shoulder of these heights is the hardest piece of mountaineering we ever accomplished. Heated, in spite of the cool breeze that is blowing, and tired, we reach at last the ultimate ridge, where the magnificence of the view burst upon us as we touched the crowning point. It is beautiful in its undulating, wooded slopes, its cultivated fields. It is grand in its mountains, huge, and black, and stately, in the distance, fading and melting in the haze, with solitary peaks jutting boldly out, breaking the ranges as far as the eye can follow. Through the valley between flows the Potomac, curving to the right, then deflecting to the left, and, with a long stretch by the base of intruding hills, lost to sight, only to reappear, for the last time, a gleaming mass in the brown, blended landscape. Loudon Heights lie on the other side of the river, and beyond is the rich Quaker settlement of Loudon County, that blessed spot, where the land drops fatness, and poverty is said to be unknown. We look down upon the broken outlines of the Short Hills, half concealed from view, in which lies Lovettsville Valley, and, on the other side, the Valley of the Shenandoah. At our feet are the fertile farms, the tree-embosomed houses, the symmetrical orchards, and the brown, harvested fields of Pleasant Valley. We are at an elevation of thirteen hundred feet. At the time of the war the whole apex was bared of its trees, and the old height lifted its head, a very monk among mountains, with a shaven crown and a narrow belt of timber midway of the summit. A dense growth of shrubbery now protects this space, save where the stone foundations and the scattered rocks that composed the walls show how the soldiers who encamped here endeavored to shelter themselves.

The broad rampart of the Old Stone Fort now forms an excellent post of observation. From it the view is unobstructed, except where the Blue Ridge, throwing out spurs here and there, mountain linked to mountain in endless variety of height and shape, rises and divides valley from valley. This Blue Ridge has another peculiarity besides the soft, enveloping, distinctive color from which it takes its name. It is not a continued line, but a series of mountain-ranges pocketed into each other. First one mountain will take up the elevation for ten or twenty miles, and then, in its turn, some detached height will continue the broken chain, only to give place to a third, and this to others, before the Susquehanna is reached. All along its course it forms the dividing-line of States and counties. From these heights we look, for instance, into seven counties—Jefferson, Loudon, Frederick, Fauquier, and Clarke, in Virginia, and Frederick and Washington Counties, in Maryland; and into three States—Virginia, West Virginia, and Maryland. Through all the scene the eye traces the Potomac, entering at the north, and flowing southeast; the white houses of the scattered towns, Martinsburg, Shepherdstown, Knoxville, Berlin, Hagerstown, and, on a clear day, following the road that winds over the hills—a yellow, wavy ribbon, now seen, now lost—Charlestown and Winchester. The horizon is bounded, to the north and west, by the Loudon and North Mountains, enveloped in a haze of smoky whiteness; and cultivated fields, checkered with square blocks of forest left for timber, lie as if in the hollow trough of two immense billows, whose crests are these swelling undulations of the land. The Potomac, coursing through sunlight or shade, adds beauty, and life, and changeableness. There would be a sombreness in the view that would detract much from its attractiveness without this beautiful river.

Reluctantly we leave our breezy station, and descend by the longer way around the shoulder of the ridge overlooking the Ferry. A few moments' rest as we join our horses; then down the steep and tortuous road at a rattling pace; along the still waters of the canal; by the Potomac; over the bridge, with thoughts full of the beauties of mountain and river, and a longing to see them all once again.

The evening falls among the mountains, calm and peaceful. The huge shadows of

OLD BRIDGE AND MILL, ANTIETAM—ANTIETAM ROLLING-MILL.—BURNSIDE BRIDGE.

the dusky heights overcast the town and river. The night darkens, and the Ferry puts on another aspect, both novel and singularly beautiful. The mountains, dimly seen, close in upon the murmuring river and the quiet town. They rise, still sombre and black, unrelieved by a single gleam of light, and shut out the sky, except immediately overhead, or where the long reach of the river has made a break in their continuity, which the eye follows, and down which the twinkling stars, reflected in the water, glitter brightly. Along the foot of the Maryland Heights, bright, golden-rayed lights creep in slow motion. They are those that show the path of the innumerable boats that convey the freight of the Chesapeake and Ohio Canal—that old, expensive, and until lately

Mill on Antietam Road.

unremunerative work of internal improvement, begun under the auspices of Washington, and laboriously pushed to completion. Occasionally a skiff crosses the Potomac, its lamp casting a long, flashing, illuminating path before it. Over the bridge, and where the perpendicular, barred, and veined rocks of the Maryland Heights come down to the river, the red signals that denote the coming of a train suddenly appear, and presently, with a rumble and jar across the bridge, the loaded cars slacken speed, stop a moment, and take up their usual hurrying, anxious, noisy crowd of passengers; the shrill whistle awakens the answering echoes from the surrounding hills, and the train carries its burden westward, its long array of shining windows flashing on the river and growing dimmer and dimmer, until, all confused and blended, they disappear beyond the rounded western hills. Again the quiet is only broken by the ceaseless ripple of the Potomac, as it frets and chafes over its obstructions, and by the weirdly-musical horns of the boatmen as they play fantastic tunes, as a warning of their approach, to the keeper of the lock.

Wandering off from the Ferry by the banks of the river, by mountain-streams, often falling in graceful cascades, or pursuing their course along the indented base of the Blue Ridge, many forest-roads present little "bits" of striking beauty dear to the eye of the artist. The road to Antietam and the battle-field of Sharpsburg is especially rich in these cabinet-pictures set in Nature's framework.

The drive is along the mountain-side from Pleasant Valley. It runs for part of the way under overhanging rocks and above deep-wooded ravines, into which foaming cascades leap, sounding in their far recesses. All along the elevated road beautiful views of mountain and valley open, ever-varying.

After the mountains are left, the Antietam gives a different scenery. Old mills border the sleepy stream—called, in the speech of the country, a "creek." Quaint stone bridges span it, and near its juncture with the Potomac stands the rambling, uneven range of buildings which form the Antietam Rolling-Mills.

On the road leading from Pleasant Valley and that from Boonsborough came the army of McClellan to the battle of Antietam, or Sharpsburg. These two roads are the only ones that cross the Antietam on stone bridges. The Burnside Bridge is on the Pleasant-Valley road, and here some of the most desperate fighting of that battle occurred. Burnside was ordered to attack and carry this bridge at all hazards. From eight in the morning until mid-day the bridge was alternately in the hands of each of the opposing forces. Finally, at four in the afternoon, the bridge was carried by the bayonet, but it was then too late. After the terrible fight that had continued throughout the day along the whole line, McClellan was too weak to reënforce Burnside. During the night Lee retreated by the Shepherdstown Ford into Virginia.

Harper's Ferry, long before the war brought it conspicuously to the attention of the world, had derived an extensive fame from Jefferson's description of it. This description the visitor of to-day is apt to believe exaggerated. But Jefferson's account was written before we were familiar with all the natural wonders of our land, and hence, while its beauties are very great, it is scarcely "one of the most stupendous scenes in Nature;" nor are we apt to believe a view of it "worth a voyage across the Atlantic."

MACKINAC.

WITH ILLUSTRATIONS BY J. DOUGLAS WOODWARD.

Arched Rock.

THE history of Mackinac (Mackinac, or Mackinaw, is an abbreviation of the full title of Michilimackinac, which, according to Lippincott's "Gazetteer," should be pronounced *Mish-il-e-mak'e-naw*) may be divided into four periods — the explorer's, the military, the fur-trading, and the tourist's. The first period embraces the early voyages of

Father Marquette; his college for the education of Indian youths, established on the straits in 1671; the death of the explorer, and the remarkable funeral procession of canoes, which, two years afterward, brought back his body, from its first burial-place on Lake Michigan, to the little mission on the Straits of Mackinac, which in life he loved so well. Here, in 1677, his grave was made by his Indian converts; its exact site was lost during the warfare that followed, but it was in the neighborhood of the little church whose foundation remains still visible, and here it is proposed to erect a monument to his memory.

In 1679 the daring explorer, Robert Cavalier de la Salle, sailed through the straits on his way to the Mississippi, in a vessel of sixty tons, called the Griffin, built by him-

View of Fort and Town.

self, on Lake Erie, during the previous spring. He stopped at old Mackinac, on the main-land; and Hennepin, the historian of the expedition, describes the astonishment of the Indians on seeing the Griffin, the first vessel that passed through the beautiful straits.

In 1688 a French officer, Baron la Houtan, visited the straits, and in his journal makes the first mention of the fur-trade: "The *courriers de bois* have a settlement here, this being the depot for the goods obtained from the south and west savages, for they cannot avoid passing this way when they go to the seats of the Illinese and the Oumamis, and to the river of Mississippi."

Arched Rock by Moonlight.

In 1695 the military period begins. At that date M. de la Motte Cadillac, who founded the city of Detroit, established a fort on the straits. Then came contests and skirmishes, not unmingled with massacres, and finally the post of Mackinac, with all the French strongholds on the lakes, was surrendered to the English, in September, 1761.

In 1763 began the conspiracy of Pontiac, wonderful for the sagacity with which it was planned and the vigor with which it was executed. Pontiac, the most remarkable Indian of all the lake-tribes, lived on Pêché Island, near Lake St. Clair. He was a firm friend of the French, and, to aid their cause, he arranged a simultaneous attack upon all the English forts in the lake-country, nine out of twelve being taken by surprise and destroyed, and among them the little post on the Straits of Mackinac.

During the War for Independence the fort was established in its present site on Mackinac Island; and the stars and stripes, superseding the cross of St. George and the lilies of the Bourbons, waved for a time peacefully over the heights; but the War of 1812 began, and the small American garrison was surprised and captured by the British, who, having landed at the point still known as the "British Landing," marched across the island to the gate of the fort and forced a surrender. After the victory of Commodore Perry, on Lake Erie, in 1813, it was determined to recapture Fort Mackinac from the British, and a little fleet was sent from Detroit for that purpose. After wandering in the persistent fogs of Lake Huron, the vessels reached the straits, and a brisk engagement began in the channel, between Round Island and Mackinac. At length the American commander decided to try a land attack, and forces were sent on shore. They landed at the "British Landing," and had begun to cross the island when the British and Indians met them, and a desperate battle ensued. The enemy had the advantage of position and numbers, and, aided by their innumerable Indian allies, they succeeded in defeating the gallant little band, who retreated to the "Landing," leaving a number killed on the field. The American fleet cruised around the island for some time, but "the stars in their courses fought against Sisera."

As far back as 1671, Marquette had noticed and described the currents of air that disturb the navigation of the straits, in the following quaint terms: "The winds: for this is the central point between the three great lakes which surround it, and which seem incessantly tossing ball at each other. For no sooner has the wind ceased blowing from Lake Michigan, than Lake Huron hurls back the gale it has received; and Lake Superior, in its turn, sends forth its blasts from another quarter; and thus the game is constantly played from one to the other." The clumsy vessels could do nothing against the winds and waves; and not until the conclusion of peace, in 1814, was the American flag again hoisted over the Gibraltar of the lakes.

Points on the Straits of Mackinac began to be stations for the fur-trade as early as 1688, but the constant warfare of the military period interfered with the business. In 1809 John Jacob Astor bought out the existing associations, and organized the American Fur Company, with a capital of two millions. For forty years this company monopolized the fur-trade, and Mackinac was the gayest and busiest post in the chain—the great central mart. Here were the supply-stores for the outgoing and incoming *voyageurs*, and the warehouses for the goods brought from New York, as well as for the furs from the interior. From here started the *bateaux* on their long journey to the Northwest, and here, once or twice a year, came the returned *voyageurs*, spending their gains in a day, with the gay prodigality of their race, laughing, singing, and dancing with the pretty half-breed girls, and then away into the wilderness again. These were

Chimney Rock.

Mackinac's palmy days; her two little streets were crowded with people, and her warehouses filled with merchandise. All the traffic of the company centred here. Again the scene changes, and the present brings us to the last period—that of the tourist. Magnificent hotels with all the attractions of a modern summer resort are here. The fur-

Fairy Arch.

traders with their lively entertainments have given way to the dancing in the ball-room and the flirtation on the veranda.

The natural scenery of Mackinac is charming. The geologist finds mysteries in the masses of calcareous rock dipping at unexpected angles; the antiquarian feasts his eyes on the remains of an earlier civilization; the invalid sits on the cliff's edge, in the vivid sunshine, and breathes in the buoyant air with delight, or rides slowly over the roads, with the spicery of cedars and juniper alternating with the fresh forest-odors of young maples and beeches. The haunted birches abound, and on the crags grow the weird larches, beckoning with their long fingers—the most human tree of all. Bluebells, on their hair-like stems, swing from the rocks, fading at a touch, and in the deep woods are the Indian pipes, but the ordinary wild-flowers are not to be found. Over toward the British Landing stand the Gothic spires of the blue-green spruces, and now and then an Indian trail crosses the road, worn deep by the feet of the red-men, when the Fairy Island was their favorite and sacred resort.

The Arch Rock, one of the curiosities of Mackinac, is a natural bridge, one hun-

Sugar-Loaf Rock—(East Side).

Sugar-Loaf Rock—(West Side).

dred and forty-five feet high by less than three feet wide, spanning the chasm with airy grace. This arch has been excavated by the action of the weather on a projecting angle of the limestone cliff. The beds forming the summit of the arch are cut off from direct connection with the main rock by a narrow gorge of no great depth. The portion supporting the arch on the north side, and the curve of the arch itself, are comparatively fragile, and cannot long resist the action of rains and frosts, which in this latitude, and on a rock thus constituted, produce great ravages every season. The arch is peculiarly beautiful when silvered with the light of the moon, and hence on moonlight nights strangers on the island always visit it.

Fairy Arch is of similar formation to Arch Rock, and lifts from the sands with a grace and beauty that justify the name bestowed upon it.

The Sugar-Loaf is a conical rock, one hundred and thirty-four feet high, standing alone in hoary majesty in the midst of a grassy plain.

The Lovers' Leap, on the western shore, is two hundred feet high, rising from the lake like a rocky column, and separated from the adjoining bank by a deep chasm. The legend is of an Indian squaw, who, standing on the rock, waiting and watching for the

Lovers' Leap.

return of her lover from battle, saw the warriors bringing his dead body to the island, and in her grief threw herself into the lake.

The cliff called "Robinson's Folly" has its legend also. This time it was a young officer who went over; indeed, there may have been half a dozen of them, for the Folly was a summer-house where cigars and wine helped to pass away the long summer days, and when at last the rock crumbled and carried them over, Robinson's folly was complete, and is still remembered, although it was finished more than a hundred years ago.

Old Fort Holmes, on the highest point of the island, was built by the British in

1812. It was then named Fort George, but, after the Americans took possession of Mackinac, it was renamed after the gallant Major Holmes, who was killed in the battle on Dousman's farm the preceding year. The ruins are still to be seen, and from the summit the view of the straits is superb.

The present Fort Mackinac was built by the British over a century ago. It stands on the cliff overlooking the town, and its stone-walls and block-houses present a bold

"Robinson's Folly."

front to the traveller wearied with the peaceful, level shores of the fresh-water seas. This ancient little fort has a long list of honored names among its records—veteran names of the War of 1812, well-known names of the Mexican contest, and loved, lamented names of the War for the Union. It has always been a favorite station among the Western posts, and many soldiers have looked back with loving regret as the boat carried them away from the beautiful island.

In 1823 a Protestant mission-school for Indian children was built upon the slope at the eastern end of Mackinac village. This was one of the most flourishing of the Indian schools in the United States, containing Indian boys and girls gathered from the lake-country as far west as the Red River of the North. The beautiful island was to be evangelized, Indian children were to be Christianized, educated, and sent back to their homes. The *voyageurs* and traders were to be gathered into the fold, and their half-forgotten religion revived and trained into a purer and more vigorous growth. The mission was continued for fifteen years, but with the extinction of the fur-trade, the removal of the agency, and the diminution of the Indians its opportunities for good waned.

The island of Mackinac was a sacred spot to the Indians of the lakes. They believed it to be the home of the giant fairies, and never passed its shores without stopping to offer tribute to the powerful genii who guarded the straits. Even now there is a vague belief among the remnants of the tribes that these mystic beings still reside under the island, and sometimes sally forth by night from the hill below the fort.

It is not often that we can obtain a specimen of the poetry of the Indian race before intercourse with the white man had corrupted its simplicity. Some years ago an Indian chieftain left his Mackinac home to visit some of his tribe elsewhere, and as he sat upon the deck of the steamer in the twilight and watched the outlines of the island growing faint in the distance, the old man's heart broke forth in the following apostrophe:

"Michilimackinac, isle of the clear, deep-water lake! how soothing it is, from amidst the smoke of my *opawgun*, to trace thy blue outlines in the distance, and to call from memory the traditions and legends of thy sacred character! How holy wast thou in the eyes of our Indian seers! How pleasant to think of the time when our fathers could see the stillness which the great Manitou shed on thy waters, and hear at evening the sound of the giant fairies, as with rapid step and giddy whirl they danced upon thy limestone battlements! Nothing then disturbed them save the chippering of birds and the rustling of the silver-barked birch. Michilimackinac, isle of the deep lake, farewell!"

Indian Hut.

SCENES IN FLORIDA.

WITH ILLUSTRATIONS BY HARRY FENN.

Fort of San Marco, St. Augustine.

THE quaint little city of St. Augustine, Florida, the oldest European settlement in the United States, is situated on the Atlantic coast, in a narrow peninsula formed by the Sebastian and Matanzas Rivers, on the west side of a harbor which is separated from the ocean by the low and narrow island of Anastasia. It lies about forty miles south of the mouth of the great river St. John's, and about one hundred and sixty miles south from Savannah, in Georgia.

St. Augustine was founded by the Spaniards under Menendez in 1565, and during its more than three centuries of existence has suffered much at the hands of foreign enemies. In 1567 its inhabitants were massacred by the French, but Menendez soon returned with new settlers to fill their places. Sir Francis Drake, in 1586, plundered and then burned the town. A party of English buccaneers appeared in 1665, and St. Augustine succumbed. In 1702 an expedition from South Carolina captured the place, but, unable to take the castle, burned the town and then retreated. Unsuccessful expeditions from the Carolinas and Georgia, led by the English, appeared before St. Augustine in 1702, 1740, and again in 1743, but San Marco was strong and saved the place from capture. Finally, in 1763, Florida was ceded to the English for Havana, but twenty years later it again passed into the hands of the Spaniards, this time in exchange for the Bahama Islands.

In 1821 Florida passed by treaty from the dominion of Spain to that of the United States, and since then there is little in the history of St. Augustine that demands particular notice.

The most conspicuous feature in the town is the old fort of San Marco, which is built of coquina, a unique conglomerate of fine shells and sand, found in large quantities on Anastasia Island, at the entrance of the harbor, and mined with great ease, though it becomes hard by exposure to the air. It is quarried in large blocks, and forms a wall well calculated to resist cannon-shot, because it does not splinter when struck.

St. Augustine Cathedral.

The Convent-Gate.

The fort stands on the sea-front at one end of the town. It was a hundred years in building, and was completed in 1756, as is attested by the following inscription, which may still be seen over the gateway, together with the arms of Spain, handsomely carved in stone: "Don Fernando being King of Spain, and the Field-Marshal Don Alonzo Fernando Herida being governor and captain-general of this place, St. Augustine of Florida and its provinces, this fort was finished in the year 1756. The works were directed by the Captain-Engineer Don Pedro de Brazos y Gareny."

While owned by the British, this was said to be the handsomest fort in the king's dominions. Its castellated battlements; its formidable bastions, with their frowning guns; its lofty and imposing sally-port, surrounded by the royal Spanish arms; its portcullis, moat, draw-bridge; its circular and ornate sentry-boxes at each principal parapet-angle; its commanding lookout tower; and its stained and moss-grown massive walls—

A Street in St. Augustine.

impress the external observer as a relic of the distant past: while a ramble through its heavy casemates—its crumbling Romish chapel, with elaborate portico and inner altar and holy-water niches; its dark passages, gloomy vaults, and more recently-discovered dungeons—brings you to ready credence of its many traditions of inquisitorial tortures; of decaying skeletons, found in the latest-opened chambers, chained to the rusty ring-bolts, and of alleged subterranean passages to the neighboring convent. Besides the

SCENE IN ST. AUGUSTINE.—THE DATE PALM.

general view of the fort, at the head of our article, we give an illustration of the quaint old watch-tower, overlooking the sea, and a glimpse of the interior, showing a stairway crumbled away out of almost all resemblance to its original form, and beneath an elliptical arch the entrance to the dungeons we have referred to. Here only a few years since, in a cavity revealed by the sinking of the parapet above, were found two skeletons hermetically walled in. The traveller curious in these old fortifications will be disposed to visit the ruins of a fort about twenty miles south of St. Augustine, on Matan-

The City Gate.

zas Inlet. Of the history of this structure nothing is known. It is different in form of construction from St. Marco, but was probably erected about the same time.

Several other buildings in the town are worthy of notice for their quaintness or antiquity. The cathedral is unique, with its belfry in the form of a section of a bell-shaped pyramid, its chime of four bells in separate niches, and its clock, together forming a cross. The oldest of these bells is marked 1682. The Catholic convent is a tasteful building of the ancient coquina, and is a suggestive relic of the days of papal rule. The United States barracks, remodelled and improved, are said to have been built as a Franciscan monastery. The old government-house, or palace, is now in use as the post-office and United States court-rooms. At its rear is a well-preserved relic of what seems to have been a fortification to protect the town from an over-the-river or inland attack. An older house than this, formerly occupied by the attorney-general, was pulled down a few years ago. Its ruins are still a curiosity, and are called (though incorrectly) the governor's house.

The "Plaza de la Constitucion" is a fine public square in the centre of the town, on which stand the ancient markets, and which is faced by the cathedral, the old palace, the convent, a modern Episcopal church, and other fine structures. In the centre of the plaza stands a monument erected in honor of the Spanish Liberal Constitution.

Watch-Tower, San Marco.

The old Huguenot burying-ground is a spot of much interest; so is the military burying-ground, where rest the remains of those who fell near here during the prolonged Seminole War. Under three pyramids of coquina, stuccoed and whitened, are the ashes of Major Dade and one hundred and seven men of his command, who were massacred by Osceola and his band. A fine sea-wall of nearly a mile in length, built of coquina, with a coping of granite, protects the entire ocean-front of the city, and furnishes a delightful promenade of a moonlight evening. In full view of this is the light-house on Anastasia Island, built in 1872–'73, and surmounted with a fine revolving lantern.

The appearance of St. Augustine to the visitor from other parts of the country is as quaint and peculiar as its history is bloody and varied. Nothing at all like it is to be seen except in the southwestern part of the United States. It resembles some of the old towns of Spain and Italy. The streets are quite narrow; one, which is nearly a mile long, being but fifteen feet wide, and another but twelve feet, while the widest of all is but twenty-five feet. An advantage of these narrow streets in this warm climate is that they give shade, and increase the draught of air through them as through a flue. Indeed, some of the streets seem almost like a flue rather than an open way;

most solid part. The age of the palm is legibly written upon its exterior surface; the age of the northern tree is concealed under a protecting bark. The northern tree, though native of a cold, inhospitable climate, is adapted to give shade; the palm, with its straight, unadorned trunk and meagre tuft of leafy limbs, gives no protection to the earth or to man from the burning tropical sun.

In "A Garden in Florida" we have, with surroundings of a more refined character, other specimens of Southern vegetation. The cactus on the right of the picture is an exceptional development of this singular plant, which is usually a humble occupier of the soil. Its habit is to push a few leaves upward, and then shed them one after another,

Coquina Quarry, Anastasia Island.

something after the fashion crabs dispose of an offending claw. Each discarded leaf, however, sets up growing for itself, and thus the cactus, in a modest way, usurps large tracts of favorable soil, forming an undergrowth more impenetrable to man and beast than walls of wood or iron. But our cactus in the garden has been led by the skilful hand of the cultivator upward, and, by removing every exuberant bud, developed into proportions quite foreign to its customary experience. At the left of the picture we have, in the banana, another phase of tropical horticulture, with its broad leaves, that unfold in a single night from a long, slender stem, and its pendent clusters of fruit.

Such was the St. Augustine of a score of years back. Then the railroad came, and the magnate with his wand of gold built palaces, and St. Augustine, with its Ponce de

A GARDEN IN FLORIDA.

Leon, its Alcazar, and its Cordova, became fashionable. With the change it has lost something of its quaintness, but there is left enough of the old Spanish flavor to still make the town a unique one.

Florida is now a great winter resort, and is easily accessible from any of the Northern capitals. During the season the entire peninsula from Jacksonville to Lake Worth on the Atlantic coast, and to Tampa on the Gulf coast, is filled with those who come to avoid the severe winters of the North. Steam navigation preceded the railway, and the rivers of Florida were long the chief highway of travel. Let us follow our artist as he makes the journey up the St. John's and then up the Ocklawaha, which were for so many years the objective points of the tourist's ambition.

Originally starting out for the avowed purpose of hunting the picturesque, we sailed for the mouth of the St. John's—a river that reaches into the very heart of the penin-

Mouth of the St. John's River—Looking in.

sula, and from the ill-defined shores of which you can branch off into the very wildest of this, in one sense, desolate region. The approach of the mouth of the harbor, as is the case with all our Southern rivers, is interrupted by a bar, over which the surf beats always more or less wildly. This entrance, once so dangerous, is now made without any difficulty, owing to the jetties that have been constructed. The scene has a strange look, for, as far as the eye can reach, a long, low reef of burning sand presents itself; the only vegetation visible being a jungle of sunburnt, wind-blasted palmettos. A little north was Fort St. George Island, the most southern of the cultivated sea-islands. Once fairly launched on the waters of the St. John's, after making a sketch of the harbor looking toward the sea, we impatiently passed all intervening places until we arrived at Palatka, a central point, from which we could easily reach the Black River, and the more famous Ocklawaha, and other small streams, only navigable for boats of miniature size.

From Palatka we made the trip in a steamer we named the "Flying Swan," a craft which, from its simplicity of construction and rude machinery, might have been the first

Mouth of the St. John's River—Looking out.

model constructed by Fulton when he was putting into practical shape the use of steam in propelling boats. Its general outline was that of an ill-shaped omnibus, with the propelling-wheel let into its rear, and, on further examination, we found the smoke-pipe, the engine, pilot-house, and all other of the usual gear of steamers, were housed, for the excellent reason of protecting them from being torn away by the overhanging limbs or protruding stumps everywhere to be met with in the narrow and difficult navigation of the swamps.

A sail of twenty miles along the St. John's brought us, a little before sunrise, to the mouth of the Ocklawaha River, looking scarcely wide enough to admit a skiff, much less a steamboat. As daylight increased, we found that we were passing through a dense cypress-swamp, and that the channel selected had no banks, but was indicated by "blazed" marks on the trunks of the towering trees. There was plenty of water, however, to float our craft, but it was a queer kind of navigation, for the hull of the steamer went bumping against one cypress-butt, then another, suggesting to the tyro in this kind of aquatic adventure that possibly he might

BAR LIGHT-HOUSE, MOUTH OF ST. JOHN'S RIVER.

6*

be wrecked, and subjected, even if he escaped a watery grave, to a miserable death, through the agency of mosquitoes, buzzards, or worse.

As we wound along through the dense vegetation, a picture of novel interest presented itself at every turn. We came occasionally to a spot a little elevated above the dead-water level, covered with a rank growth of lofty palmetto, the very opposite, in every respect, to those stunted, storm-blown specimens which greeted us at the mouth of the St. John's River. Here they shot up tall and slender, bearing aloft innumerable

Green Cove Springs.

parasites, often surprising the eye with patches, of a half-mile in length, of the convolvulus, in a solid mass of beautiful blossoms.

Another sharp turn, and the wreck of an old dead cypress is discovered, its huge limbs covered with innumerable turkey-buzzards, which are waiting patiently for the decomposition of an animal that some successful sportsman has shot, and left for the prey of these useful but disgusting birds. The sunshine sparkles in the spray which our awkward yet efficient craft drives from its prow, and then we enter what seems to be a

A Florida Swamp.

cavern, where the sun never penetrates. The tree-tops interlace, and the tangled vines and innumerable parasites have made an impenetrable mass overhead.

The swamps of Florida are as rich in birds as in vegetation. It is no wonder that Audubon here found one of the finest fields from which to enrich his great works of

Waiting for Decomposition.

natural history. A minute list of the varieties we sometimes saw in a single day would fill a page. One of the most attractive was the water-turkey, or snake-bird, which was everywhere to be met with, sitting upon some projecting limb overlooking the water, the body as carefully as possible concealed from view, its head and long neck projecting out, and moving constantly like a black snake in search of its prey. Your curiosity is excited; you would examine the creature more critically, and you fire, at what seems a short, point-blank shot. The bird falls, apparently helpless, in the water; you row rapidly to secure your prize, when, a hundred yards ahead, you suddenly see the snaky head of the "darter" just protruding above the surface of the water. In an instant its lungs are filled with air, and, disappearing again, it reaches a place of safety.

Ascending the Ocklawaha River at Night.

Another conspicuous bird is the large white crane. It is a very effective object in the deep shadows of the cypress, as it proudly stalks about, eying with fantastic look the finny tribes it hunts for prey. Especially is it of service in seizing upon the young of the innumerable water-snakes which everywhere abound. With commendable taste, it seems to pay especial áttention to the slimy, juvenile moccasins, which have a fondness for sunning themselves on harsh dried leaves of the stunted palmetto.

Formerly the alligator was the prominent living object in the Florida swamps. Fashion claimed his hide, and now the alligator has almost entirely disappeared. Those who have escaped the sportsman find here a climate suited to their delicate constitutions; and while those in the Louisiana swamps find it necessary to retire into the mud to escape the cold of winter, the Florida representatives of the tribe are happy in the enjoyment of the upper world the year round. It was a comical and a provoking sight to see these creatures, when indisposed to get out of our way, turn up their piggish eyes in speculative mood at the sudden interruption of a rifle-ball against their mailed sides, but all the while seemingly unconscious that any harm against their persons was intended. Like Achilles, they possess a vulnerable point, which is just in front of the spot where the huge head works upon the spinal column. There is of necessity at this place a joint in the armor, and a successful hunter, after much experience, seldom lets one of the reptiles escape. On one occasion we fired into a herd of alligators, and the noise of two or three shots caused all but one to finally disappear. For some reason it seemed difficult to get the remaining one to move, the creature lying with its head exposed to our gaze, looking as demoniac as possible. A bullet, which struck somewhere in the vicinity of its jaws, touched its feelings, and then, with a grunt not unlike that of a hog, buried itself in the muddy water. This unwillingness to move was then explained by the appearance of a large number of young alligators, which, in the confusion, came to the surface like so many chips. We had, without being aware of it, attacked the mother while she was protecting her nest.

In the vicinity of the alligator's nest we came upon a primitive post-office, consisting of a cigar-box, bearing the magic letters "U. S. M.," nailed upon the face of an old cypress-tree. It was a sort of central point for the swampers, where they left their soiled notes and crooked writing to be conveyed to the places of destination by "whomever came along." We, desiring to act the part of a volunteer mail-carrier for the neighborhood, peeped into the post-office, but there were no signs of letters; so our good intentions were of no practical effect.

Our little craft bumps along from one cypress-stump to another, and fetches up against a *cypress-knee*, as it is termed—sharp-pointed lances which grow up from the roots of the trees, seemingly to protect the trunk from too much outside concussion; glancing off, it runs into a roosting-place of innumerable cranes, or scatters the wild-ducks and huge snakes over the surface of the water. A clear patch of the sky is seen, and the bright light of a summer evening is tossing the feathery crowns of the old cypress-trees into a nimbus of glory, while innumerable paroquets, alarmed at our intrusion, scream out their fierce indignation, and then, flying away, flash upon our admiring eyes their green and golden plumage. It now begins to grow dark in earnest, and we become curious to know how our attentive pilot will safely navigate this mysterious channel in what is literally Egyptian darkness. While thus speculating, there flashes across the landscape a bright, clear light. From the very intense blackness we have a fierce, lurid glare, presenting the most extravagantly-picturesque groups of overhanging palmettos, draped with parasites and vines of all descriptions; prominent among the latter is the scarlet trumpet-creeper, overburdened with wreaths of blossoms, and interwined again with chaplets of purple and white convolvulus, the most minute details of the objects near being brought out in a sharp red light against the deep tone of the forest's depths. But no imagination can conceive the grotesque and weird forms which constantly force themselves on your notice as the light partially illuminates the limbs of wrecked or half-destroyed trees, which, covered with moss, or wrapped in decayed vegetation as a winding-sheet, seem huge unburied monsters, which, though dead, still throw about their arms in agony, and gaze through unmeaning eyes upon the intrusions of active, living men.

The Lookout.

Another run of a half-mile brings us into the cypress again, the firelight giving new ideas of the picturesque. The tall shafts, more than ever shrouded in the hanging moss, looked as if they had been draped in sad habiliments, while the wind sighed through the limbs; and when the sonorous sounds of the alligators were heard, groaning and complaining, the sad, dismal picture of desolation was complete.

A sharp contact with a palmetto-knee throws around the head of our nondescript steamer, and we enter what appears to be an endless colonnade of beautifully-proportioned shafts, running upward a hundred feet, roofed by pendent ornaments, suggesting the highest possible effect of Gothic architecture. The delusion was increased by the waving stream-

A Post-office on the Ocklawaha.

ers of the Spanish moss, which here and there, in great festoons of fifty feet in length, hung down like tattered but gigantic banners, worm-eaten and mouldy, sad evidences of the hopes and passions of the distant past. So absorbing were these wonderful effects of a brilliant light upon the vegetable productions of these Florida swamps, that we had forgotten to look for the cause of this artificial glare, but, when we did, we found a faithful negro had suspended from cranes two iron cages, one on each side of the boat, into which he constantly placed unctuous pine-knots, that blazed and crackled, and turned

A Slight Obstruction in the Ocklawaha.

what would otherwise have been unmeaning darkness into the most novel and exciting views of Nature that ever met our experienced eyes.

The morning came, and the theatrical display of the swamp by torchlight ended, when we were destined to be introduced to a new feature of this singular navigation. A huge water-oak, seemingly in the very pride of its matured existence, had fallen directly across the channel. Its wood was only a little less hard than iron, and the labor to be performed to get this obstruction out of the way was contemplated with anger by the captain of our craft, and in sadness by the "hands," to whose lot fell the labor of clearing the obstruction away. However, the order was given, and no inhabitant of the swamp is inexperienced in the use of the axe. The sturdy blows fell thick and fast, as one limb after another broke loose from the parent trunk and floated slowly away. The great butt was then assailed, and, by a judicious choice in the assault, the weight of the huge structure was made to assist in breaking it in twain. While this work was going on, which consumed some hours, we waded—we won't say ashore—but from one precarious foothold to another, until, after various unpleasant experiences—the least of which was getting wet to our waist in the black water of the swamp—we reached land, which was a few inches above the surface of the prevailing flood.

We were, however, rewarded for our enterprise by suddenly coming upon two "Florida crackers," who had established a camp in a grove of the finest cypress-trees we ever

Cypress-shingle Yard.

saw, and were appropriating the valuable timber to the manufacture of shingles, which shingles, we were informed, are almost as indestructible as slate. These men were civil, full of character, and in their way not wanting in intelligence. How they manage to survive the discomforts of their situation is difficult to imagine, but they do exist, the mosquitoes drawing from their bodies every useless drop of blood, the low swamp malaria making the accumulation of fat an impossibility, while the dull surroundings of their life, to them most monotonous, cramp the intellect until they are almost as taciturn as the trees with which they are associated. But their hut was a very model of the picturesque, and the smouldering fire, over which their dinner-pot was cooking, sent up a wreath of blue smoke against the dark openings of the deep forest that gave a quiet charm, and a contrast of colors, difficult to sufficiently admire, and impossible to be conceived of in the mere speculations of studio-life.

One of our strangest experiences in these mysterious regions was forced upon us one morning, when, thrusting our head through the hole that gave air to our "sleeping-shelf," we saw a sight which caused us to rub our eyes, and gather up our senses, to be certain we were positively awake. Our rude craft was in a basin, possibly a quarter of a mile in diameter, entirely surrounded by gigantic forest-trees, which repeated themselves with the most minute fidelity in the perfectly translucent water. For a hundred feet downward we could look, and at this great depth see duplicated the scene of the upper

A Sudden Turn in the Ocklawaha.

world, the clearness of the water assisting rather than interfering with the vision. The bottom of this basin was white sand, studded with eccentric formations of crystals of a pale emerald tint. This we soon learned was the wonderful silver spring of which we had heard so much, which every moment throws out its thousands of gallons of water without making a bubble on the surface. The transparency of the water was marvellous. A little pearly-white shell, dropped from our hand, worked its zigzag way downward, deepening in its descent from a pale green to a rich emerald, until, finding the bottom, it seemed a gem destined forever to glisten in its silver setting. Procuring a "dug-out," we proceeded to inform ourself of the mysteries of the spot. Noticing the faintest possible movement on the surface of the basin at a certain point, we concluded that it must be over the place where the great body of the water entered the spring.

So, paddling to the spot, we dropped a stone, wrapped in a piece of white paper, into the water at the place where the movement was visible. The stone went down for some twenty-five feet, until it reached a slight projection of limestone rock, when it was suddenly, as if a feather in weight, forced upward in a curving line some fifteen feet, showing the tremendous power of the water that rushes out from the rock. The most novel and startling feature was when our craft came from the shade into the sunshine, for then it seemed as if we were, by some miraculous power, suspended seventy feet or more in the mid air, while down on the sanded bottom was a sharp, clear *silhouette* of man, boat, and paddle. A deep river a hundred feet wide is created by the water of this spring, which in the course of seven miles forms a junction with the Ocklawaha.

Silver Spring.

NIAGARA.

WITH ILLUSTRATIONS BY HARRY FENN.

Stairway at Whirlpool.

NIAGARA! Who has not heard of this peerless cataract, which is among waterfalls what the Himalayas are among mountain-ranges, not only the grandest, but so greatly preëminent as to be without rivalry? The essential quality of Niagara is its sublimity. Other falls are dashed from more stupendous heights, and lost amid chasms of rocks of wilder and more savage formation. But none of them even approach this cataract in that first essential of magnificence. Nor can we be surprised at this when we consider that over the ledge of limestone at this point the accumulated waters of four vast inland seas hurl themselves madly on their way to the ocean, and that, during the last half-mile before the wild descent, there is a decline so great as to produce the most superb rapids. The territory that these lakes drain is equal to that of the entire Continent of Europe, many of the streams that feed Lake Superior being fully two thousand miles away. Hence the volume of water precipitated is so enormous as to produce the most majestic effects, and it may well be doubted if Niagara would gain much by an increase in the height of the fall. At present the height is, on the American side, one hundred and sixty-seven feet, and, on the Canadian, one hundred and fifty-eight feet.

The Brink of the Horseshoe.

Niagara has no advantages of striking character in its surroundings. All that it boasts of the sublime and the beautiful is contained within the rock-walls of its stupendous chasm. All its approaches are plain, dull, and tedious. The country around is almost absolutely flat, divided into fields that wave pleasantly with bearded grain, and dotted with white-painted wooden houses, ugly churches, homely factories, and mills. It may be well said that Niagara resembles a superb diamond set in lead. The stone is perfect, but the setting is lamentably vile and destitute of beauty. And, even in the chasm to which its glories are confined, there is but little loveliness or majesty in the configuration of the rocks, though they are deeply interesting because the river has laid them bare from top to bottom, and exhibits their stratification as clearly as if they were shown on a geologic chart.

The first discoverer of Niagara Falls quickly perceived that Table Rock, on the

Canada side, was the best point of view for the ordinary spectator, though for the artist there are several other spots which bring into prominence various interesting features. It is so still, even after the fall of Table Rock itself, all that remains of that famous slab of limestone being a mere shelving rock. Here a rustic seat is arranged for the accommodation of visitors, who from this point can take in at one glance the whole of

Barnett's Stair, under Table Rock.

the falls. Immediately in front of them is the Horseshoe Fall, where, from the extreme depth of the channel, the water has a deep-emerald tinge of exquisite beauty. Next to it come shelving down the shores of Goat Island, with which, by bridges of frail aspect, on the right hand, is connected Terrapin Rock, where formerly the tower shown in the illustration stood, and, on the left hand, Luna Island. Beyond Luna Island stretches the American Fall. The whole width of the river here is four thousand five hundred feet, of which the American Fall occupies eleven hundred feet, Goat and Luna Islands fourteen hundred, and the remaining two thousand feet belong to the Horseshoe Fall. But it must be remembered that this measurement is from point to point; for, in reality, the curvilinear shape of the Horseshoe has a much more extensive line—over one-third more. These details one learns afterward, the first gaze of the visitor being

Barnett's Stair, in Winter.

too productive of the stupefaction of extreme awe to allow him to notice individual details. One sees the extraordinary volume of water, and its deep, rich color; one sees the clouds of smoke-like spray rising at the base; one hears the roar of the cataract; and that is all. At the time being, nobody can estimate what is seen. The mind is stunned, and what the eye sees produces no effect upon the imagination. Sit as long as you will on the scanty remnant of Table Rock, and, after all, you have not seized

UNDER THE FALLS, CANADA SIDE.

the idea of Niagara, and you will not be able for some time. Best is it, therefore, to glance from Table Rock last of all, and to examine first the details which the enterprise of individuals has placed within our power to study and observe. The great feat, of course, is to descend the stairs underneath the Table Rock, and to penetrate under the Horseshoe Falls as far as one's courage will permit. For this purpose we have to procure oil-skin suits, and caps like those worn in former times by coal-heavers, and known as fantails. The feet have also to be encased in India-rubber shoes, and, if the descent is made in winter, iron spurs are fastened under-

Horseshoe Fall.

neath, so as to give us a firm footing on the ice and snow. The wooden stairways are narrow and steep, but perfectly safe; and a couple of minutes brings us to the bottom. Here we are in spray-land, indeed; for we have hardly begun to traverse the pathway of broken bits of shale when, with a mischievous swoop, the wind sends a baby cataract in our direction, and fairly inundates us. The mysterious gloom, with the thundering noises of the falling waters, impresses every one; but, as the pathway is broad, and the walking easy, new-comers are apt to think that there is nothing in it. More and more arched do the rocks become as we proceed. The top part is of hard limestone, and the lower part of shale, which has been so battered away by the fury of the waters that there is an arched passage behind the entire Horseshoe Fall, which can easily be traversed, if the currents of air would let us pass. But, as we proceed, we begin to notice that it blows a trifle, and from every one of the thirty-two points of the compass. At first, however, we get them separately. A gust at a time inundates us with spray; but, the farther we march, the more unruly is the Prince of Air. First, like single spies, come his winds; but soon they advance like skirmishers; and at last, when a column of thin water falls across the path, they oppose a solid phalanx to our efforts. It is a point of honor with visitors to see who can go the farthest through these corridors of Æolus.

It is on record that a man, with an herculean effort, once burst through the column of water, but was immediately thrown to the ground, and only rejoined his comrades by crawling face downward, and digging his hands into the loose shale of the pathway. Professor Tyndall has gone as far as mortal man, and he speaks of the buffeting of the air as indescribable, the effect being like actual blows with the fist.

As we return along the narrow path, we have leisure to examine the rock-wall, and to discern in it masses of white sulphur, mixed with limestone-lumps lying among the shale. We find also masses of pure, white quartz, sparkling like sugar, and pieces of selenite, or crystallized gypsum, which has a faint resemblance to asbestos, but is translucent. There are ferns growing in patches here and there, and a kind of water-cress. Of moss—long, fine, green moss—there is an abundance, and it grows so delicately that it forces cries of admiration and thrills of delight from the ladies of our party. When we remount the stairs, and find ourselves once more on the upper earth, we are divided in our minds whether to turn and examine the falls, which now begin to be comprehensible, or to doff our dripping oversuit and remove our drenched shoes. Romance and æstheticism suggest the former, but prudence thunders out that the latter is imperative; so prudence carries it.

A pleasant occupation is to stroll along the ramble of Queen Victoria Park and glance at the rapids above the Horseshoe Falls. At Dufferin Island the view is, indeed, transcendent. Immediately below us is the river, which, from our point of observation, we are forced to look up to rather than across, so that it appears like a raging,

THE WHIRLPOOL.

shoreless sea. To the left are Goat Island and the Three Sisters; midway, in the distance, the green slopes of the Grass Islands; and, beyond, the wooded bluffs of Navy Island complete the view. The rapids extend from the verge of the falls for half a mile; and so furious is the rapidity of the current that the centre is heaped up in a ridge-like form, and the waves on every side suddenly leap up in the air, like great fish, and fall down with a sullen sough. The wind comes sweeping over them, and drives their crests along the surface in showers of spray. Great logs and trees burdened with all the glory of their branches, with their greenery still untorn, come swooping down, taking leaps like greyhounds, and giving us the idea of independent life and motion. Here is a great hemlock, bearded with age, and with an abundant spread of branches. How he darts along, showing only portions of his length, like a great brown-and-green serpent! He nears us, he passes us; we turn—we must do it—and we see him, in an instant, shoot with tremendous speed over the brow of Niagara. Another and another still succeed; and, as we gaze, the instinct of cruelty arises in us, despite ourselves, and we long to see a bear or a deer driven down the rapids, and disappear over the abyss, uttering hoarse cries of fear and anguish.

Returning, the next objects on the regular line of observation are the Whirlpool and its rapids. These are more than a mile below the falls, and the best point of observation is from the American side, and we have again to cross the Suspension Bridge, and to pass the lower bridge, which is justly considered one of the marvels of modern engineering. It was built in 1855, and it combines the advantages of the tubular form of construction and the principle of suspension.

The width of the chasm at the rapids, immediately above the Whirlpool, is only eight hundred feet—this contraction being caused by the compact, hard nature of the sandstone rocks through which the river here had to cut its way. The depth of the Niagara here must be very great, and the rapidity of the current, combining with the volume of the stream, actually heaps up the centre in a broken ridge, from which waves are perpetually forced into the air. The Whirlpool is not exactly a whirlpool. It is a vast and furious eddy, which, meeting with a very faint resistance from the shale and gravel of the hills at that precise point, cuts out, by its force, a huge, semicircular curve, and would, no doubt, have cut its way, but was suddenly arrested by hard rocks, which forced it to make a sudden turn to the right hand. We traverse painfully the downward road, and find ourselves at the bottom, among huge masses of fragments, principally of gypsum, of a very hard character, whose edges, however, have been strangely and fantastically worn by the action of the water. These blocks are of all sizes, from slabs weighing many tons to pieces no larger than one's hand. Mingled with them are granite bowlders, whose pink hue makes them conspicuous among the gray gypsum. All are partially covered with a thick, velvety moss, of an intensely dark color. Seating ourselves on these rock-fragments, we discover that we are at the head of the

Whirlpool, in the very fullest frenzy of its rushing fury. The chasm is still contracted here, and the rocks on the Canada side are sandstone, but on the American side limestone. There is on both sides a fine growth of deciduous trees, and the cedars come sloping daintily down to the line of broken rocks. The water fairly hisses as it undulates, seethes, and boils. The waves seem to have a life of their own, and to be animated with human passions. Here, at this exact point, comes the reaction of the eddy; and here one of a series of small whirlpools is formed, which suck down trees headforemost in an instant, and vomit them out in a few minutes, with every vestige of branches and bark completely gone, and great splinters riven out of the hard wood. Even after this they do not escape, for they are borne into the semicircular eddy, and go wandering round and round for days together. At the points where the whirlpools are, the scene is fairly terrific; the waters battle and rage and foam. Current opposes

Rapids above American Fall.

THE CAVE OF THE WINDS

current, wave fights wave, with hideous uproar. Sometimes a wave is forced into the air by fierce collision with another from an opposing side, and is broken into masses of boiling foam, which the wind, as it comes bellowing down the gorge, drives in sheets of spray along the surface of the struggling eddies, and upon the cedars at the brink, which in winter-time become masses of icicles, and sparkle, when the sun falls on them, with a radiance greater than any chandelier of banquet-hall or ballroom.

Ascending the elevator, we turn our steps to the bridge above the Cataract House, which connects the American side with Bath Island, and thence again with Goat Island. From the first-named the view of the rapids, above the falls, is immeasurably finer than on the Canadian side, and this for two reasons: the first, because the point of observation is not much above the level of the rapids; whereas, from the Canada side, you see them from a greater elevation; and, secondly, because the water is contrasted at this point with numerous small islets, which are crowned with cedars, growing at every possible angle. These give an immense relief to the current, and exhibit its

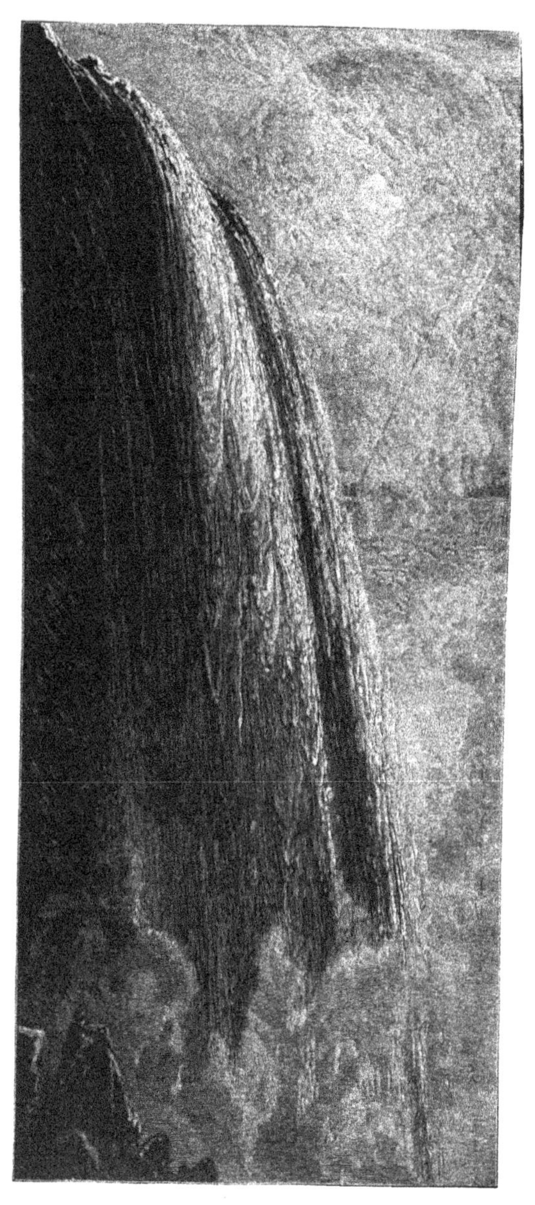

Below American Fall.

Luna Island in Winter.

rapidity in the strongest possible manner. Where all is moving, the motion seems less fierce; but these stationary islets act as a foil. This is a good place to study the lines of waves, for the appearance of these rapids is exactly that of a tempestuous sea, whose billows are heaved and tossed in every direction, and yet, at the same time, are forced forward by an irresistible current. The time to visit this spot is at night, for then the moon, rising slowly in the heavens, sends its light through the very verge of the cataract, shining through the extreme edge. Rising higher, it casts its beams over the angry rapids, turning the dark waves into moving ebony, and the foam into molten silver. But we cannot delay

Ice-Forms.

here, for we must hasten over the farther bridge to Goat Island, where we land amid all the smiling glories of a garden, and inhale with satisfaction the perfumes of roses and heliotropes. The soil of the island is fertile; fine cedars grow in every direction, and there are elms and basswoods. But the fate of Goat Island is doomed. Sooner or later it will be all carried away by the remorseless water, which bears away, year after year, yards upon yards of its circumscribed and narrow bounds. It is a pleasant and a smiling spot in summer-time, much beloved by the white gulls that hover with impunity close over the falls, and seize their finny prey in the very shadow of the great cloud of smoky spray that rises from the Horseshoe.

On the left side there is a bridge which connects with Terrapin Rock, on which for many years stood the famous tower, right upon the verge of the precipitous cataract.

We go across, and from the rock we catch the sublimest view of the falls which can be found. We see nothing but the Horseshoe Falls, it is true; but we see all of that, and we discern the full fury of the torrent, and catch the utmost glory of the rainbow. From Termination Point we go down Biddle's Stairway, having donned oil-skin suits, and make our way painfully to the bottom of the rocks, which are even more arched in formation than those on the Canada side, and the pathway is still broader. Here we come to the famous Cave of the Winds, which is almost underneath the American Fall. Nature has been assisted here by the hand of man, for bridges have been built from rock to rock, under the very cataract itself, and amid all its vapory spray and turmoil and deafening roar. The sun shines down upon the seething waters, and its slanting arrows of light are seized upon by the mist, and broken into myriad scintillations of prismatic hues, into fragmentary rainbows, and globes and bubbles of crimson and green. The cataract shrieks and groans and howls and bellows in fifty different accents at once, while over all dominates the deep, booming roar of the distant Horseshoe Fall.

Close here is the bridge which leads over to Luna Island, a small grain of dry land in the very curve of the fall. It is pleasant enough in summer, but its great glory is in the winter, when all the vegetation is incrusted with frozen spray. The grasses are no longer massed in tufts, but each particular blade is sheathed in a scabbard of diamonds, and flashes radiantly at every motion of the wind. Every tree, according to its foliage, receives the frozen masses differently. In some, especially evergreens with pinnatifid leaves, each separate needle is covered with a fine coating of dazzling white. In others, where the boughs and branches are bare, the spray lodges upon the twigs, and gives to the eye cubes of ice, that greatly resemble the uncouth joints of the cactus. In some evergreens the spray, being rejected by the oleaginous particles, forms in apple-like balls at the extremities of the twigs and the nooks of the branches. Those close to the verge of the fall are loaded so completely with dazzling heaps of collected frozen spray, that the branches often give way, and the whole glittering heap comes flashing down in crumbling ruin. On the ground, the spray falls in granulated circular drops of opaque white; but, wherever there is a stone or a bowlder, ice is massed about it in a thousand varying shapes. Let us peep down from the verge, and, regardless of the noise and the smoke of the water-fall, give our attention solely to the ice. It stretches in great columns from the top to the bottom of the falls, forming a colonnade. The frozen spray, descending upon these, covers them with a delicate tracery of flowers and ferns. In winter-time we may not descend on the American side; but, if we might, surely we should discern the most wondrous ice-configurations along the verge of the pathway. The descent can be made at this time under Table Rock; and the visitor passes from the stairways into a defile such as Dante dreamed of in his frozen Bolgia. Along the side of the rock-walls are rows of stalactites, about the size of the human body, to which they bear a quaint resemblance. Upon the other side, massed along the verge of the bank, are ice-heaps that mount up fifty feet into the troubled air, some of them partially columnar in shape, but the majority looking like coils of serpents, that have been changed, by the rod of an enchanter, into sullen ice.

If winter gives much, it also takes much away. If it covers the trees and the grass with diamonds, and heaps up ice-serpents, and builds colonnades and spires and obelisks, it takes away a great part of the volume of the water, for the thousand rills

Tree crushed by Frozen Spray.

that feed the great lakes have been rent from the hills by the fierce hand of the Frost-giant, and clank around his waist as a girdle. Those who love color and light, and majesty of sound, will do well to come in the summer; those who like the strange, the fantastic, and the fearful, must come in the winter. The true lover of the picturesque in Nature will come at both times. Each has its charm; each has some things which the other lacks; but in both are features of transcendent beauty.

ON THE SAVANNAH.

THE CITIES OF SAVANNAH AND AUGUSTA.

THE SAVANNAH, the largest river of Georgia, and forming the boundary between that State and South Carolina, rises by two head-streams, the Tugaloo and Keowee, in the Appalachian chain, and near the sources of the Tennessee and Hiawassee on the one side, and the Chattahoochee on the other. From the junction of these confluents at

Andersonville the river has a course of four hundred and fifty miles to the sea. Savannah and Augusta, two of the largest cities in the State, are situated upon its banks, the former eighteen miles from where it empties into the Atlantic, the latter at the head of navigation, two hundred and thirty miles from its mouth. The river between these points glides between richly-wooded banks, with occasional glimpses of cotton-plantations in the upper portion and of rice-plantations below. The wild swamp-wastes that mark its lower shores are full of a strange, weird beauty, and the groves of massive live-oaks, hung with their mossy banners, that shadow and conceal the mansions of the planters, have a noble grace that is very captivating.

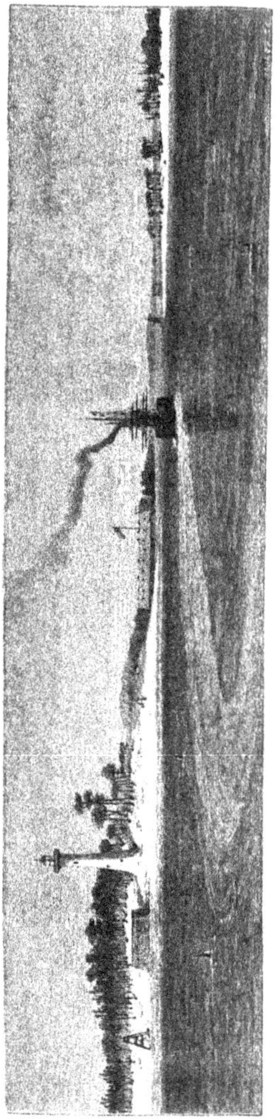

Mouth of the Savannah River.

The site for the city of Savannah was selected by General Oglethorpe, the founder of the colony of Georgia, who made his first settlement at this point in February, 1733. The city occupies a promontory of land, rising in a bold bluff, about forty feet in height, close to the river, extending along its south bank for about a mile, and backward, widening as it recedes, about six miles. The river making a gentle curve around Hutchinson's Island, the water-front of the city is in the form of an elongated crescent, about three miles in length. The present corporate limits extend back on the elevated plateau about one and a half miles, and is three and one-third miles square. Beyond the city limits, to the south, suburban settlements are fast growing up, and the adjacent grounds are both eligible and available to an unlimited extent.

In its general plan, Savannah is one of the handsomest of the American cities; and in view of its antiquity, and the fact that its founders were for the most part poor refugees, seeking a home in the wilderness among savages, it is a matter of surprise that they should have adopted a system at once so unique, practical, and tasteful. The streets—running nearly east and west and north and south, and crossing at right angles—are of various widths; the very wide streets, which run east and west, being alternated with parallel narrower streets, and each block intersected with lanes twenty-two and a half feet in width. The streets running

View of Savannah from the River.

north and south are of nearly uniform width, every alternate street passing on either side of small public squares, or plazas, varying from one and a half to three acres in extent, which are bounded on the north and south by the narrower streets, and intersected in the centre also by a wide street.

These plazas—twenty-four in number, located at equal distances through the city, handsomely enclosed, laid out in walks, and planted with the evergreen and ornamental trees of the South—are among the distinguishing features of Savannah, and in the spring and summer months, when they are carpeted with grass, and the trees and shrubbery are in full flower and foliage, afford delightful shady walks and playgrounds, while they are not only ornamental, but conducive to the general health by the free ventilation which they afford. They have well been called the lungs of the city.

In addition to these squares, a public park, comprising some thirty acres, called Forsyth Place, was, when laid out, a considerable distance south of the city limits. It is, however, now being rapidly enclosed by buildings, and is almost in the centre of one of the finest and most populous portions of the city. Many of the origi-

Fountain in Forsyth Park.

nal pine-trees were left standing on the grounds, which are laid out in serpentine walks, and ornamented with evergreen and flowering trees and shrubbery. In the centre is a handsome fountain, after the model of that in the Place de la Concorde in Paris. The lofty pines still standing, with the ornamental trees, afford a grateful shade; while the beautiful shelled walks, the luxuriant grass, the fragrant flowers, and the plashing fountain, make Forsyth Place a delightful retreat from the noise, bustle, dust, and heat of the city.

Among the peculiar features of Savannah which command the admiration of strangers are the wideness of its principal streets, abounding with shade-trees, and the flower-gardens which, in the portions of the city allotted to private residences, are attached to almost every house. Ornamental trees of various species, mostly evergreens, occupy the public squares, and stud the sidewalks in all the principal thoroughfares; while the gardens abound with ornamental shrubbery and flowers of every variety. Conspicuous among the former are the orange-tree, with its fragrant blossoms and golden fruit in their season, the banana, which also bears its fruit, the magnolia, the bay, the cape-myrtle, the stately palmetto, the olive, the *arbor-vitæ*, the flowering oleander, and the pomegranate. At all seasons, Savannah is literally embowered in shrubbery, and in the early spring months, when the annuals resume their foliage, and the evergreens shed their darker winter dress for the delicate green of the new growth, the aspect of the city is truly novel and beautiful, justly entitling it to the appropriate name by which it has long been known, far and wide, of the "Forest City."

The old city of Oglethorpe's time was located on the brow of the bluff, about mid-way between the present eastern and western suburbs, and its boundaries are still defined by the Bay and East, West, and South Broad Streets. Upon the river-front, a wide

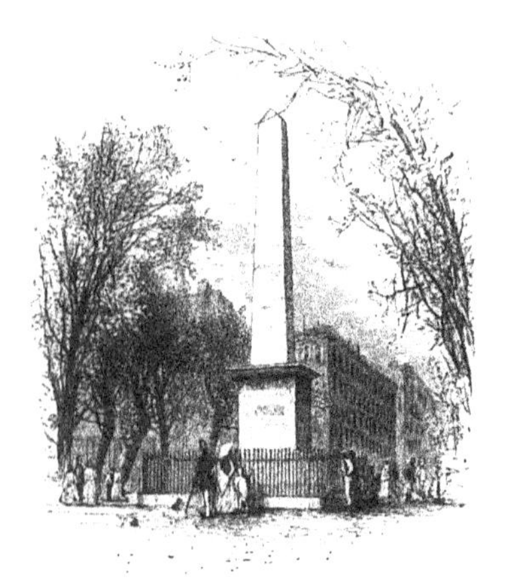

Monument to General Greene.

Church, Bull Street.

esplanade, about two hundred feet in width, extending back from the brink of the bluff, was preserved for public purposes. This is called the Bay, and is now the great commercial mart of Savannah. As commerce grew up, warehouses and shipping-offices were built by the first settlers, under the bluff between it and the river. In time these were replaced by substantial brick and stone structures, rising four and five stories high on the river-front, with one or two stories on the front facing the Bay, connecting with the top of the bluff by wooden platforms, which spanned the narrow roadway beneath, passing between the buildings and the hill-side. Some of these buildings, spared by the great fire of 1820, which consumed the larger portion of the old town, are interesting for their antique and quaint architecture. A range of them, opposite the foot of Bull Street—the fashionable thoroughfare of the city—is made the subject of a sketch by our artist. These relics of old Savannah, and a few others, hold their place in the line of stately modern buildings, which now extend along the larger portion of the city-front under the bluff. Platforms still connect the upper stories of the stores under the bluff with the Bay; and at the foot of the principal cross-streets walled roadways lead to the quay, which is wide, and occupied at intervals with large sheds for the protection of goods in the process of shipping and discharging. Along the quay, in close proximity to the wharves, are also located the cotton-presses and rice-mills.

While Savannah makes no special pretensions to architectural beauty, nevertheless the city contains many fine public and private buildings, and the good taste which characterizes her modern improvements evinces a progressive spirit and liberality worthy of

Bull Street.

her rapidly-increasing wealth and commercial importance. Some of her church edifices are models of architectural beauty; and among the new buildings, many of which have been erected within the past twenty years, are some substantial and imposing structures. In Monument Square there is a fine marble obelisk, erected to the memory of General Greene, of Revolutionary fame, the corner-stone of which was laid by Lafayette, during his visit to America in 1825. The shaft is fifty-three feet in height. Another and very elegant structure was erected in 1853, to the memory of General Pulaski, who fell, it will be remembered, during an attack upon the city by the British, in the year 1779.

Owing to the crescent form of the city-front, its elevation, and the absence of any eligible point of observation on the opposite side of the river, it is difficult to obtain a view that will convey a correct impression of its size and appearance. This difficulty our artist experienced, as the best position which he could obtain, on Fig Island, pre-

Bonaventure Cemetery.

sented but a meagre profile of the city-front and its eastern environs. He has, however, made a sketch of the city from that point. The view takes in the line of Hutchinson's Island, on the opposite side of the river, which extends the length of the city.

The view of the mouth of the Savannah River conveys a very correct idea of the appearance of the entrance to the harbor, which is capacious and well protected, Tybee

Island being the head-land on the left, and the extreme southern point of Dawfuskie Island defining the entrance to the river on the right. Passing up the river, the stranger is struck with the peculiar aspect of the wide expanse of salt-marsh through which it meanders, forming many islands, but preserving at all times ample width for the navigation of vessels.

Old Houses in Savannah.

Until the construction of the Central Railroad, half a century ago, Savannah was comparatively isolated from the internal commercial world, her only communication with the interior of the State being by the Savannah River to Augusta, the head of steamboat-navigation—the wilderness and the great swamps of the Altamaha interposing an impassable barrier to the vast and fertile regions of the Southwest.

The temperature of Savannah has all the mildness of the tropics in winter, without the intense heat in summer, the mean temperature being very nearly the same as that of Bermuda. The sultriness of the "heated term" in Savannah is less oppressive than in New York or Boston, mitigated as it is by a soft, humid atmosphere, and the never-failing breath of the trade-winds, so grateful at that season.

Savannah is not without suburban attractions, there being several places in its vicinity of historical interest, whose sylvan character and picturesque beauty are in keeping with the "Forest City" itself. Thunderbolt, White Bluff, and Isle of Hope, are all rural retreats on "the salts," within short drives of the city, where, in the summer months, the bracing sea-breeze and salt-water bathing are enjoyed.

Our artist presents a sketch of Bonaventure, which is located on Warsaw River, a branch of the Savannah, about four miles from the city. A hundred years ago, the seat of a wealthy English gentleman, the grounds around the mansion, of which only a dim outline of its foundations remain, were laid out in wide avenues, and flanked with native live-oaks. These

The Savannah, near Augusta.

trees, long since fully grown, stand like massive columns on either side, while their far-reaching branches, interlacing overhead like the fretted roof of some vast cathedral, the deep shade of their evergreen foliage shutting out the sky above, and the long, gray moss-drapery depending from the leafy canopy, silent and still, or gently moving in the breeze, give to the scene a weird and strangely-sombre aspect at once picturesque and grandly solemn. Many years ago Bonaventure became the burial-place of many of the prominent families of Savannah, whose memorial

THE SAVANNAH, AT AUGUSTA.

monuments add to its solemn beauty. It was incorporated in 1869 and called Evergreen Cemetery.

Augusta, which lies at the other extreme of the navigable waters of the Savannah, was settled only two years later than its seaward rival. Like Savannah, it was laid out under the personal supervision of General Oglethorpe, to whom it is indebted for its name, given in honor of one of the English princesses. It is situated on a broad plain. The wooded and winding Savannah waters one of its sides; villa-crowned hills environ it on others. The taste which has made Savannah one of the handsomest of cities is apparent here also in its broad avenues richly shaded with antique trees.

The most beautiful of its avenues is Greene Street, which is lined with fine mansions. Tall, spreading trees not only grace the sidewalks, but a double row, with grassy spaces between, run down the centre of the ample roadway. Here stands the City Hall, a fine building of venerable age, set in an ample green amid trees, and having about it an air of dignity and repose. A tall granite column standing before the hall in the green of the roadway, commemorating the signers of the Declaration of Independence from Georgia, adds dignity and finish to the picture.

Augusta is an important cotton-market, its situation at the head of navigation on the Savannah giving it good facilities for shipping. Hence cotton centres here from all the surrounding country; it comes in the shipping-season in vast abundance, both by rail and by wagons. Active scenes are witnessed on the banks of the river, where small stern-wheel steamers come up and bury themselves to their smoke-pipes in cotton-bales. The groups of boats shown in the engraving illustrating this scene are just below the bridge that connects Augusta with the town of Hamburg, on the South-Carolina side of the river. Along the high banks upon which Augusta is situated are rows of old mulberry-trees, the trunks of which are covered with warts and knobs, and their gnarled, fantastic roots exposed by the washings of many freshets.

We give a view of Augusta from Summerville, a suburban town of handsome villas, on high hills two or three miles from the city. A line of electric-cars runs from the town to the summit of the range. Here are many villas and cottages, embowered in trees, with verandas, gardens, and many signs of wealth and culture. The scene is more Northern in its general features than Southern; the houses are like the Northern suburban villas, and the gardens not essentially different, although the Spanish-bayonet—that queer horticultural caprice, with its bristling head of pikes—shows a proximity to tropical vegetation. The United States arsenal, built in 1827, is on the Summerville hills. These heights form a part of the famous red sand-hills of Georgia, and a characteristic feature are the rich red tints of the roadways.

VIEW OF AUGUSTA, FROM

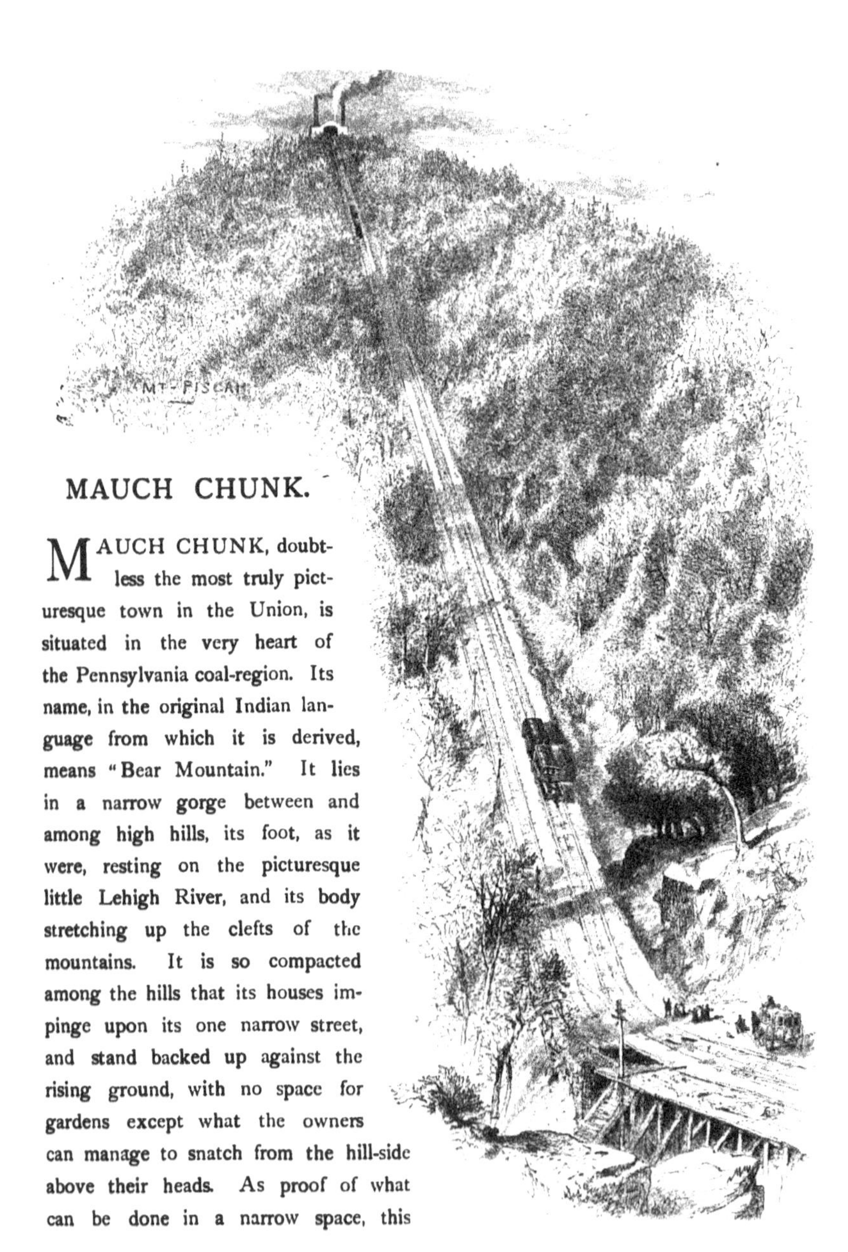

MAUCH CHUNK.

MAUCH CHUNK, doubtless the most truly picturesque town in the Union, is situated in the very heart of the Pennsylvania coal-region. Its name, in the original Indian language from which it is derived, means "Bear Mountain." It lies in a narrow gorge between and among high hills, its foot, as it were, resting on the picturesque little Lehigh River, and its body stretching up the clefts of the mountains. It is so compacted among the hills that its houses impinge upon its one narrow street, and stand backed up against the rising ground, with no space for gardens except what the owners can manage to snatch from the hill-side above their heads. As proof of what can be done in a narrow space, this quaint Swiss-like village affords an example. In one portion, just where the turbulent Lehigh sweeps around, as if to give the town a salute, and then rushes merrily off again, one sees the river, a canal, the railways, a road, and a street, all in a space scarcely more than a stone's-throw wide.

There is a great deal in knowing how to find the picturesque, and our artist, in his large drawings, has selected views that present the hills and the town in their best aspect. The first of these is taken from the road that runs along the side of the high hill just below the town. In the second illustration, one can discern the road, faintly marked, ascending obliquely the distant hill. From this road the picture gives just a glimpse of the receding town to the left; shows in the distance Mount Pisgah, which is not a volcano, notwithstanding the smoke that seems to issue from its apex; and gathers at the feet of the spectator hurrying river, busy canal, railways, and highway, as they lie crowded between the steep hills. Here there is always the stir of a great traffic. Ceaselessly day and night the long, black coal-trains come winding round the base of the hills, like so many huge anacondas, often with both head and tail lost to the eye, the locomotive reaching out of sight before the last car comes swinging round the curve. So continuous is their coming and going, sweeping now around the foot of the hill opposite, and now around the base of the hill on which we stand, that usually several trains are visible at the same time. These trains are almost exclusively employed in freighting coal; and this immense traffic in black diamonds becomes still more surprising when it is remembered that, in addition to the trains, canal-boats similarly freighted ceaselessly pass the town with the regularity, order, and succession of a procession. Up here on the hill-side the scene before us is certainly novel and picturesque. We may watch the stirring traffic, the quiet canal, the swift Lehigh—sometimes only the small thread of a river barely covering its rocky bed, but occasionally a roaring flood bringing ruins upon its surface and carrying ruin before it—or we may study the tints and forms of the receding hills, or note a singular locomotion far up on the sides of the distant Mount Pisgah.

On the highest part of this mountain are two tall chimneys, ascending to which is the line of a railway. The chimneys and the building thereto give note of a stationary engine at this crowning apex of the height, and the line up the mountain-side shows us where the famous Mount-Pisgah inclined plane ascends to its top. The line crossing the hill half-way down, and just below Upper Mauch Chunk, marks the course of the Gravity Railway, one of the marvels of the place. We will descend our mountain-highway, picturesque and beautiful every step of it, with beetling cliffs above and precipitous reaches below, and prepare for an odd sort of journey to the top of Mount Pisgah, and, by the Gravity Road, to the coal-mines beyond.

The mines which supply the principal traffic of Mauch Chunk are situated nine miles back from the river, on Sharp and Black Mountains, and in Panther-Creek Val-

MAUCH CHUNK AND MOUNT PISGAH

MAUCH CHUNK, FROM FOOT OF MOUNT PISGAH.

ley, lying between. The first anthracite coal was discovered on Sharp Mountain, sometimes known as Summit Hill, by Philip Ginter, in 1791. It was not, however, until 1820 that shipments became at all regular or noteworthy. Coal was brought from the mines, slowly and wearisomely, by wagons, until 1827, when a track was constructed, with a falling grade, from Summit Hill to the Lehigh, by which cars were run down by their own gravity—hence the name Gravity Road. The cars were drawn back by mules, which, of course, had to be sent down on cars with each train. This method continued for a long time; but the traffic at last so increased that a more expeditious return of the cars to the mines was needed, and in 1844 the plan of a back-track was arranged.

Canal-boats receiving Coal.

An inclined plane was laid to the top of Mount Pisgah, up which the empty cars were elevated by means of a stationary engine; the track, then, by a downward grade, the cars moving by force of their own weight, reached the foot of Mount Jefferson, up which they ascended by another plane—the power a stationary engine—and then, by another downward grade, reached Summit Hill. From Summit Hill the cars descended to the mines in the valley, by what was called the Switch-back, a term now often given to the entire road. The cars are returned to Summit Hill by means of inclined planes and stationary engines; and from the Summit to the Lehigh, a distance of nine miles, the gravity-impelled cars dash at a rapid rate with their spoils from the heart of the mountain.

In the first of our larger illustrations, the Mount-Pisgah inclined plane and a portion of the Gravity Road, as already mentioned, are shown. The cars which are seen on the grade may be discovered at their terminus in the engraving given on the preceding page. Here they rattle down into huge coal-boxes, into which their contents are dumped and shot into the canal-boats, which are always gathered here by hundreds.

A Mauch-Chunk Highway.

But we must hasten to the station for a trip on the famous plane, which, if our description has not adequately depicted to the mind's-eye of the reader, the initial illustration will bring before him accurately and clearly. It may be mentioned here that the length of this plane is twenty-three hundred and twenty-two feet, and its elevation sixteen hundred and sixty feet above tide-water. At its foot we find a very small passenger-car—designed to hold ten passengers—in which we may enter. The plane appears, when standing at its foot, to reach almost perpendicularly up into the air; and when at last the ascent begins, one feels as if he were drawn up into the clouds, and naturally commences to speculate with what terrible swiftness the car would shoot down the plane if it should get loose. Behind the miniature carriage is what is called a safety-car. From this car extends an arm over a ratchet-rail, laid between the tracks. Should an accident occur either to the car or to the gearing, this arm, the moment a downward movement begins, inevitably falls into the notch of the ratchet-rail, and, being too strong to break, the train is at once brought to a stand-still. It is frightful-looking, notwithstanding this assurance, and one discovers that his imagination takes a strange pleasure in depicting the terrible whirl through space and the horrible splintering upon the rocks, should it please Fate to give the pleasure-trip a tragical turn. As the car ascends, the prospect enlarges; and, when the height is reached, a splendid view presents itself.

What follows may now easily be conceived, by means of the description of the road already given. The car runs easily and swiftly along, without other force than its own weight, the road being through beautiful woodland-scenery. As we draw near the mines, large villages appear, occupied principally by the miners, and at Summit Hill is a hotel, church, and other evidences of civilization. The huge structures, called coal-breakers, at the mouths of the mines, form new, striking, and picturesque objects, and immense piles of *débris*, accumulated in excavating for the black wealth below, look like small mountains. Near abandoned mines, these vast heaps give indications of a new soil gathering on their surface. Bushes and small evergreen trees have already managed to find sufficient nurture amid the slate and coal-dust for their roots, and in time there can be no doubt that what are now unsightly masses of *débris*, will be covered with grass and trees.

The circuit completed, we leave the car well up Mount Pisgah, and descend the mountain-road to the village. The roofs show far down below us among the trees, and the houses, hugged in close by the hills, are grouped in most picturesque form. As we near the houses they seem so directly beneath that we wonder if a slip would not precipitate us down a chimney, or impale us on a steeple. The second of our larger illustrations shows the scene as we near the town from this approach. There is a church-roof below the point of view, and a row of houses in the middle ground on the hill-side, and a picturesque church, built just where it would add most to the beauty and effectiveness of the scene.

The street-scenes in Mauch Chunk are quaint enough. They are literally highways. As there is no room for gardens or out-buildings back of the houses, they are built up above them, and are reached by ladders. It is not uncommon, in the ruder parts of the town, to see a pigsty, up above the house-top, reached by a ladder; another ladder extending above this to a potato or cabbage patch, and another leading to the family oven, presiding over the strange group with suitable honor and dignity.

A visit to Mauch Chunk makes a pleasant summer-trip; but it is in October, when all the superb hills that encircle the quaint town are in the full glow of their autumn tints, that the innumerable mountain-excursions which then may be taken, mostly enhance the pleasure of the visit.

LOOKOUT MOUNTAIN AND THE TENNESSEE.

WITH ILLUSTRATIONS BY HARRY FENN.

The Tennessee.

THE first sensation of the prospect from the top of Lookout Mountain is simply of immensity. The eye sweeps the vast spaces that are bounded only by the haze of distance. On three sides no obstacles intervene between your altitude and the utmost reaches of the vision. To your right, stretch successive ranges of hills and mountains that seem to rise one above another until they dispute form and character with the clouds. Your vision extends, you are told, to the great Smoky

LOOKOUT MOUNTAIN.—VIEW FROM THE "POINT."

Mountains of North Carolina, which lie nearly a hundred miles distant. The whole vast space between is packed with huge undulations of hills, which seem to come rolling in upon your mountain-shore, like giant waves. It is, indeed, a very sea of space, and your stand of rocks and cliffs juts up in strange isolation amid the gray waste of blending hills. Directly before you the undulations are repeated, fading away in the far distance where the Cumberland Hills of Kentucky hide their tops in the mists of the horizon. Your eye covers the entire width of Tennessee; it reaches, so it is said, even to Virginia, and embraces within its scope territory of seven States. These are Georgia, Tennessee, Alabama, Virginia, Kentucky, North and South Carolina. If the view does in truth extend to Virginia, then it reaches to a point fully one hundred and fifty miles distant. To your left, the picture gains a delicious charm in the windings of the Tennessee, which makes a sharp curve directly at the base of the mountain, and then sweeps away, soon disappearing among its hills, but at intervals reappearing, glancing white and silvery in the distance, like mirrors let into the landscape.

Lookout Mountain presents an abrupt precipice to the plain it overlooks. Its cliffs are, for half-way down the mountain, splendid palisades, or escarpments, the character of which can be altogether better conceived by the study of our illustrations than by the most skilful description. The mountain-top is almost a plateau, and one may wander at his ease for hours along the rugged, broken, seamed, tree-crowned cliffs, surveying the superb panorama stretched out before him in all its different aspects. The favorite post of view is called the "Point," a plateau on a projecting angle of the cliff, being almost directly above the Tennessee, and commanding to the right and left a breadth of view which no other situation enjoys. Beneath the cliff, the rock-strewed slope that stretches to the valley was once heavily wooded, but during the war the Confederates denuded it of its trees, in order that the approaches to their encampment might be watched. It was under cover of a dense mist that the Federal troops on the day of the famous battle skirted this open space and reached the cover of the rocks beyond, up which they were to climb. The "battle above the clouds" is picturesque and poetical in the descriptions of our historians, but the survey of the ground from the escarpments of the mountain thrills one with admiration. It is not surprising that the Confederate leader believed himself secure in his rocky eyrie, and the wonder must always remain that these towering palisades did not prove an impregnable barrier to his enemy.

On the summit of Lookout Mountain the northwest corner of Georgia and the northeast extremity of Alabama meet on the southern boundary of Tennessee. The mountain lifts abruptly from the valley to a height of fifteen hundred feet. It is the summit overhanging the plain of Chattanooga that is usually connected in the popular imagination with the title of Lookout, but the mountain really extends for fifty miles in a southwesterly direction into Alabama. The surface of the mountain is well wooded, it has numerous springs, and is susceptible of cultivation. There is a small settlement

CHATTANOOGA AND THE TENNESSEE FROM LOOKOUT MOUNTAIN.

on the crest of the mountain, which is reached by a switch-back railway, and thither tourists come both in winter and in summer to view the scenery.

The sunshine that seduced us from Chattanooga only kept our company until we reached the mountain-top, when clouds began to obscure the scene, and winds to chill the air. There were glimpses of sunshine, and the clouds would lift and give us superb vanishing pictures of the valley and distant hills, touched in spots with sunlight; but the cold winds of early spring and the ever-recurring showers made sketching out-of-doors cold and dismal work.

The majority of visitors go to Lookout only for an hour or two, and hence miss some very striking characteristics of the mountain. A lake and a cascade of uncommon beauty are about six miles distant from the "Point," and a singular grouping of rocks, known by the name of "Rock City." The City of Rocks would be a somewhat more correct appellation. Vast rocks of the most varied and fantastic shape are arranged into avenues almost as regular as the streets of a city. Names have been given to some of the main thoroughfares, through which one may travel between great masses of the oddest architecture conceivable. Sometimes these structures are nearly square, and front the avenue with all the imposing dignity of a palatial residence. But others exhibit a perfect license in capricious variety of form. Some are scooped out at the lower portion, and overhang their base in ponderous balconies of rock. Others stand balanced on small pivots of rock, and apparently defy the law of gravitation. I know of nothing more quaint and strange than the aspects of this mock city--silent, shadowy, deserted, and suggestive, some way, of a strange life once within its borders. One expects to hear a foot-fall, to see the ponderous rocks open and give forth life, and awaken the sleep that hushes the dumb city in a repose so profound.

Lookout Mountain is remarkable generally for its quaint and fantastic rocks. Near the "Point" are two eccentric specimens that are pointed out to every visitor. The "Devil's Pulpit," almost at the extreme end of the "Point," consists of a number of large slabs of rocks, piled in strange form one upon the other, and apparently in immediate danger of toppling over. The reader will readily discover this strange formation if he consults Mr. Fenn's drawing showing the view from the "Point." Another odd mass is called "Saddle Rock," from a fancied resemblance to a saddle. It consists of a great pile of limestone, that has crumbled and broken away in small particles, like scales, until in texture one may discover a likeness to an oyster-shell, and in form something of the contour of a saddle-tree. With queer rock-forms, Lookout Mountain is certainly abundantly supplied. It is supposed that these rocks, jutting so far above the level of the Palisades, are remains of a higher escarpment, which, during uncounted centuries, has gradually worn away.

The lake and cascade to which reference has been made are known as "Luluh Lake" and "Luluh Falls," *Luluh* being a corruption of the Indian name of Tullulah.

The cascade is one of uncommon beauty. It is nearly as high as Niagara, and far more picturesque in its setting. Lookout Mountain, indeed, is very imperfectly seen by those who make a hurried jaunt to its Palisades, glance at the prospect so superbly spread out before them, and then hurry back again. There is no mountain and no landscape that does not require its acquaintance to be cultivated somewhat. The supreme

Rock City, Lookout Mountain.

beauty, the varied features, the changing aspects, the subtle sentiments of the "rock-ribbed hills," enter the soul by many doors, and only after a complete surrender on our part to their influences. One may comfortably house himself on the great plateau of Lookout, and there give many days to wandering along its Palisades, or in search of the thousand picturesque charms that pertain to its wooded and rocky retreats.

The Tennessee River comes sweeping down upon Lookout Mountain as if it confidently expected to break through this rocky barrier and reach the Gulf by an easy course through the pleasant lowlands of Alabama. The flood reaches the base of Lookout's tall abutments, and, finding them impenetrable, sweeps abruptly to the right, breaking through the barrier of hills that lie in its course, and, as if with a new purpose at heart, abandons its hope of the Gulf, to eventually reach it, however, after a double marriage with the Ohio and the Mississippi.

The Tennessee is formed by the Union of the Clinch and the Holston Rivers, at Kingston, and, together with its principal affluent, attains a length of eleven hundred miles. Steamers navigate different portions, but a succession of shallows and rapids in Alabama, known as "Muscle Shoals," bar vessels from its lower waters to the upper;

Rock-Forms on Lookout Mountain.

and below Chattanooga the "Suck" and the "Pot" at certain times form serious obstacles to navigation.

The "Pot" is some twenty miles below Chattanooga; it is a maelstrom which, at certain depths of water, is wild and beautiful. The swift current is impinged sharply upon a high bluff, and turns to escape, at an angle so acute, that a perfect whirl of waters ensues. Vast trees have been seen caught in its fierce turmoil and swept out of sight; and, in the time of freshets, houses, carried off by the flood, have plunged into the gulf, to reappear none knew where or how. The "Suck" is thirteen miles from Chattanooga. This phenomenon is caused by a fierce little mountain-current, called "Suck Creek," which, in times of high water, brings from its rocky fastnesses such masses of *débris* that the river-bed is strewed with bowlders, and a bar formed, which compresses the channel into a narrow, swift, and dangerous current. In former years it was no uncommon sight to see a steamer struggling up against an adverse current by means of a windlass on the bank, with the songs of the deck-hands.

To visit this famous "Suck," and get a sketch or two of the shore, was the purpose of our journey along the Tennessee. There was once a fine bridge across the Tennessee, at Chattanooga, but it fell a victim to a great flood. The Chattanoogians have been so busy since erecting new warehouses, new railway-stations, and new hotels, that they have forgotten the piers of masonry in the river-bed, which in grim solitude seem to utter a protest against their neglect. Not that we, searchers for the picturesque, would have had it otherwise—for a bridge would have deprived us of a sketch of one

Rocks, Rock City.

of the quaintest ferries in the country. It is a rope-ferry, having for motive-power the river-current, which it masters for its purpose by a very simple application of a law in physics. A long rope from the ferry-boat, supported at regular intervals on poles resting on small flat-boats, is attached, several hundred feet up-stream, to an island in mid-water. The boat thus secured is pushed from the shore, when it begins to catch the force of the current, a greater surface of pressure being secured by a board, like the centre-board of a sail-boat, which is dropped down deep into the water on the upper side. The current sweeping against the boat would carry it down-stream, but the attached rope retains the vessel in place, and we have, as a result of the sum of the forces, the boat swiftly propelled on the arc of a circle across the stream. A very odd effect in the scene is the fleet of small flat-boats, upholding the long and heavy rope, which start in company with the large vessel in the order and with the precision of a column of cavalry. Moving in obedience to no visible sign or force, they impress one as being the intelligent directors of the movement, and are watched, when first seen, with lively interest.

The method adopted at this ferry is occasionally found in the South, but, ordinarily, ferry-boats are carried from one side of the stream to the other by means of a suspended rope from shore to shore. The Chattanooga ferry is very picturesque, apart from the method of progression. Our road to the "Suck," one of the most attractive and charming we had ever travelled, certainly outdid in roughness of surface any previous experience. It led through superb woods; under high banks; over rocks and bowlders; into swift-running streams; up steep hills, and down declivities. We were pitched into the bottom of the wagon one moment, tossed against the top at another, now precipitated affectionately into each other's arms, now hurled discordantly apart against the wagon-sides—all of which, however, while trying to one's bones, added to the relish of the journey, or rather, to the relish of our recollections of it.

The Tennessee, as already said, runs between high hills, mountains even, being the continuation of the Cumberland range. Spreads of table-land, with intervening dips of the forest, mark one side of the river, while on the other the rocky hills rise abruptly from the water's edge. The river is very winding, and the road sometimes runs along its course, sometimes loses sight of its silvery waters altogether; but the appearing and reappearing surface of the stream affords continual changes to the picture. Between the bluff and the river are narrow strips of arable bottom-land; and these, which sometimes are only narrow ribbons bordering the stream, and at others wide fields, are very rich in soil and carefully cultivated. But the owners, almost without exception, live in rude log-cabins. Although the cabins are rude, the grounds limited, the means scanty, the residents are a proud, intelligent set, who should be classed as hunters and woodsmen rather than as husbandmen. Their delight is the woods and the mountains, and they almost live on horseback. Their needs are a gun, a dog, a horse, a cottage, a wife, and a cow—and pretty much in the order enumerated. They are semi-sportsmen, accomplished in woodcraft, who delight in all kinds of hunting, but exhibit very little energy in developing the resources of the country. It would be a mistake to accuse them of a lack of intelligence. We met many people on the road that day whose faces were refined and handsome. With their sloping *sombreros*, their gray shawls or army coats, their picturesque saddles, and their general air of graceful dilapidation, they looked like

FERRY AT CHATTANOOGA.

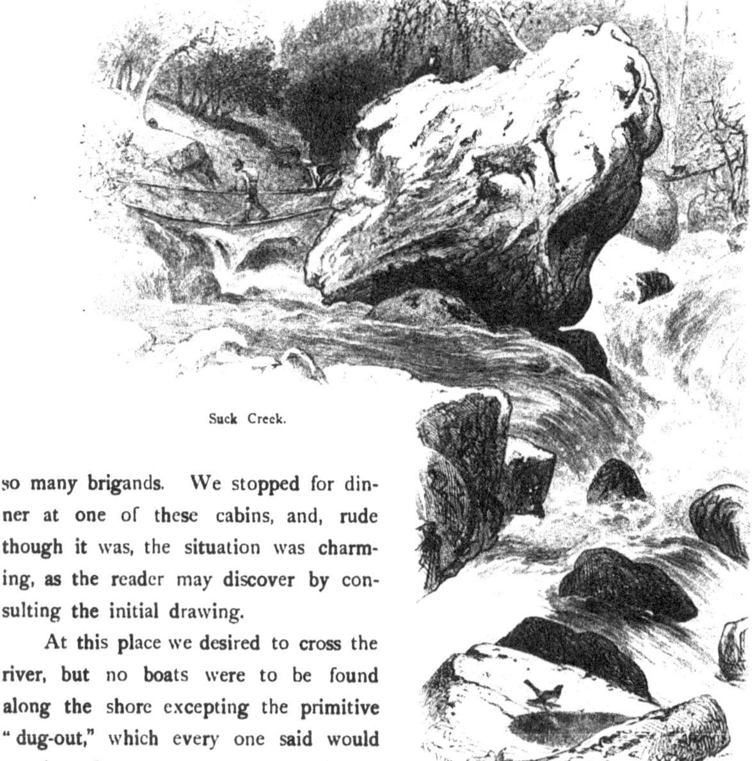

Suck Creek.

so many brigands. We stopped for dinner at one of these cabins, and, rude though it was, the situation was charming, as the reader may discover by consulting the initial drawing.

At this place we desired to cross the river, but no boats were to be found along the shore excepting the primitive "dug-out," which every one said would not be safe on account of the swiftness and turbulence of the current. The rudest savage tribes of the Pacific build canoes that can sail far out at sea in high winds and rough water, but the boats of the Tennessee can only be employed in the smoothest of water. The skill of our Tennessee men is equal, no doubt, to many emergencies of the mountains, but their resources for the water are certainly very limited. As we could not get on the other side of the river, we started in search of the most eligible points on this side. In order to reach the shore, we had a wild and picturesque walk, reaching in due time the romantic stream which ignobly rests under the title of "Suck Creek." This stream is a mountain-torrent; it comes tumbling through rocky crevices above with all the flash and splendor of the "waters of Lodore," and pours with turbulent energy into the Tennessee. In freshets it comes from its mountain-home with tremendous volume and force, burying far under water even the high rocks shown in the illustration, and sweeping into the river a score or so of smaller impediments. We crossed this torrent on a round and very small tree-trunk, and, not having the skill of the natives, crept along it on our hands and knees. But, shortly after, seeing one to the manner born, with a pack on his back, and a load in each hand, quietly and confidently walk the shaking and unsteadfast bridge, we on our return plucked up courage and performed the feat in an upright position. The picture here was very charming—mountains closing us in all around, a canopy of noble forest-trees, and the music of the mountain-stream as it plunged over its bed of rocks.

If the morning drive was charming, the return was enhanced by the beauty of the setting sun. The river, the trees, the hills, gained new beauties from the rays of the

Steamer on the Tennessee warped through the "Suck."

level light; and Lookout Mountain, whose high top would occasionally reveal itself, towered superbly, purpling in the evening air. Arriving at the ferry near sunset, we experienced some amusing incidents in crossing the stream. It is one feature of this method of ferrying a river that the exact place of landing cannot be controlled, the rise or fall of the stream varying it considerably. On our return we found the nose of the boat thrust into a bank, and some apprehensions prevailing as to how the waiting cargo was to be got on board. Our horses were unharnessed, and the vehicle, by the strenuous effort of half a dozen negroes, lifted on the boat. Then the horses, our own and several others, without much difficulty, jumped the space; but the cattle struggled, and

THE TENNESSEE AT CHATTANOOGA.

backed, and plunged, with the most incorrigible perversity. Some charged back, and tried to escape up the hill; others plunged into the water; and one fine heifer was with difficulty saved from drowning. Immense numbers of live-stock constantly traverse the road along the Tennessee, and cross by this ferry into Chattanooga.

The Tennessee road is of historic interest, as being the principal avenue during the civil war by which supplies were sent for the army in East Tennessee. Ceaseless trains of army-wagons wound over the rough, devious, and picturesque road. The Confederate sharp-shooters hung along the southern bank, and it was not uncommon for a sudden fusillade from the opposite hills to send death and consternation among the draught-animals and their drivers.

There is one feature of the Tennessee at Chattanooga that remains to be described. Under a high cliff near the ferry-landing may, at suitable season, be seen a number of flat-boats unloading their cargoes of grain or other produce from the upper waters of the Tennessee. Here is often a very stirring picture. Crowds of vehicles are receiving grain; there is the bustle of loading and unloading, the clamor of many voices, the noisy vociferation of the negro drivers, altogether making up a scene of great animation. These flat-boats come mainly far up through the Clinch or the Powell River, from the northern border of Tennessee and the southern counties of Virginia, bringing corn, wheat, and bacon. A striking characteristic of their construction is their ponderous stern-oars, which often reach a hundred feet in length. Floating with the current, these oars are only needed as rudders, and the necessity of their great length is not obvious. The flat-boatmen of the Tennessee are not, like those of the Mississippi, notorious as "hard characters." They are mostly farmers, who, once a year possibly, bring down their harvests, and perhaps those of their neighbors, to market.

Chattanooga is a great railroad centre; it is on the main line of travel between the North and the South, and is rapidly developing into an important place. The influx of capital, with fresh energy and a more varied industry, has given a marked impulse in the development of a section rich in so many natural resources.

have been torn apart by the waves, and which stand either isolated or in groups in the midst of the waters. All along the coast from Eureka, on Humboldt Bay, to Sonoma County, the shore is rendered interesting by these gigantic fragments, around which the wind howls with fruitless fury, and where the wild birds of the ocean congregate in myriads, deafening the visitor with their tumultuous cries.

The tourist will find little scenery of sufficient grandeur to interest him until he approaches Cape Mendocino. Here the mountains, which previously were low down upon the line of the horizon, come right up to the sea. After crossing the Eel River the road winds along the skirts of Mount Pierce, a huge mountain, which terminates a long range of high hills, running parallel to and not far from the sea-coast. The sides of Mount Pierce are covered with the famous red-wood; and the eye ranges over miles of this magnificent tree without detecting any other kind. Many of them are three hundred feet high and twelve feet in diameter, and the magnificence of these mountain-forests can well be imagined. In the early morning the eye delightedly watches the mist slowly departing from the tall tops of these giants. A thick veil lies upon the cliffs and the sea, also unillumined by the sun. To the left, however, slanting arrows of red light come up beside the crags and fall upon the columnar trunks of the red-woods. The deep-green leaves seem gilded at the edges, and the bark of cinnamon-color glows under the red rays. Half-way up the trees the mist wreathes and circles about under the influence of the sun, and this movement communicates itself slowly, very slowly, to the deep bank of mist above, where the grays are pure. The sky above is wonderfully clear, tinged a little with saffron back of the mountain, and a few stars tremble lazily over the deep, dark pall of gray fog that overhangs the ocean. We can hear the slow, solemn pulsing of the waves and the roar of the breakers as they beat upon the rocks. A few light, wandering clouds suddenly become visible overhead, a tongue of fire licks the topmost crag of Mount Pierce, and warms its barrenness. The cloudlets become a pale red, the mist upon the trees creeps up higher, and more and more of the dense foliage becomes visible.

The cliffs are not high, but along them are the fragments that the sea in its fury has overwhelmed after centuries of never-ending warfare. In a kind of inlet, standing like the monument of some great one in a market-place, is an isolated rock of fantastic shape. It is of basalt, seamed and scarred very strangely. The sea has worn a passage through the base, through which the waters plash and rage unceasingly. The height of the arch thus made gives us an idea of the fury of the storms that have beat upon this tower of the sea-birds. If this did not exist, we might infer it from the difference of color in the rock. Above, the tones are pure gray; but below, where the tempest reaches, of a dark-brown. The crest is of a dazzling white, from the guano of the wild-fowl that make their nests there.

Nothing can be more tumultuous or less pacific than the waters of the Pacific

SCENE AT THE MOUTH OF RUSSIAN RIVER.

Ocean along the Mendocino coast. Where there is a sandy beach, which is not often, it is pleasant to watch the incoming waves, and to compare them with those of the Atlantic. On the Eastern coast the surf is seldom more than six feet high, and the serried line of waters that comes dashing onward is rarely more than two hundred yards long. In fact, gazing at the sea that breaks upon the Atlantic shore, one can see without difficulty ten or a dozen waves breaking on the shore or advancing in line, all within the field of vision afforded by one glance. It is not so here. The waves on the Pacific coast are less frequent, but they are twelve feet high and a mile in length. The curves described by the falling crests of such waves are infinitely finer than any thing which the Atlantic presents; and the boiling fury with which they crash upon the beach and churn the sands is, at first sight, appalling. All along the Mendocino coast the waves have worn the cliffs into strange and wondrous forms, beating out caverns where the lower part is conglomerate rock, and series of arched cellars,

Basaltic Formation near Russian River.

COAST SCENE, MARIN COUNTY

into which heaps of sea-weed and *débris* are thrown. The basalt, which is the leading character of the crust, is not uniform in texture, some parts being very much softer than others. Wherever this occurs in the proximity of the waters, they have invariably scooped out the soft rock, making all kinds of mystic arches.

Back of the hills that hem in Mendocino there is a stretch of level country, known as the Long Valley, which is watered by the main fork of the Eel River. The mountains press closer and closer to the sea until the mouth of Russian River is reached. This is a small stream, and is only navigable for about twelve miles from its mouth; but there are many saw-mills on its banks, and Bodega, the nearest town, does quite a lumber-business. The entrance to the mouth of Russian River is quite picturesque. The northern side of the little bay is very bold. The promontory is of the most striking character, coming down from the mountain-peak in a succession of grand, sweeping terraces. On the flanks there is the inevitable red-wood forest, which, in places, ascends almost to the summit. In other places the mountain is bare and rugged, showing huge masses of grayish granite verging on purple. The cliffs at the extremity of the promontory have been torn and rent by some dreadful convulsion until they are almost separated from the main-land; and their jagged summit bears a quaint resemblance to the spires and minarets of a cathedral. At the entrance to the mouth of the river are huge detached cliffs of basalt, which form two groups, called by the boatmen the Brothers and Sisters. On the southern side of the mouth of Russian River there are broad sweeps of fine pasturage, from which, however, the basalt crops up occasionally in isolated peaks. One of these peculiar formations—shown in the engraving—is more than three hundred feet high, and affords a pleasant shadow in the hot noons for the flocks of sheep and their shepherds. It is nearly square, and the sides are so steep that no one has ever succeeded in climbing, though many have tried.

The town of Bodega was once a Russian station, and in the vicinity there are still the frail and fading remains of a stockade and fort, with an old church, built in 1787. There is an old hotel, built of adobe, in the regular Spanish style, with a garden attached. From Bodega it is a short distance to Valley Ford or Bodega Corners on the railroad, but for the tourist who wishes to see the coast this route is inadmissible, since it is a diversion inland, and a turning one's back upon the scenery and the difficulties of the shore-line. A stage will take him to the town of Two Rocks, which derives its name from the configuration of the coast. One of these rocks is shown in the illustration. The height is about two hundred and sixty feet, but the mass is enormous Detached rocks, like needles worn to a point by the eternal blustering of the winds and waves, surround it on every side, and on the flanks there are broad, flat masses, which are the favorite resorts of seals. Here also come innumerable birds to make their nests upon the broad, flat summits of the rocks.

For the lover of natural scenery, these enormous, isolated rocks have a grand fasci-

MOUNT TAMALPAIS AND RED PORCH.

nation. The grayish tones of the upper part, melting into the deepest brown, with the glowing white of the summit produced by guano, and the broad yellow of the withered grasses, delight the artistic eye. The shadows vary from pale violet to deep purple, according to the hues of the lights. The lines are also most picturesque, Nature having contrasted all varieties of lines—perpendiculars, diagonals, horizontals, vanishing curves in the rocks; while below, in the swash and foaming of the tumultuous seas, there are other curves of a totally diverse character, and other tones, which contrast strangely with the colors of the rocks. The seas are deep-green, like emerald, or muddy-green, according to depth and other conditions; and there is great variety, also, in the white tints of the foaming crests, according to their volume. The sky above is not a very deep blue. It is a softer, milder cerulean than that which arches over our heads in the Atlantic States; it is not so splendid, but it is more tender.

Alcatraz Island.

THE "CLIFF HOUSE."

Persisting in our resolution not to be diverted from the coast, we must, now that we have arrived in Marin County, take a schooner from the pretty little harbor of Olema, only fifteen miles from San Francisco, and enter the famous bay in this way rather than yield to the seduction of stages or railways. As we approach the entrance, the hills on the left loom up through the deep haze like giants, and are, indeed, more than two thousand feet high. To the right, they are by no means so lofty. As the mist clears off, they are bare and sandy, and are not very picturesque, though on the left the peak of Tamalpais shows grandly. The view opens, and the splendid strait called the Golden Gate appears. Through it we can see the island-rock of Alcatraz, with its fortifications gleaming in the distance. The enormous mass of Tamalpais, which showed at first boldly in our front, seems still behind Alcatraz. Between the last and the shore is Angel Island, very high, and covered with rich green vegetation. Goat Island, with its fort, is on the left of Alcatraz. To our right hand is Fort Point, where the United States flag floats, and, a little beyond it, the old Presidio. Then comes the city—the glorious city that leaped full-born into existence. It rises up with numberless towers and spires, and great warehouses, as the schooner, with her sails filled to bursting with the fresh sea-breeze, staggers on. Little craft and big craft, steamers from the ocean, tugs, and every variety of floating thing, are spread upon the gleaming waters, whose green waves dash into white foam upon the three islands ahead. Beyond the city one can catch momentary glimpses of shipping, which grow fuller and fuller until we get abreast of Alcatraz, when all the glory of the bay bursts upon the sight. Far on the other side are Benicia and the glittering waves of Carquinez Straits. Beyond we catch a glimpse of the peak of Mount Diablo, at the base of which seems to crouch the town of Oakland, though it is really a very large place. But the air is so pure, so serene, that one can see the scarred ravines on the sides very far, and almost fancy that Stockton is in sight. It is not from the bay itself, however, that the finest view can be obtained. To survey all the beauty of the Golden Gate it is necessary to climb Telegraph Hill. From that elevated position, with roofs and buildings lying peacefully below one's feet, and stretching far out to one's right hand, the prospect of the Golden Gate is, indeed, exceedingly beautiful. The portals of the "Gate" seem but a mile apart, and, through the mist that hangs upon the farther side, the giant Tamalpais looms with tremendous force. The steamers, with their crowds of passengers swarming along the bulwarks, move majestically through the heaving tide, which makes the white-sailed schooners dance, and rocks the three-masted merchantmen that have traversed wild wastes of water around Cape Horn. The islands show plainly, and the fortifications gleam brightly, under the full glare of the sun. To the inhabitants themselves there is no pleasure equal to the drive through the sand-hills, over a fine, hard road, to the Cliff House. This is emphatically the most picturesque part of San Francisco, both in its surroundings and in its seal-cliffs, where the sea-lions bark and whine and roar, with

SEAL ROCKS.

none to make them afraid. The Cliff House is built, as its name imports, upon frowning basalt; and the road that winds from it to the ocean hence has been cut through solid rock. The bluff of the hotel is about one hundred and thirty feet in perpendicular height, of a gray color, verging into the deepest brown. Detached bowlders lie at its base, and are tormented by the fierce rollers. Beyond, at some distance, are the rocks on which the sea-lions congregate.

Point San Pedro, near San Francisco.

From the Cliff House, a road extends along the beach for about five miles, which has much to recommend it in the scenery by which it is surrounded, for from this sandy strand one can see the Pacific Ocean roll in with uninterrupted grandeur. Nothing can be conceived more majestic than this sight, especially at that part of the beach which gives a fair view of Point San Pedro. The length of the wave-walls is fully a mile, and the height of the rollers twelve feet. The enormous mass of water comes onward with a solemn grandeur which appalls.

The Sacramento River.

There is no hesitation, no tremor, along the whole line; and it looks like the charge of an army of cavaliers galloping with perfect regularity and even line upon the foe. Solemnly it advances, with the crest just flecked with foam; and everything seems hushed, as if in expectation of the onset. Suddenly, as it nears the shore, there is a trembling all along the mile of sea, and the crests begin to curve slightly over. The line halts; the crests curve more and more; and suddenly the immense length pours down like a cataract upon the shore. Everything has a throb; the solid earth seems to tremble, and the great rocks to oscillate. The white rime that was poured over the strand rushes back with incredible velocity. As it retreats it meets still another oncoming wave, and, striking its base, hurtles it down with crashing fury; and then there is a hush. The sea is silent. The birds and the insects, taking courage, begin to sing and to chirp until there comes another solemn booming, and the roar of another broken, rolling wave. And this eternal symphony takes place in a kind of bay, where the mountains, rushing to battle with the sea, have advanced far into the waters, and their outposts have been terribly mangled. The great promontory has been severed from the mountain; and between them are three square, isolated crags, with shallow water around them. Here the sea rages and bellows like a wild thing; and the waters seem to lose themselves in eddies and whirlpools, and to be unable to find their way back to the sea, so that they might charge in line with the great, solemn roll-

Marysville Buttes.

LASSENS BUTTE, SACRAMENTO VALLEY.

ers. The old promontory, now become an isolated crag, is covered with sea-birds. Seals sometimes come here, but not very often, as they are not protected. Blocks of granite show themselves occasionally peeping up from the sand, and probably are bowlders deposited there in by-gone ages, which the sands have covered. The sea is diversified with the sails of fishing-boats, for fish are abundant in these waters, and the birds are busy all day long in the neighborhood of their stronghold.

Let us turn our faces northward and make our way toward Oregon overland, for the wonders of the inland scenery rival, in their wildness and sublimity, the sights that we have just described.

The valley of the Sacramento, at first with an undulating surface, and diversified with earth-waves crowned with noble oaks, as we ascend northward spreads out into treeless prairies, as flat as those of Illinois. The first break in the monotony of the expanse is made by the Marysville Buttes—a short range of low, volcanic hills, which rise suddenly up from the eastern bank of the Sacramento, which is lined with alders and cottonwood-trees. The latter are broad and umbrageous, but not high, and their dark-green foliage contrasts pleasantly with the gloomy brown hues of the fire-born hills. Native grape-vines trail along the ground, and cling around the trunks of the trees, making picturesque bowers. Between these one catches glimpses of the river, reflecting placidly the masses of green foliage and the trailing vines. Deep pools here and there give back the blue of the cloudless sky; and, as a bass accompaniment, come in the dark shadows of the Buttes, with their sharply-drawn angles and their truncated cones. The slopes that rise from the banks have a very gradual ascent, and are dotted with ranches, pleasantly hid by orchards and vineyards. Looking from the cones of these hills to the right and left, the eye glances over miles upon miles of flat plains, where fields of wheat succeed to vineyards and to groves of oak, broken only by the homes of the settlers. In the far distance can be faintly discerned the undulations of the foot-hills on either side—the first indications of the Coast Range to the left and the Sierra Nevada to the right.

The Lassens Buttes are the first mountains which are capped by perpetual snows. The river winds at the feet of these giants of stone, and its banks begin to show traces of the higher ground, in the degeneration of the cottonwood-trees, and the improvement in the appearance of the oaks, which here are very lordly. The trunks are huge, though the height is not remarkable, and the spread of the boughs is enormous. Not only are the banks crowned with groups of these trees, but the low, brown hills at the foot of these snow-clad summits are fairly embowered in their luxuriant groves. Seen in the early morning, when enveloped in sombre mist, they give a mysterious beauty to the bases of the mountains at this point. Above is the vault of intense blue, with the stars glittering like "patens of bright gold." A few faint, rosy clouds fleck the heavens, and reveal the iridescent splendor of the snowy crests, which glow with all the colors

LIMESTONE FORMATION, ON PIT RIVER.

of the opal or of an iceberg. The shadows from these summits are of the most brilliant purple, and blend slowly and gradually with the dull brown of the mountain-sides.

On the crest that surmounts the valley of the Pit River fine pines are reached, which are grouped in masses considerable enough to be styled a forest. This extends along the line of the river, which, cutting its way through the Sierra, falls toward the west in a series of white, tumultuous rapids. Rising directly from this pine-laden crest is a range of granite and limestone rocks, which attains an elevation from the plain of three thousand feet, and is broken into a multitude of ragged forms. The granite is a bluish gray, which relieves the dazzling white of the limestone. When the sun shines upon the latter the observer can hardly tell it from marble, so brilliant is its snowy hue. The line of these singular hills is of considerable length, extending, indeed, along the whole valley of the Pit. When the crests are of granite the forms are of that bold, bluff character so peculiar to crystalline rocks, but where they are of limestone their appearance is finely castellated. The peaks have been wrought, by the cunning hand of Nature, into the guise of gigantic castles of the El Dorado land. Battlement and bartizan and huge donjon-keep present their sharp angles and clearly-defined walls against the brilliant blue of the California sky. Here and there coniferous trees have attained a root-hold on the almost perpendicular walls, and flaunt their branches like the banners of a proud castellan.

The eye turns, with a sense of exquisite relief, to the wooded crest below, where the sugar-pines stand in glorious phalanxes. These trees grow to an immense height, often not less than three hundred feet, though their diameter is only eight. This gives them an appearance of slenderness and grace resembling the effect produced by Saracenic columns—an effect heightened to the utmost pitch of idealism by the character of the trees. For fully one hundred and fifty feet these lovely trunks are branchless, and as symmetrically rounded as even a Neo-Greekish architect could desire. The hue is a bluish purple, delicately marked with a net-work of scorings. From the point where the branches commence they stretch out with nearly level poise straight from the shaft, and their leaves are dark-green spiculæ, to which the noonday sun gives a yellow tinting. Lying down at the bases of these regal pines and gazing upward, one sees the foliage massed with fairy-like grace against the white walls of the limestone, and above these three thousand feet of blinding glare is the sky, like blue fire, into whose depths the eye seems to pierce. But the sugar-pines are not alone. Often they are mixed with firs of feathery, bluish-green foliage, hiding by its mass the dark-brown trunks. And then, more rarely, is found the big tree, the *Sequoia gigantea*. These, however, are generally to be met with in open glades, green with herbage, and bright with the blossoms of many flowers. Some have a circumference at the base of one hundred and thirty feet, and rise to an altitude of three hundred feet. The bark is excessively thick, scored with deep, regular grooves, and of bright-brown color, mottled with purple and yellow.

PILOT ROCK.

The foliage of these huge trees is delicate beyond description, and has been aptly compared to a pale mist of apple-green hue. Old, very old, are these trees. Many of the sugar-pines are a thousand years in being, and the giant *sequoias* more than double that tremendous age.

Though the granite and limestone hills extend with unbroken grandeur along the line of the Pit River, the pine-grown crest is broken here and there where the valley broadens, and from the opposite bank the land stretches out into breadths of prairie. There are not wanting trees, for manzanitas and sugar-maples grow in clumps upon the plain; and on the opposing slopes, wherever the limestone is not too precipitous, there are dense, serried ranks of firs, and occasionally in the ravines a stunted growth of cottonwood. And, wherever there are foot-hills, the oaks abound, displaying their far-reaching branches and their lustrous leaves. The river is quite rapid, but neither broad nor deep, and abounds in salmon.

The region of the Pit River is an absolute wilderness. Lake and field and fell, naked crag and towering pine-clad crest, succeed each other with a savage grandeur similar to, but far greater than, that of the wilderness of the Adirondacks. To the south and east are the iron-hills, enfolding in their rocky clasp millions of treasure, that will be brought out by the future generations.

From the valley of the Pit the traveller rises, continually traversing woods covered with fair mountain-pines, until, through a notch to the northward, a glimpse can be caught of the huge summit of Shasta, which may be ascended from Sisson. Rising from the plain are hundreds of small volcanic hills, built up out of the lava, the mud, and *scoriæ*, thrown out from the crater above in other times. Beyond, there is what may be termed the base of the mountain, attaining an altitude of some two thousand feet, and throwing out spurs in every direction. Above this the cone of the mountain rises in one tremendous sweep to a sheer height of eleven thousand feet. The stupendous proportions of this great snow-peak would alone be sufficient to rivet the attention of every traveller. But to these must be added a most wonderful play of color. The lava forming the body of the mountain, which penetrates often through the snow-part, is of a pale rosy hue, and, when the sun shines on this, it has a splendor which words are too weak to render adequately. The snow, with its pure, white, fleecy fields, is in many places diversified by great glaciers of ice and yawning crevasses, in whose depths are shadows of the most intense blue. Upon the veins of the ice the sunbeams fall with refracted glory, giving forth the most wonderful opalescent tints. Here, in some places, the hues are green as emerald; then, in others, there is a lurid purple, interstriated with a tender pink. In other spots, the prevailing tone is a rich cream-color, perfectly translucent. The snow, too, has its colors, but generally glows with an incandescent fire under the welcoming kisses of the solar rays. So beautiful, so varied, are the effects produced by the mingling colors of lava, of snow, and of ice-enamelling, that the be-

UMPQUA CAÑON.

holder cannot consider other things. His eyes are ever strained upon the peak, and bent admiringly upon its lustrous hues and the deep, violet shadows that contrast them. He has but one thought—to watch the radiation of color at sunrisings and settings, and see the fiery rays slant and shoot across the great mass, working its parts up from the still white and steely gray of night to all the splendors of the northern lights.

When the eye has been satiated with the radiant colors of Shasta, the mind begins to be impressed with its vast proportions. Its total elevation above the sea is fourteen thousand four hundred and forty-two feet. From the base the cone rises upward in one tremendous sweep of lava and ice. There was no cumulative series of effects of Nature in building up this mountain, for it is a gigantic peak set simply upon a broad base that sweeps out far and wide in every direction. Very sheer and precipitous is Shasta to the north and south, but east and west there are two grand slopes, from the plain right up to the rim of the crater. These are the buttresses of Nature's great chimney. Beyond a well-defined line the ascent toils upward without a tree or shrub to cheer it on the way, retaining nothing save a little stunted herbage. This is soon replaced by the pale, roseate lava, and above that comes the deep blue of the snow in shadow. The road winds through Strawberry Valley, over a soil entirely of pumice-stone; and it is odd to see great sugar-pines, whose roots are firmly embedded in masses of this substance. Around Shasta this tree produces its most enormous cones, some of them being fully eighteen inches in length.

The road passes, on its left, Black Butte, now called Muir's Peak, another extinct volcano of considerable size, which appears dimly in the far distance. After leaving behind Lower Klamath Lake, one sees rising in the blue air the singular form of Pilot Rock, an elevation of the Siskiyou Mountains. This is a great mass of black volcanic substance, which rises perpendicularly from the mountain-crest. The Siskiyou Range has here an elevation of twenty-five hundred feet, and the knob is about five hundred feet higher.

The volcanic origin of the mountains all through this region accounts for their singular lack of beauty. The angles are so sharp that the earth which covers their skeletons cannot adhere, and comes off in great land-slides, leaving the mountain-sides bare and exposed. But the trees which skirt the base of the hills are very beautiful. Every step toward Oregon from this point seems to increase the size of the forests. The trees grow thicker together, and the firs and pines are larger. There is no lack of oaks through these valleys, and the trails often wind through groves that are park-like in the beauty of their natural arrangement.

Beyond Pilot Rock the old immigrant-trail crosses the Rogue River, a beautiful stream, full of salmon, and then entering the Umpqua Mountains is soon involved in the terrors and the gloom of the Umpqua Cañon. This is a pass through the mountains, eleven miles in length, with sides twenty-five hundred feet in height. So high are

THE THREE SISTERS.

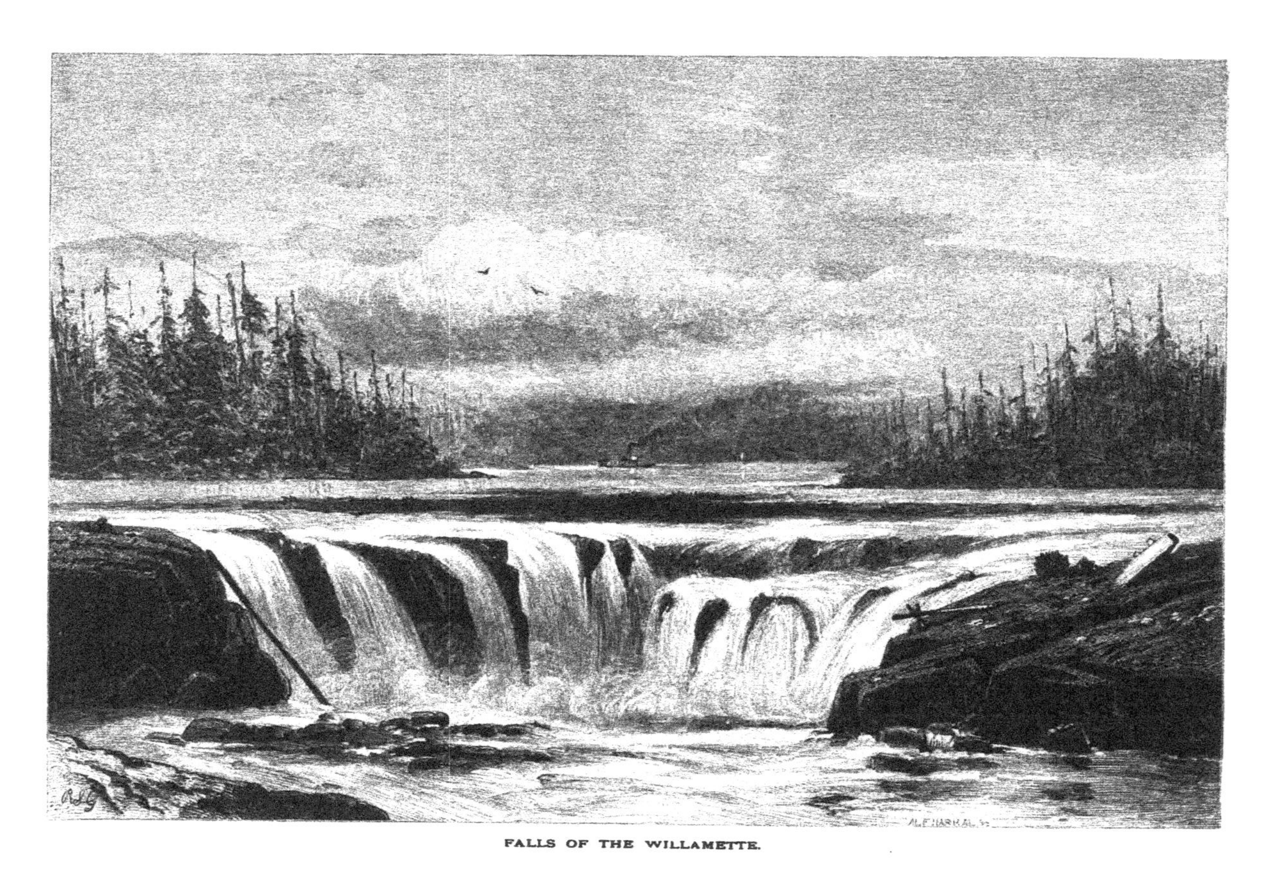

FALLS OF THE WILLAMETTE.

these tremendous walls that it is with difficulty one can discern the blue sky. If you listen attentively you will hear a feeble trickling below the road, which, the walls re-echoing, swell into a low murmur. This is the Umpqua River, a narrow thread of water, which, in the summer and autumn, hides itself under the vegetation of the pass, and steals away unknown and unnoticed. But in the winter and spring, flushed with rains and melting snows, it becomes a torrent, occupying the whole width of the cañon, and sweeping away every thing that obstructs its course.

On emerging from the cañon the traveller is in Oregon, and in the valley between the Coast Range of mountains on the west and the Cascade Range on the east. Striking the forks of the Mackenzie River near Eugene, the snow-clad summits of the Three Sisters loom up into the pleasant air. They rise from a range of volcanic hills, of moderate height, to a considerable elevation, being capped with perpetual snows. The Mackenzie River flows along the edge of the plain; and, from the eastern side, bluffs of basalt rise perpendicularly to a great height, which then spread out into a fair table-land, reaching, by a fine, gradual ascent, to the base of the mountain-range. The sides of the Three Sisters are finely zoned with a broad belt of forest, which mounts to an altitude of eight thousand feet. The angles of the Sisters are less acute than those of other snow-mountains in this region, and consequently there are fewer slides, and the peaks are always covered with the glistening folds. The clouds rest continually upon the Sisters, adding their contributions of vapor, to be turned into tiny snow-flakes; and of mornings, ofttimes, the haze wraps them round in mazy folds, producing vague, fantastic images. The Indians believe that these three peaks were three female giants, who had been wives of Manitou, and, having rebelled against him, were turned into stone.

Not far from them, in the same range, rises the Willamette River, on which Portland, the chief city of Oregon, is situated. The falls of this stream are justly celebrated for their beauty. The river, which is generally about a mile wide, narrows suddenly near Oregon City, as if preparing for its tremendous leap. The rocks on each side are of frowning basalt, of a deep black, rendered more intense by the foaming waters. By the action of the stream—the current being strongest in the centre—the falls have been worn into a horseshoe form, the two sides being so close that one can throw a stone to the other shore. The water, rushing with a very swift current, precipitates itself down a sheer fall of seventy feet, rising, in smoke-like mist, from the mysterious depths below, where it issues in great swaths of turbid water, streaked with green and curved like glass, jostling with the bowlders of basalt, and roaring in rage at the contest. There is not so much mist as in some less grand cataracts, but there is enough to hide the fortunes of the fallen river, and the confusions of its lines, as they beat against the masses of rock which they have detached through long successions of heroic charging.

SOUTHERN SIDE OF WILLAMETTE FALLS.

THE DELAWARE WATER-GAP

WITH ILLUSTRATIONS BY GRANVILLE PERKINS.

THE Indians called the Kittatinny the Endless Mountain; and, disregarding all the discussions of modern science, we may still say that the great range stretching from Maine to Georgia was the strong backbone of the thirteen colonies, that made them stand erect among the nations. In such union, indeed, there is strength; and grandeur and beauty invest the whole—whether, as the Green Mountains, giving its euphonic name to Vermont, or when the snow-capped peaks become the White Mountains of New Hampshire; whether Dutched into Kaatskill, or when, in Pennsylvania and the more Southern States, the even tinting of the forest-clad sides renames them, as, melting softly into the atmosphere, they are as blue as the circumambient air.

Pretty streams rise on the western declivity of the Catskills, and, quitting their mountain birthplace, wander toward the southwest until near the line of Pennsylvania they unite, and thence, as the mighty Delaware, move on in constantly-increasing volume, the fitting boundary of majestic commonwealths.

Near the junction of the three States of New York, New Jersey, and Pennsylvania, the river again approaches the mountains, and follows their western side through a succession of magnificent scenes, which, increasing in grandeur, find a sublime culmination where the river turns abruptly into the mountain, which opens to give it passage into a defile, or cañon, called the Delaware Water-Gap. Thenceforward the forms soften from grandeur into grace, and the river, escaping from bluff, precipice, and rock, pursues its way through rolling lands to the sea.

To the first settlers the mountains proved a troublesome barrier, and all intercourse to the southward necessarily passed through the natural gate-ways of the gaps; but the Delaware writhed its way through its cavernous passage with contortions too like those of the rattlesnakes that thronged upon the banks, and the dangerous pass was long avoided for the easier road through the Lehigh Gap, where the water-course of a pretty stream led to the head-waters of Cherry Creek, and a pleasant road followed its bank through the beautiful Cherry Valley, full of dimpling hills and fine orchards. This Cherry Creek, running toward the north along the western side of the mountain, to join the Delaware just above the Gap, formed a natural road to Philadelphia, which by reason of its pleasantness long maintained its popularity. Nearly midway between the two rivers, Nature had also provided another gate-way in the Wind Gap, called, by the early Dutch settlers, "Die Wind Kaft," a sharp notch, which, descending almost to the base of the mountain, but not low enough for a water-passage, was only a pass for the winds. This route was the road cut by General Sullivan and his army in 1779.

The earliest history of the region is involved in obscurity; but, shortly after Hendrick Hudson, in his little Half-Moon, passed up the river that was thenceforth to bear his name, his enterprising countrymen founded settlements at Orange, afterward to be known as Albany, and at Esopus, since the historic city of Kingston. The pretty valleys leading to the southwest wooed these colonists to travel, and the Dutch, certainly at an early date, traversed the valleys of the Mamakating and Neversink to the land of the Minisink. Near the Gap were found mines of copper and iron, and "the mine-road" was soon opened, proving so available that, even as late as the year 1800, it was chosen by John Adams as the best route from Boston to Philadelphia.

The two grand mountains which form the mighty chasm of the Gap have been fittingly named. The one on the Pennsylvania side is Minsi, in memory of the Indians, who made the Minisink their hunting-ground. The opposing more rugged and rocky cliff in New Jersey bears the name of Tammany, the chief of chiefs, who clasped hands in solemn covenant with William Penn under the elm-tree of Shackamaxon.

DELAWARE WATER-GAP.

The ruggedness of the narrow defile is seen in the sketch of the entrance. The bold face of Tammany exhibits vast, frowning masses of naked rock, while the densely-wooded Minsi displays a thicket of evergreen, with the railway-track skirting it down by the water's edge. Mount Tammany defies ascent except by a vigorous climber, but the bold and distinct stratification shown in the great rocky mass called the Indian Ladder adds to the grand abruptness of the outlines, and from the narrow mountain-top is best beheld the wide, extended view of the scenery above the Gap.

Mount Minsi owes its sweeter beauty to the lovely streams of water that descend

Distant View of the Gap.

its sides beneath a dense foliage, which veils the mossy pools and fern-draped cascades from the sunlight into the cool twilight that enraptures the summer tourist.

Successive ledges, or geological steps, mark the face of Minsi, and upon the lowest of these, at nearly two hundred feet above the river, stands the modern hotel, on the site of old and well-known

> "Kittatinny House, that on a rock is founded,
> So, when floods come, the folks won't be drownded."

The stream that issues beneath the hotel, to fall in a cascade into the river, has come down the mountain-side through a dark ravine. The densest bordering of rhododendrons fringes its sides with dark foliage and lovely blossoms, while tall trees complete the shade. Far up the ascent it takes its rise in the Hunter's Spring, whose cool margin has long been known as a welcome resting-place to the sportsmen who sought deer along the range. Under the name of Caldeno Creek, it continues its downward course by cascade and water-fall, and, to those who have once followed its devious way through the shaded ravine, the lovely glens and fairy grottos must return in dreams, for to dream-land does their witching, twilight beauty seem to belong.

Along the face of Minsi, about five hundred feet above the river, runs a grand horizontal plateau of red shale. Extending for several miles along the mountain, it

Cherry Valley.

makes one of its most remarkable features, and is known as the Table Rock. Over the slope of this ledge, at an angle of forty-five degrees, the lovely Caldeno flows in a charming succession of miniature falls or rapids. The rocky strata beneath are densely covered with moss, which, kept ever verdant by the passing streamlet, is still further fostered in its growth by the thick shade of towering trees, and gives the spot its claim to the name of Moss Cataract.

Lower down, Caldeno, stilling its wavelets into temporary repose, rests a while in the cool confines of a rocky basin. Shade even more dense makes a twilight at midday, and, dark, silent, and secure, a happy fancy has made it Diana's Bath.

At a still lower range, or ledge, the stream dashes at Caldeno Falls over a rugged, rocky precipice, in which the singular regularity of the formation is exposed in the broken surface of the falling water.

One of the loveliest aspects of the varied beauties of the Gap is under the early morning light, when—

"The mountain-mists uprolling let the waiting sunlight down—"

dense clouds of vapor break the contours of the peaks, causing uncertainty of vision, increasing or diminishing the apparent height, at times making the tops suddenly appear to bend forward as if threatening to fall, or as suddenly recede into vast distance, while softly-tinted masses of veiling vapor are wafted hither and thither by the wind at its own sweet will to catch the morning splendors, and wreathe in many-colored scarfs around rock, and crag, and lofty pine.

Poetry and romance have familiarized us with the legend that, as a forerunner of storm, Pontius Pilate still appears above the mountain that bears his name, and, bending in cloudy presence, wrings his hands in remorse for his evil deed. A cloud-phenomenon somewhat similar occurs upon these heights. A narrow space between two jutting peaks foretells by clearness or cloud the fortunes of the morrow, but no legend lingers around the summits, and prosaic Americans call it the Rain-Hole.

But, wild and wonderful as is the interior of the Gap, it is outside its limits that the grand scenery of the region must be sought. From the mountain-peaks on every hand open magnificent vistas, and from the river, both below and above the chasm, the views are of marvellous extent. Spurs jutting out from the main range give endless variety to the landscape, while hollows, gaps, and ravines add their countless beauties.

Several miles above the Gap the Delaware is joined by the mountain-stream called the Bushkill. This creek was long regarded as the extreme limit of civilization in this direction, all beyond being a howling wilderness too often full of howling savages. In this neighborhood were the copper-mines which at an early date attracted the Dutch settlers from the Hudson, and induced them to open the famous Mine Road, which became the thoroughfare from Albany to Philadelphia, following the Esopus and Neversink Creeks to the Delaware, crossing that river to reach the western side of the Blue Ridge, and passing near the Gap of the Delaware to find a passage at the Lehigh Gap, and thence a southerly course to Philadelphia.

Upon the Bushkill is one of the most beautiful water-falls of the district. A chasm one hundred feet in height is surrounded upon three sides by an almost perpendicular wall of rock, over which the water falls. From a point below, the scene is grand in its sombre magnificence, as the swift torrent, striking midway upon a projecting ledge in the rock, rebounds in snowy foam-flakes, which, after the momentary interruption, continue to fall into the dark chamber of rock below. On the walls of the chasm, at a level with the summit of the water-fall, there is still another scene of equal beauty, as the rapid stream emerges from the dark shades of the forest to make the sudden plunge from the precipice.

Another small mountain-torrent near by frets its way through a tortuous channel of dark fossiliferous limestone, until, in a sheet of foam, it leaps over a precipice in a shower of dazzling whiteness, this singularly beautiful cascade has been called But-

Delaware Water-Gap, from the South.

termilk Falls. Upon the same stream the Marshall Falls deserve special note for their picturesqueness. The dark surrounding rock is crowded with fossil impressions, which fill the stone with irregular fissures; through this ledge the waters have torn and gnawed their way down a chasm fifty feet in depth, leaving a veil of overhanging rock in front, through which the spectator gazes at the gloomy cataract as through a curtained casement.

DELAWARE WATER-GAP, LOOKING SOUTH FROM SHAWNEE.

That the Minisink was a favorite abode of the red-men is proved at almost every step. The plough turns up quantities of spear-heads and arrow-points, as well as hammers, and tomahawks of stone, and rude cutting instruments fashioned out of flint; stone mortars and pestles have been found, with bowls and jars of earthenware.

Upon commanding elevations, where small plateaus permit at once a kind of seclusion as well as an extensive outlook over the mountains and the river, there are many Indian burial-grounds, always chosen for the beauty of the position. In the graves almost invariably are found articles of personal adornment, with warlike weapons, and frequently vessels of clay. Glass beads, bells, and trinkets of metal, are supposed to prove some intercourse with the white race, but beads of bone, bowls of baked earthenware composed of pounded shell and clay, and the ruder instruments made out of stone, mark Indian workmanship, and may belong to more remote generations.

As one of the wonders of the Gap, must be counted the marvellous lake upon Tammany—a lake so singular that popular superstition has been tempted to add a final touch to its surpassing strangeness, and declare that it has no bottom. As if in quaint climax to her wild work, Nature, after riving the mountain to its very base, here places beside the rude chasm, on the very apex of the lofty peak, a peaceful lake. Masses of bare gray sandstone stand about its margin, and within the stern encirclement the pure water reflects alone the swift-darting birds or the slowly-moving clouds, for naught else comes between it and the sky. In this unbroken solitude, beside the lonely lake, is a single Indian grave in a narrow cleft of rock. On a lower level, near at hand, many graves were gathered into one place of sepulture, as if to make the loneliness of this solitary tomb even more marvellous; and fanciful conjecture can but gather round the grave to ascribe to its tenant some strange history, and imagine him to be a king who disdained companionship in death with those he had ruled when living; or a poet who sought a resting-place beneath the clouds, or a prophet entombed by his devout followers beneath the skies in which he had beheld visions.

Throughout the whole Minisink single bodies are occasionally exhumed by the plough, or washed out from the river-banks, but it has been conjectured that these have been enemies, or those whose fate was unknown or not regarded, for the numerous burial-grounds attest that even the wild wanderer of the forest craved to find his last resting-place in companionship with his kind. In these ancient cities of the dead each tenement is a low mound surrounded by a clearly-marked trench, and frequently several mounds are connected into a single group by a ditch encircling the whole. In the graves that have been examined in the plateaus consisting of coarse gravel and clay, the bodies are found embedded in the river-sand, which must necessarily have been carried a considerable distance expressly for the purpose.

The last lingerer of the primitive people was Tatamy, veritably the last of the Mohicans. He had long served as interpreter to the travelling Moravian ministers, and his sympathies bound him so closely to the region of the Minisink that he voluntarily remained behind when his tribe moved to the West. An iconoclastic generation has degraded the name of his lonely home into the wretched diminutive of Tat's Gap. In this wild spot he remained, and a touching picture is drawn of the solitary man sitting alone at the door of his wigwam, hunting alone upon the mountain, singing in the forest wilds the songs of his departed nation, and striving feebly to preserve the habits of his old life amid the encroachments of civilization.

Moss Cataract.

The story of the relations between the aboriginal races of America and their European conquerors has always been a sad one, and it is especially so in the land of the Minisink. Here a singularly mild and cultivated tribe, the Lenni-Lenape, welcomed the early settlers with unusual kindness, a feeling which seems to have been quite heartily reciprocated by the French and Dutch. This friendly intercourse was preserved unbroken for a long period, and promised to remain so, when it was utterly destroyed by the incidents of the disastrous "Walk" of the year 1737.

Diana's Bath.

The Indians had apparently been perfectly satisfied with the terms of the purchases made by William Penn. According to the native custom, the territory sold was always measured by distances to be walked within specified times. In the first walk, William Penn had taken part in person, and the affair had been conducted in true Indian fashion, the walkers loitering, resting, or smoking, by the way. But the successors of Penn had determined upon a

Moss Grotto.

different policy, and prepared a scheme for driving a sharp bargain.

The boundaries of the territory were to be determined by the point reached by walking for a day and a half from a certain chestnut-tree at Wrightstown Meeting-house, and the proprietors were undoubtedly determined to make, what in modern phrase is termed, a "good thing of it."

By the terms of the agreement, the governor was to select three persons for the task, and the Indians to furnish a like number from their own nation. The men engaged, as particularly fitted for the purpose on the part of the province, were Edward Marshall, James Yates, and Solomon Jennings.

Also, according to Indian usage, the measurement was to be decided when the days and nights were equal, marking precisely twelve hours between sunrise and sunset. Therefore, attended by a large number of curious spectators, belonging to both of the interested parties, the six walkers met before sunrise on the 20th of September.

By established custom, a day's walk was, with the Indians, a well-ascertained distance, and the day and a half from Wrightstown was expected by them to end at the Blue Ridge, the savages never intending, or even supposing, that the boundaries of the purchase could by any possibility intrude into, much less include, their favorite hunting-grounds of the Minisink.

The previous arrangements had, however, been made with care; the direction of the route had been distinctly marked, and a line run to the greatest advantage of the purchasers. That no time should be lost, relays of horsemen attended the walkers with liquors, and refreshments awaited them at places along the route.

Marshall fulfilled his part of the contract, walking with great rapidity and without pause. This infringement, at least of the spirit of the bargain, provoked incessant complaints and protests from all the Indians, not only those who belonged to the party, but those who were assembled as spectators, the savages exclaiming again and again, in angry expostulation: "No sit down to smoke—no shoot squirrel; but *lun, lun, lun,* all day!"

Caldeno Falls.

Before the first day ended, one of the white men and two of the Indians had given out; and when, before sunset, Marshall and Yates reached the Blue Ridge, they met there assembled a great number of the savages gathered to witness the expected ratification of the boundary. When it was discovered that not even the first day's walk was yet accomplished, the manifestation of anger became general; the Indians loudly proclaiming the whole affair a cheat, by which all the good land would be taken from them, indignantly refusing their assent to the purchase.

By sunset Marshall and Yates had passed the mountains, and started afresh at sunrise the next day; but Yates soon turned faint and fell from exhaustion, while Marshall pursued his course, and at noon reached the Pocono Mountain, having walked about eighty-six miles.

Bushkill Falls.

The indignant Indians immediately inaugurated a systematic retaliation, and the purchasers, who began to move upon the land in considerable numbers, found the savages arrayed in armed hostility. The warfare in this case did not consist of the usual occasional skirmishing, but was much more formidable, as being a part of a determined attempt of the Indians to regain the lost territory, which they believed had been taken by fraud, and which they never relinquished until the year 1764.

RICHMOND, SCENIC AND HISTORIC.

WITH ILLUSTRATIONS BY HARRY FENN.

IN the Century Club, New York, there may be seen a painting of a quaint old mansion of red brick, architecturally of the reign of Queen Anne, one wing of which stands only in its charred timbers and blackened walls. This mansion is situated on the left bank of the James River, and, a century and a half ago, was the stately dwelling of the "Hon. William Byrd, of Westover, Esquire."

There were three William Byrds, of Westover, grandfather, father, and son, each one of whom makes a figure in the colonial history of Virginia, but it was the second of the name and title to whom reference is made above—a man of many shining traits of character and of imposing personal appearance, as we know from contemporary records and from the full-length portrait of him, in flowing periwig and lace ruffles, after the manner of Vandyck, which is still preserved at Lower Brandon. He had an immense estate, and at his death left the Westover Manuscripts, from which we learn—a fact not mentioned in his epitaph—that he was the founder of Richmond.

On the 19th day of September, in the year 1733, he says, on their return from the boundary expedition, one Peter Jones and himself laid out two towns or cities, one on the Appomattox and the other on the James River, twenty-two miles apart. The one they called Petersburg, from the baptismal name of the Jones of the period; and the other they called Richmond, from a resemblance, in its site with soft hills, and far-stretching meadows, and curving sweep of river, lost to view at last behind glimmering woods, to the beautiful English town in Surrey.

Colonel Byrd did not live to see Richmond attain unto any considerable size, for the town was not established by law until 1742, and he died only two years later. A few warehouses for the storage and shipment of tobacco were built first of all; then an irregular and scattering collection of houses for trade grew up around them; and on the hills overlooking the settlement arose the dwellings of a few rich planters and the thriving Scotch and English merchants who had established themselves at the place. Richmond, indeed, had no importance until it supplanted Williamsburg as the seat of the State government in 1779, and so little prepared was it for defence in war that it was given up to the British troops, in Arnold's descent upon Virginia, without the firing of a gun.

Immediately after the War of the Revolution, sanguine expectations were entertained that Richmond would soon become, not only the seat of a large trade, but a centre of learning and science. Commercial relations were established with London, and vessels of small tonnage made passages of sixty days from the wharves of Richmond to the pool of the Thames. Before many years an India-house was built, with the vague idea that the fabrics and spices of the East would be brought from Bombay and Calcutta direct to the capital of Virginia.

The point from which the most commanding and comprehensive view of Richmond is visible bears the name of Hollywood Cemetery, a picturesque elevation in the north-western suburbs, where rest the remains of many illustrious men, and of thousands who died in the civil war. Far away from the noises of city-life, curtained by Nature with the luxuriant foliage of tree and flower, and presenting at every turn of hill and dell patches of beauty which art cannot improve, there is perhaps no spot in America more suggestive of the solemn associations that attach to the sacred circle of the dead. At the southern extremity may be seen the monument erected to the memory of President Monroe, whose remains were removed from New York under the escort of the Seventh Regiment of that city.

The scene from President's Hill, in Hollywood, is one that never tires the eye, because it embraces a picture which somewhere among its lights and shadows presents features that constantly appeal to imagination and refined taste. In the great perspective which bounds the horizon the distant hills and forests take new color from the changing clouds; while nearer—almost at your feet—the James River, brawling over the rocks, and chanting its perpetual requiem to the dead who lie around, catches from the sunshine playing on its ruffled breast kaleidoscopic hues. Hundreds of willowy islets impede its flow, diversifying the picture with patches of green, and the brown-backed rocks and ledges peeping out are marked by silvery trains of foam.

RICHMOND, FROM HOLLYWOOD.

St. John's Church.

That, however, which attracts the attention of the visitor above all other objects as he views the broad prospect, is the city itself, with its bold yet broken outline of roofs and spires.

The ground on which Richmond is built is a succession of hills and valleys. Indeed, it is sometimes called, like Rome, "the seven-hilled city," and, in approaching from almost any direction, it produces upon the stranger the imposing effect of a large and populous capital. Nor will he be disappointed by his subsequent experience, for he will still find the city a place of interest as the social and political centre of Virginia.

From the period of the Revolution down to the present time the flower of the country-people have been in the habit of spending here a considerable portion of the year, while the sessions of the Legislature and the courts drew together many of the most brilliant intellects of the land. In 1861 still greater prominence was given to Richmond by its selection as the capital of the Southern Confederacy. It became the home of the Southern leaders and the resort of the officers of its armies, while the net-work of intrenchments that almost encircled the city and the battles fought in the neighborhood tell of the obstinacy with which it was defended as the key-stone of the cause. In April, 1865, when the Confederate forces evacuated their positions, nearly one thousand houses, including property to the value of eight million dollars, were destroyed by fire. Since then, however, Richmond has recovered from her misfortune, and there are now visible but few traces of the great conflagration.

Chief among the public buildings, and one that may be said to belong to the post-Revolutionary period, is the Capitol, a structure which lifts itself above all other buildings as from an Acropolis, and has, indeed, an imposing effect, which is not wholly lost when one gets near enough to see the meanness of its architectural details and the poverty of its materials. The Maison Carrée at Nismes, in France, was selected as the model for the structure, but so many alterations were made in this model that the Capitol resembles it about as much as the Hall of Records in New York City resembles the Temple of Wingless Victory. For purposes of the picturesque, the Capitol serves as well in the prospect from Hollywood as if it were the Parthenon restored.

The James above Richmond.

The James, from Mayo's Bridge.

It stands on the brow of what is known as Shockoe Hill, in the centre of a public square of about eight acres, which, being beautifully laid out, is a favorite place of resort for both citizens and strangers, who find in its shady recesses and the music of its fountains a grateful contrast to the dust and bustle of the streets. The building is of the Græco-composite order, adorned with a portico of Ionic columns, and the view from it is extensive, varied, and beautiful. The entrances are on the two longer sides, and lead to a square hall in the centre of the building, surmounted by a dome. In the centre of this hall is the famous marble statue of Washington, bearing this inscription:

"*Fait par Houdon, Citoyen Français*, 1788."

The statue is clothed in the uniform of an American general during the Revolution, and is of the size of life. In one of the niches of the wall is a marble bust of Lafayette. Among other objects of interest here is an antique English stove covered with ornamental castings and inscriptions, and dating far back beyond the Revolution. It was used to warm the old Virginia House of Burgesses at Williamsburg, in colonial times, and still holds its place in the present hall as the centre of legislative discussion and gossip, as it no doubt was more than a hundred years ago. The library contains many historic relics and valuable old pictures, and indeed the entire building is rich in associations which make the place seem almost sacred. Here Aaron Burr was tried for treason before John Marshall; here Lafayette was received by his old companions in the cabinet and the field; here the memorable Convention of 1829–'30 held its sessions, among whose members were Madison, Monroe, Marshall, John Randolph, Leigh, and many other men of national fame; and here, at a later period, Stonewall Jackson lay in his coffin, with the flag of the Confederate States (then first used) for his pall.

A few rods distant from the Capitol stands the celebrated equestrian statue of Washington, by Crawford. It consists of a bronze horse and rider of gigantic size, artistically poised upon a pedestal of granite, and surrounded by immense bronze figures of Patrick Henry, Thomas Jefferson, John Marshall, George Mason, Thomas Nelson, and Andrew Lewis. Each of these statues is a study in itself, as a specimen of the sculptor's genius, and as an almost "speaking likeness" of the original. Henry is represented in the act of delivering an impassioned address; Jefferson, with pen in hand, and thoughtful brow, appears the statesman; Marshall wears the dignity and firmness of the great judge; while the noble form of General Andrew Lewis, arrayed in the hunting-costume of the pioneer, recalls the romance and daring of early days. On smaller pedestals are civic and military allegorical illustrations, also in bronze; and, altogether, the monument is perhaps the most imposing in America.

In another portion of the Capitol grounds is a life-size marble statue of Henry Clay, and elsewhere is one in bronze of Stonewall Jackson.

The prominent public buildings of Richmond are substantial, and in most instances handsome specimens of architecture. The new City Hall, Custom-House, Governor's Mansion, Penitentiary, Medical College, and State Armory, are severally worthy of a visit; while, among the many churches, that which occupies the site of the ill-fated theatre destroyed by fire in 1811, when the Governor of the State and sixty others perished in the flames, is the most notable. The "Old Stone House" is cherished in the affections of the citizens of Richmond as the first dwelling erected within the city limits. Washington, Jefferson, Madison, and Monroe, have all been beneath its roof; and, says Lossing, in his "Field-Book of the Revolution," "Mrs. Welsh informed me that she well remembers the fact that Monroe boarded with her mother while attending the Virginia Convention in 1788."

At the remotest point of the landscape in the drawing of "Richmond from Hollywood" may be seen a white spire on the summit of a hill. This is the old parish-church of St. John's, Henrico. At what exact period this church was built the local historians do not inform us; but there are tombs in the burial-ground bearing date 1751, and probably no interment was made there until after parish services were regularly performed in the building itself. Originally, it was without architectural pretensions of any kind; but many years ago it was modernized by the erection of a tower, and enlarged by an addition joining the ancient part at right angles. During the civil war the tower fell in a high wind, and has been replaced by the spire which is seen in the drawing.

The associations of the building are of the most stirring and interesting character. Here assembled, on the 20th of March, 1775, the Second Convention of Virginia, which was called to determine the question of peace or war between the colony and the crown, and which gave to the Old Dominion the honor of organizing the first plan of resistance to British tyranny. The body contained a large number of men who were destined to become illustrious in the annals of the Commonwealth and the country. The delegate from Albemarle was Thomas Jefferson, and the delegate from Fairfax was George Washington. But the leading spirit of the convention was Patrick Henry, and the walls of this old church gave back the animating strains of his eloquence, as, rising to the full height of his argument, he uttered the war-cry of the Revolution: "I know not what course others may take, but, as for me, give me liberty or give me death!"

From the hill on which the church stands the James River is in view for several miles of its course, and lends much to the attractiveness of the prospect. Above the city, in the rapids which for six miles tumble over a rocky bed, we see whence is derived the water-power that animates the mills, and how art has overcome the obstructions of Nature by means of a canal which opens the navigation of the river above the falls. Below the bridge the scene is more peaceful, and the tran-

Rapids in the James.

quil surface of the water reflects the steadily-increasing commerce of the capital. The canal, which is seen in the last of our illustrations, is connected with tide-water by a series of locks, with an aggregate lift of ninety-six feet. Two of these locks on the highest level constitute the central part of a sketch which, at first glance, looks as if it

Scene on the Canal.

were designed to set before us a quaint, old, tumble-down nook or corner of some European city.

Richmond is indeed fortunate in the beauty of its site and in the charm of its landscape, so that, when the whole river-margin along the rapids shall be lined with factories, and Richmond shall become a great manufacturing city, even then it will tempt the wandering artist to take out his portfolio and sketch the outlines of its hills and the tumult of its leaping waters.

LAKE SUPERIOR.

WITH ILLUSTRATIONS BY WILLIAM HART.

Grand Portal.

SUPERIOR is three hundred and sixty miles long, one hundred and sixty broad, and eight hundred feet deep. Its general shape was best described by the French fathers, more than two hundred years ago, as "a bended bow, the northern shore being the arc, the southern shore the cord, and the long point the arrow." This long point is Keweenaw, a copper arm thrust out seventy miles into the lake from the south shore. Passing the Sault Sainte-Marie—commonly called the "Soo"—Point Iroquois is seen on the west, and opposite the Gros Cap of Canada, six hundred feet high. Beyond Iroquois stretches long White-Fish Point, and, this turned, the Sables come into view—sand-dunes hundreds of feet high, golden by day, crimson at sunset, and silver by night.

Beyond the Sables lie the Pictured Rocks. These stretch from Munesing Harbor eastward along the coast, rising, in some places, to the height of two hundred feet from the water, in sheer precipices, without beach at their bases. It is impossible to enumerate all the rock-pictures, for they succeed each other in a bewildering series, varying

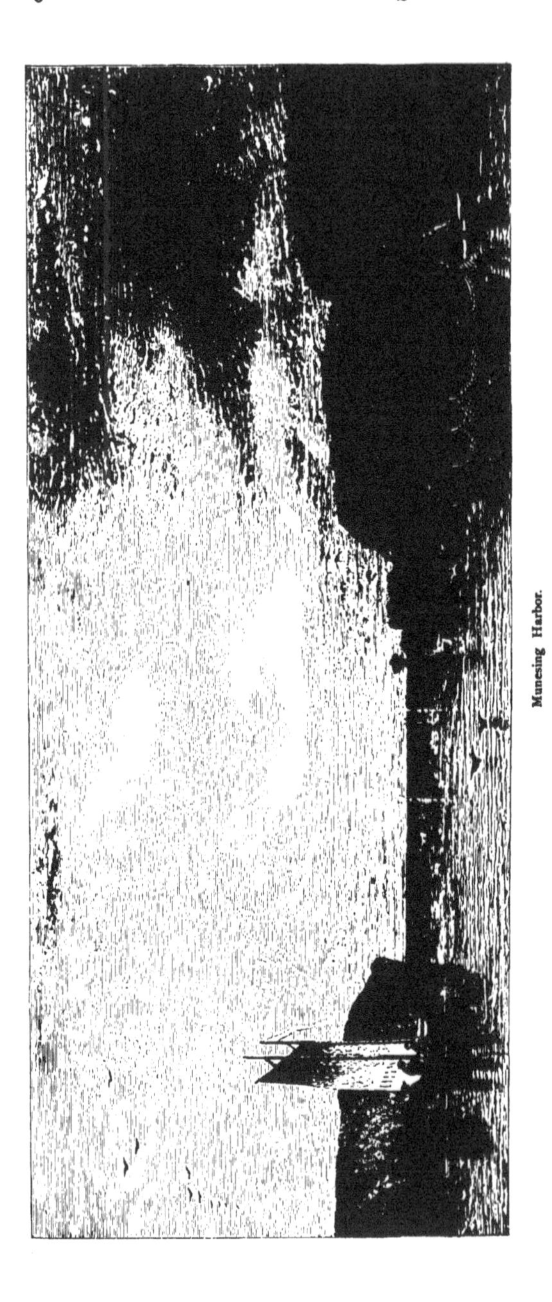

Munesing Harbor.

from different points of view, and sweeping, like a panorama, from curve to curve, mile after mile. Passing the Chimneys and the Miner's Castle, a detached mass, called the Sail Rock, comes into view; and so striking is its resemblance to a sloop, with the jib and mainsail spread, that, at a short distance out at sea, any one would suppose it a real boat at anchor near the beach. Two headlands beyond this is Le Grand Portail, so named by the *voyageurs*—a race now gone, whose unwritten history, hanging in fragments on the points of Lake Superior, and fast fading away, belongs to what will soon be the mythic days of the fur-trade. The Grand Portal is one hundred feet high by one hundred and sixty-eight feet broad at the water-level; and the cliff in which it is cut rises above the arch, making the whole height one hundred and eighty-five feet. The great cave (of which our artist has made so excellent a sketch), whose door is the Portal, stretches back in the shape of a vaulted room, the arches of the roof built of yellow sandstone, and the sides fretted

CHAPEL BEACH

SILVER CASCADE.

into fantastic shapes by the waves driving in during storms, and dashing up a hundred feet toward the reverberating roof with a hollow boom. Floating under the Portal, on a summer day, voices echo back and forth, a single word is repeated, and naturally the mind reverts to the Indian belief in grotesque imps who haunted the cavern, and played their pranks upon rash intruders—pranks they still play, and dangerous ones, too, for the whole coast of pictures is dreaded by the lake-captains, and not a few craft have gone down close to the shore, lost in treacherous fogs.

Farther toward the east is La Chapelle of the *voyageurs.* This rock-chapel is forty feet above the lake—a temple, with an arched roof of sandstone, resting partly on the cliff behind, and partly on massive columns, as perfect as the columned ruins of Egypt. Within, the rocks form an altar and a pulpit; and the cliff in front is worn into rough steps upward from the water, so that all stands ready for the minister and his congregation. The colors of the rock are the fresco; mosses and lichens are the stained glass; and, from below, the continuous wash of the water in and out through holes in the sides is like the low, opening swell of an organ voluntary. A manitou dwelt in this chapel—not a mischievous imp, like the sprites of the Portal, but a grand god of the storm, who, with his fellow-god on Thunder Cape, of the north shore, commanded the winds and waves of the whole lake, from the Sault to Fond-du-Lac. On the chapel-beach the Indians performed their rites to appease him; and here, at a later day, the merry *voyageurs* initiated the tyros of the fur-trade into the mysteries of their craft, by plunging them into the water-fall that dashes over the rocks near by—a northern crossing-the-line.

The Silver Cascade falls from an overhanging cliff, one hundred and seventy-five feet, into the lake below. The fall of Niagara is one hundred and sixty-five feet, ten feet less than the Silver, which, however, is but a ribbon in breadth, compared to the "Thunder of Waters." The Silver is a beautiful fall, and the largest among the Pictures; but the whole coast of Superior is spangled with the spray of innumerable cascades and rapids, as all the little rivers, instead of running through the gorges and ravines of the lower-lake country, spring boldly over the cliffs without waiting to make a bed for themselves. Undine would have loved their wild, sparkling waters.

The coast of Pictures is not yet half explored, nor its beauties half discovered; they vary in the light and in the shade; they show one outline in the sunshine and another in the moonlight; battlements and arches, foliage and vines, cities with their spires and towers, processions of animals, and even the great sea-serpent himself, who, at last, although still invisible in his own person, has given us a kind of rock-photograph of his mysterious self. In one place, there stands a majestic profile looking toward the north—a woman's face, the Empress of the Lake. It is the pleasure of her royal highness to visit the rock only by night, a Diana of the New World. In the daytime, search is vain, she will not reveal herself; but, when the low-down moon shines across the water, behold, she appears! She looks to the north, not sadly, not sternly, like the Old Man of the White Mountains, but benign of aspect, and so beautiful in her rounded, womanly curves, that the late watcher on the beach falls into the dream of Endymion; but, when he wakes in the gray dawn, he finds her gone, and only a shapeless rock glistens in the rays of the rising sun.

Leaving the Pictures, and going westward past the Temples of Au-Train and the Laughing-Fish Point, Marquette comes into view, a picturesque harbor, with a little rock

Empress of the Lake.

islet, the outlet for the Iron Mountain lying back twelve miles in the interior, a ridge of ore eight hundred feet high, which sends its thousands of tons year after year down to the iron-mills of Cleveland, Pittsburg, and Cincinnati, and scarcely misses them from its massive sides. A fleet of hundreds of vessels belongs to this iron-bound coast; their sails whiten the lakes from the opening of navigation to its close; they are the first to start when, in the early spring, word comes that the ice is moving, and the last to leave when, in the late fall, word comes that the ice is making. Perilous voyages are theirs, in the midst of grinding ice; and sometimes they are caught in the fierce storms of Superior, going down with all on board off the harborless coast of the Pictured Rocks or the Sables.

The iron shoulders passed, next comes the copper arm of Keweenaw, the arrow in the bow; the name signifies a portage; and the Indians, by crossing the base of the

Great Palisade.

point through Portage Lake and its streams, saved the long ninety miles around it. This copper arm has its history. Centuries ago its hills were mined, and the first white explorers found the ancient works and tools, and wondered over them; when they were tired of wondering, they ascribed them to the extinct Mound-Builders, whoever they were, a most convenient race, who come in for all the riddles of the Western country.

On Keweenaw are several lakes, among them the lovely Lac-la-Belle of the *voyageurs;* the north shore of the Point is bold with beautiful rock-harbors, and beyond Ontonagon, the western end of the copper-region, rise the Porcupine Mountains. At Montreal River Michigan ends, and Wisconsin pushes forward to share a part of the rich coast. Farther to the west is the beautiful group of the Apostles; this name

Cliff near Beaver Bay.

brings up again the memory of the early missionaries, who came to these islands as far back as 1669, Father Marquette himself, the central figure of the lake-country history, having spent some time here at La Pointe, on Madeline Island. It was while attending to this mission that he first heard of the Mississippi, or Great Water, from the Illinois tribes, who were attracted to La Pointe by the trinkets distributed by the French.

The idea of seeking out this wonderful river dwelt in his mind from that time, but he was not permitted to go until several years later, entering its waters at last in June, 1673, with, as he writes in his journal, "a joy I am not able to express."

At the head of Lake Superior is the St. Louis River; here Wisconsin ends and Minnesota begins. The town of Duluth, named after a French explorer who visited its site in 1680, is called the Chicago of Lake Superior; it has upward of thirty thousand inhabitants, and stands at the extreme western end of the Great Chain. Quebec stands at the eastern end, for the St. Lawrence beyond is but an arm of the sea, and seventeen hundred and fifty miles lie between. Beyond Duluth begins the

Baptism Bay.

North Shore; and these words call up visions of grandeur, of gold and silver, of adventure and danger, not unlike the dreams of the first white men on the shores of Mexico. The long coast, the arc of the bow, is even now but vaguely known, for, although a few settlements have been made where silver exists, they are but dots on the line, and the map-makers are obliged either to leave their paper blank, or fill it up from imagination and the vague stories of the hunters. The veil of mystery adds, no doubt, a charm; but, nevertheless, the surveys, as far as they have gone, verify the visions, and the silver sent down to the lower-lake towns fairly exceeds the descriptions of the discoverers.

Until within a few years the north shore has been traversed only by the hunters, trappers, and *voyageurs* of the Hudson's Bay Company; more than half of its length is their rightful territory, and scattered along its line are several of their forts, with their motley inhabitants. This company was formed in London in 1669, under the leadership of Prince Rupert, and afterward obtained a charter from Charles II., granting "the sole right of trading in all the country watered by rivers flowing into Hudson's Bay;" this right was soon stretched until it included the whole of British America, and as much of the United States as the hunters found convenient. There were four departments:

Temperance Harbor.

the Northern, which embraced the icy region near the arctic circle; the Eastern, along the St. Lawrence and its tributaries; the Southern, lying on the shore of Lake Superior; and the Western, which took in the immense country west of the Saskatchewan, as far as the Pacific Ocean and the Columbia River, where John Jacob Astor made his brave fight, single-handed, with the vast corporation, and failed, solely on account of the incapacity or infidelity of his agents.

All through the north coasts of Superior roamed the company hunters; along the hundreds of little lakes and streams the *voyageurs* paddled their canoes, trading with

SPIRIT HARBOR.

the Indians and gathering together the furs, which, packed in bales, were sent eastward from post to post, until they reached the ocean, where the company vessels carried them to England.

The shore of Superior, north of Duluth, rises into grand cliffs of greenstone and porphyry, eight hundred to one thousand feet high. The Great Palisade, a remarkable rock-formation, is moulded in columns up and down, more regular than the Palisades of the Hudson. The rock is red, and the minute quartz-crystals scattered over its face cause it to gleam in the sunshine like a wall of diamonds. It stands almost entirely detached from the land, and, at a short distance out, might be said to resemble a row

Island No. 1.

of plants growing upward, side by side, from the water, like giant lily-stalks. The cliffs of Beaver Bay are wild and rugged; and yet, dangerous as they appear, here is one of the good harbors of the north shore.

Baptism, or Baptême, River, beyond the Great Palisade, comes dashing down to the lake in a series of wild water-falls, with a wall of rocks on one side, through which it has cut a gate-way for itself when the storms build up a sand-bar across its natural mouth. The Indians called the stream the "River of Standing Stones;" but the *voyageurs* named it "Baptême," probably from some mission or work of conversion on its banks, although the sailors of to-day declare that it was so called because a persistent

LA CROSSE HARBOR.

scoffer fell in accidentally, and, as a priest was standing by, he baptized the man in spite of his objections. It is attached to the beautiful harbor of Temperance River, which is said to have been so called because there was no bar at the mouth!

The portion of Minnesota lying back of this coast is a wilderness, with vague rumors of precious metals hidden in its recesses. At Pigeon River is the boundary-line between the United States and Canada; and here begins the Grand Portage, where, through a series of lakes and streams, the names of which have a wild sound—Rainy Lake, Lake of the Woods, and Winnipeg—the *voyageurs* were able, with short portages, to take their canoes through to the Saskatchewan and the Red River of the North.

The whole Canadian shore is grandly beautiful in every variety of point, bay, island, and isolated cliff. Passing Fort William, Thunder Cape is seen—a basaltic cliff, thirteen hundred and fifty feet high, upon the summit of which rest the dark thunder-clouds, supposed by the Indians to be giant birds brooding upon their nests. At the foot of this cliff, near the shore, is Silver Island, whose low surface, over which the waves have dashed at will, is now diked and protected in every precious inch.

Nepigon Bay, or the Bay of Clear Waters, is forty miles long by fifteen broad, and contains a number of islands. The river which flows into this bay comes thirty miles from a lake of the same name, which was long believed to equal Lake Erie in size. Beyond Nepigon Bay, eastward, the coast is studded with water-falls the hues of which are sometimes a bright-claret color of varying shades.

At Pic River the shore-line bends to the south, and the lake begins to narrow toward the Sault. At Otter-Head the cliff rises, in a sheer precipice, one thousand feet from the water, and on its summit stands a rock like a monument, which on one side shows the profile of a man, and on the other the distinct outline of an otter's head. Farther south is the broad bay of Michipicoten, or the "Bay of Hills." There are a number of islands in the lake, worthy of description on canvas, so full are they of wild beauty. Among the large islands are "Michipicoten," "Saint-Ignace," and rugged "Pie;" and farther west is Isle Royale, the largest in the lake, forty-five miles in length, and belonging to Michigan. Royale may be called an ancient settlement; for, as far back as 1847, miners, geologists, capitalists, and vessels, were there; fortunes shone in the air, and the whole lake-country rang with the name and praises of the wonderful island. Royal, however, it did not prove, in spite of its name; and the capitalists came back to civilization with empty purses, and all faith in Lake Superior gone forever. Isle Royale is again solitary, and its very light-house abandoned.

The storms of Lake Superior are often violent, but not so dangerous as those of the lower lakes, for Superior has more sea-room; the waves, although of great height and force, are regular and united when compared with the short, chopping seas of Erie and Michigan. On shore the storms of Lake Superior seem terrific; and the ocean itself cannot show a more stormy expanse than the great lake in a September gale.

CUMBERLAND GAP.

WITH ILLUSTRATIONS BY HARRY FENN.

A Glimpse of Kentucky, from Cumberland Gap.

THE tourist may be familiar with the fastnesses of Alpine scenery, the heights of Mont Blanc, the cone of Vesuvius, the bald summit of Washington, or the gigantic outlines of the Western cañons, and yet, memories and associations attached to all of these localities will be recalled by a visit to that region of America in which the Cumberland Mountains trend obliquely across the States of Kentucky and Tennessee; because, somewhere in the four thousand four hundred miles of territory occupied by these "everlasting hills," they present to the eye almost every variety of picturesque expression that elsewhere has excited wonder or admiration.

Great ridges—now roofed over with thickets of evergreen, now padded with moss and ferns, or, again, crowned with huge bowlders that seem to have been tumbled about in wild disorder by some convulsive spasm of the monster beneath—shoot suddenly upward, from two thousand to six thousand feet, and become, as it were, landmarks in

CUMBERLAND GAP, FROM THE EAST.

the skies, that are visible at such distances as to appear like a part of the clouds. Here and there a broad table-land, on which a city might be built, terminates abruptly in sharp escarpments and vertical sheets of rock, seamed and ragged, like the front of a stupendous fortress that has been raised by giant hands to protect the men of the mountains from the encroachments of the lowlanders.

The "ridges" referred to are among the curiosities of the Cumberland region. Aside from the fact that they observe a species of parallelism to each other, they contain numerous "breaks," or depressions, which, in the peculiar configuration of the country, appear to the traveller who is at the foot of the mountain to be distant only a few hundred rods; yet he must frequently ride for miles through a labyrinth of hills, blind roads, and winding paths, before he can reach the entrance and pursue his journey.

The most celebrated of these great fissures, or hall-ways, through the range, is "Cumberland Gap," which is in East Tennessee, near the Kentucky border, about one hundred and fifty miles southeast from Lexington, and may be regarded as the only practical opening, for a distance of eighty miles, that deserves the name of a "gap."

The gap depicted by our artist is about six miles in length, but so narrow in many places that there is scarcely room for the roadway. It is five hundred feet in depth. The mountains on either side rise to an altitude of twelve hundred feet; and when their precipitous faces have been scaled by the tourist, and he stands upon the summit, the view, beneath a cloudless sky, is one of the most beautiful in America. Southward, there stretch away the lovely valleys of Tennessee, carpeted in summer with every shade of green, and in autumn with every rainbow tint—the rolling surface resembling in the distance a vast plain, written all over with the handiwork of human enterprise; while, looking to the north, the vision is lost among a series of billowy-backed mountains, rising barrier-like to hide the luxuriant fields of Kentucky.

The gap delineated in the accompanying sketches is a great highway between Southwestern Virginia and her sister States adjoining. Hence, during the late war, the position was early deemed important, and was occupied and strongly fortified by the Confederate Government. Cannon bristled from the neighboring heights, and a comparatively small force held the pass for months, defending in that secluded mountain-recess the railroad connections between Richmond, North Alabama, Mississippi, Nashville, and Memphis, on the integrity of which so much depended.

The approach to the range from the northeast side, after leaving Abingdon, Virginia, is over a rough, broken country; and the only compensation to the traveller, as he saunters along on horseback, is in the enjoyment of bits of scenery wherein rocks and running streams, mountain-ferries, quaint old-fashioned mills, farm-houses and cabins perched like birds among the clefts of hills, lovely perspectives, wild-flowers and waving grain, and a homely but hospitable people, combine in charming confusion to keep the attention ever on the alert.

CUMBERLAND GAP, FROM EAGLE CLIFF.

The road through the gap, winding like a huge ribbon, to take advantage of every foot of rugged soil, up, down, and around the mountains, is but the enlarged war-trail of the ancient Cherokees and other tribes, who made incursions from one State to the other. You are following the path pursued by Boone and the early settlers of the West. Passing through the scenes of bloody ambuscades, legends, and traditions, it would seem almost a part of the romance of the place if now an Indian should suddenly break the reigning silence with a war-whoop, and its dying echoes be answered by the rifle-shot of a pioneer.

Of residents in the gap there are but few. One of these has been enterprising enough to establish, near an old bridge, which is shown in the picture, a grocery-store, and obtains his livelihood by trading in a small way with the teamsters of the passing trains, and exchanging clothing and the like for the produce of his neighbors. Similar establishments will be found at intervals of five, ten, or fifteen miles; sometimes they are half hidden from view in the coves, or "pockets," of the mountains. But they absorb much of the small "truck" that finds its way to market from this section. The commodities thus purchased and shipped in the mountain-wagons through the gap, *en route* to Baltimore and elsewhere, consist of dried apples, peaches, chestnuts, butter, lard, flaxseed, bacon, etc. Horse and mule trading is likewise carried on to a considerable extent; and sharp-witted, indeed, must be that man who can buy or sell more shrewdly than these self-same mountaineers, whose lives have been hammered out on the anvil in Nature's own workshop.

As a class, they are a large-bodied, large-hearted, large-handed people, rude in speech, brave in act, and honest in their friendships. They may know nothing of the conventionalities of society, but they will exhibit the "small, sweet courtesies of life"—as they understand them—with a readiness of generosity that makes one "feel at home." They may have but a single room in their cabin, yet you will be invited to enjoy the night's hospitality like one of the family, and may go to bed with "he, she, and it," on the family floor, with the manifestation of no more curiosity or concern, on the part of the individual members thereof, than if they had been born without eyes. And in the morning, after a "pull" at the "peach-and-honey" and a breakfast of hog and hominy, a long stride by your horse's side for three or four miles will tell you that the mountaineer knows how to "speed the parting guest," in his simple fashion, with a grace and hospitality that come straight from the heart.

The road through a portion of the gap, and one of the caravans which are frequently passing, may be seen in one of the accompanying pictures; while in another sketch is a view of a primitive old mill, now almost in ruins, where grain is ground for the neighbors; but it is situated in a spot so picturesque that, if money could buy the beauty of Nature, long ago it would have been transplanted to become the site of a rural palace.

Newport, from the Bay.

NEWPORT AND PROVIDENCE.

OF all the States in the Union, little Rhode Island alone has two capitals —Newport and Providence. As they are linked together politically, and geographically are in close proximity to each other, it is manifest that they should be considered together. Bishop Berkeley wrote, in 1728, that "Newport is the most thriving place in all America for bigness. I was never more agreeably surprised than at the sight of the town and harbor."

Then the Revolution came, and it ruined Newport beyond redemption. Fashion made it the summer home of the wealthy, and its old-time supremacy was restored. Such, in brief, is the *three-fold* aspect which Newport has assumed during the last hundred and fifty years. We now turn to the special points of attraction, as indicated by our artist.

Entering the harbor, on the left your eye rests upon a small, oval fort, gray, time-worn, and dilapidated, stand-

OLD FORT DUMPLING

ing on the island of Conanicut, and known by the somewhat impressive name of "Dumpling." The fort has been left for many years to the corroding wear and tear of the elements, but, while the interior works have been gradually destroyed, the outer walls remain as complete and firm as they ever were. As a means of defence it would be of little service in these days, for one of our modern shells dropped into the centre would blow the whole affair to fragments. Compared with Fort Adams, one of the largest and most completely equipped defences on our shores, which, with its massive walls and long rows of guns, frowns upon Dumpling from the opposite side of the bay, this little tower looks somewhat insignificant; but, as a picturesque *ruin*, it has its charms. For a century the winds have beat upon the old fort; the Cross of St. George has waved over it; the French fleet swept around it as the vessels moved up to their winter-anchorage in the harbor; the Stripes and the Stars long ago supplanted the British ensign; it is more venerable than the Republic; and we trust that it will be left undisturbed for ages—one of the few memorials in existence of our early history.

Brenton's Cove is approached by a causeway leading to Fort Adams, and affords one of the finest views that can be obtained of Newport. "The tall and delicate spires of the churches cut sharp against the blue sky; the public buildings stand out in noble relief; and the line of houses, as they rise one above another on the hill-side, is broken by open grounds and clusters of shade-trees. Each spot on which the eye may chance to rest recalls some event that happened there in earlier times." In the illustration all is placid and serene; but, when the breakers dash upon that fatal reef, and the strong waves whiten its jagged ridge, it is a place of terror. Many a vessel has been wrecked there; and the mouldering gravestones along the edge of the ocean show where the bodies of the drowned sailors were once buried. Why they should have been deposited there, where the winds and the waves sound a perpetual dirge, and the spray of the ocean always dampens the sods which cover them, instead of being taken to some rural ground, where the birds sing and flowers bloom, we do not know.

Following the southern shore, we come to what is known as the Spouting-Rock. After a southeasterly storm, the apparatus is in working-order; and, during the "season," multitudes assemble there to see the intermittent fountain play. The construction of the opening beneath is such that, when it is nearly filled and a heavy wave comes rolling in, the pent-up waters can find relief only by discharging themselves through a sort of funnel into the air. It is, however, a somewhat treacherous operator: for a long time there may be no spouting done; and, even when the waves roll in from the right quarter, it is not easy to tell just when the horn intends to blow. But the ocean-view is, at this spot, so indescribably grand after a storm, that the temptation to linger as near the edge of the rocks as possible is almost irresistible, and we have seen many a gay company pay the watery penalty.

At the west end of Sachuest bathing-beach rise the precipitous rocks, with the deep

BRENTON'S COVE.

and sharp-lined fissure, known as "The Purgatory." A little beyond this chasm there is a pleasant spot, shaded by trees, and commanding a beautiful view, which is known as "Paradise"—so that, when a stranger in that region asks the way, he is likely to be told that he must pass by Purgatory to Paradise. The opening in the cliff extends one hundred and sixty feet, and is fifty feet deep at the outer edge It is from eight to fourteen feet wide at the top, and from two to twenty at the bottom. It was once supposed that the water at the base was unfathomable; but at low tide it is actually not more than ten feet in depth. It was formerly the prevailing theory that this fissure was occasioned by a sudden upheaving of the rock; but, after careful examination, it has been decided that it was probably formed by the gradual eating away of the softer portions of the stone at a very early period.

Like most places of the kind, Purgatory has its legends.

Some little time after the settlement of the country by the whites, an Indian woman murdered one of the colonists, in revenge for certain wrongs inflicted upon her people. Walking, one day, near Purgatory, she was accosted by a person appearing to be a well-dressed Englishman, who proposed to fight with her. The stout squaw was not unwilling to accept the challenge, and in the struggle she was gradually dragged toward the edge of the chasm, when her opponent seized her in his arms and leaped into the abyss. At this moment the cloven foot appeared, his goodly garments fell off, and he was revealed in his true Satanic personality. Why the devil should have felt himself called upon to interfere in this way to punish the woman for the wrong that she had done to the English settlers, does not appear; but, as the print of his feet and marks of blood are still visible on the stones, it is not for us to gainsay the story. At any rate, it is easy to see that such a belief on the part of the Indians might have tended to promote general security.

Another legend pertaining to this spot is not quite so tragical, and perhaps can be better authenticated. A beautiful but giddy girl, heiress to a large estate, had for some time received special attentions from a young man in all respects her equal, and whose affection, notwithstanding appearances to the contrary, she warmly reciprocated in her heart. But the passion for coquetry was so strong with her, that she could never resist the temptation to torment her admirer; and, one day, as they stood together on the brink of Purgatory, and he was pleading, with impassioned eloquence, for some pledge or token of love from her, she said, "I will be your wife if you will show the earnestness of your devotion to me, and your readiness to obey all my wishes, by leaping across this abyss." Without a moment's hesitation the young man sprang to the other side of the rock, and then, politely lifting his hat, he complimented the beautiful girl upon her charms, told her candidly what he thought of her character, bade her final adieu, and she saw his face no more. After this, as the tale runs, she went mourning all her days.

THE SPOUTING CAVE.

"Berkeley's Seat" is in Paradise, within easy walking-distance of the house which he built and occupied nearly a century and a half ago. Out of regard to the memory of Charles I., to whom he was indebted for certain favors, he called his place Whitehall, one of the palaces occupied by the king. It is still standing, and in good repair. There is the room which he occupied as a study, with its tiled fire-jambs, and low ceiling, and undulating floor, and the little chamber where he slept; and it is pleasant to think that, in the sunny court-yard adjoining, he once walked—perhaps discussing with his friends the state policy of Walpole, or the probable future of the new Western land, "whither the course of empire" had already begun "to take its way," or the medical virtues of tar-water, or it may be some of the profounder problems of the soul which occupied his thoughts. When the weather was favorable he betook himself to the sheltered opening in Paradise Rocks, which is now consecrated by his name. This he is said to have fitted up with chairs and a table; and tradition says that it was in this rocky cave he wrote his "Minute Philosopher." With the broad expanse of ocean before him, and its monotonous roll sounding in his ear, it may be that he was able to give his thoughts a wider range, and fix them more intently upon the subtile questions which he was so fond of contemplating, than was possible in the pent-up little room where he kept his books; and it may have been easier for him to bring his mind to the conclusion that there is nothing in the universe but *soul* and *force*—no organic substance, no gross matter, nothing but phenomena and relations and impressions—than it would be if he were shut in by doors and walls, and nearer to his kitchen.

In the following strains Mr. Longfellow tells how "the Viking old set a[illegible]d his way from "the wild Baltic's strand" to our strange shores, and built here "t[illegible]ory. [illegible]ty tower" by the sea, commonly known as "the old stone mill:"

"Three weeks we westward bore,
And, when the storm was o'er,
Cloud-like we saw the shore
Stretching to leeward;
There for my lady's bower
Built I the lofty tower,
Which, to this very hour,
Stands looking seaward."

We wish that we could believe in our having so respectable a piece of antiquity in Rhode Island. Inasmuch as this interesting and unique structure dates back to the prehistoric times of the colony, no record of its construction being in existence, and, still further, as it has a close resemblance to certain edifices still existing in Northern Europe, many have been willing to accept the tradition that it must be of Danish origin.

The first authentic notice of the edifice is found in the will of a Mr. Benedict Arnold, dated 1677, in which he bequeaths his "stone-built windmill" to his heirs. About the middle of the last century it was surmounted by a circular roof; and one

PURGATORY.

of the old inhabitants, in a deposition signed in 1734, says, "It is even remembered that, when the change of wind required that the wings, with the top, should be turned round, it took a yoke of oxen to do it." There is abundant tradition to show that it has been used for various purposes; and a hundred and fifty years ago it was known as the Powder-Mill—the boys, as late as 1764, sometimes finding powder in the crevices; and at a later period it was used as a hay-mow. It is somewhat singular that such a substantial

Distant View of Purgatory.

and peculiar structure should have been erected simply as a windmill, but this may be explained by the facts that the first wooden mill was blown down in a great storm that occurred in 1675; that Governor Arnold was unpopular with the Indians, and would be likely to build a mill that would withstand both storm and fire, and *look* like a fort at least; and, still further, he may have seen old mills in England of the same style—there being an engraving in the *Penny Magazine*, of 1836, of one near Leamington, which is the very counterpart of the Newport mill. The various traditions connected with this old relic impart to it a special interest; and, unless it is upheaved by the earthquake or demolished by lightning, it is likely to stand for many generations.

At a little distance from the old Stone Mill, on the easterly side of the public square, stands the statue of Commodore Matthew Calbraith Perry. The material is bronze; and the accurate proportions, the graceful attitude, the well-disposed drapery, and the speaking likeness, combine to give this statue a high place as a work of art.

We have now glanced at Newport as it was a hundred years ago, as it was fifty years ago, and as it is to-day. What will be its appearance fifty years hence? The streets of the older part of the town may continue to be as narrow as ever; and, unless a wide-spread conflagration should sweep them away, the ancient wooden houses may crowd upon the gutters, as they have always done; the venerable stone mill will stand in its place, a monument of the prehistoric ages of Newport; Trinity Church, we trust, will be undisturbed, whether the congregation abide by its courts or not; the Jewish Synagogue is secured from ruin by a perpetual endowment; the port-holes of Fort Adams may still show their iron teeth, unless, indeed, the advance of military science should have made all such stone fortresses unserviceable, or the universal dominion of the doctrines of peace—which God, in his mercy, grant!—have swept them all away.

The natural features of the region will remain unchanged; the same rocks will frown upon the sea; the same purple haze rest at eventide upon the land-locked harbor; the same veil of ocean-mist temper the brightness of the noontide sun, and tide rise and fall on the sandy beach with the same rhythmical flow; the storm thunder with the same loud turbulence; but, meanwhile, what changes will the hand of man have wrought? Within the last twenty years miles upon miles of barren pasture have been converted into lawns and gardens and verdant groves; millions have been expended in the erection of beautiful villas and stately palaces; the tide of population has set in like a flood; and such are the peculiar advantages which Nature has bestowed upon this lovely spot, that no caprice of fashion can ever turn back or arrest the flow of its prosperity. Regions now unoccupied will soon be covered with habitations; the summer population will spread itself all over the southern portion of the island, from east to west, and then crowd back into the interior, until the whole area from south to north is made a garden of beauty. Newport will never again become a busy mart of traffic; its ancient commerce will never return there; the manufactures which have made "the Providence Plantations" so rich will never flourish in "the Isle of Peace," for the soft and somewhat enervating climate is not conducive to enterprise and activity; but those who need relief from the high-strung excitement of American life, the merchant who wants rest from his cares, statesmen and writers who would give their brains repose, will find it here. The men of our land, above all others, require some such place of resort, to allay the feverish activity of their lives—a place where they may come together periodically,

"BERKELEY'S SEAT."

not for debate, and controversy, and labor, and traffic, but for pleasant talk, and rational recreation, and chastened conviviality. They need to dwell where, for a part of the year, they can see the sun rise and set, and scent the flowers, and look out upon the waters. This green island seems to have been made by a kind Providence for such uses as these, where men may forget their cares and cease from their toils, and behold the wondrous works of God, and give him thanks.

Commodore Perry's Statue and the "Old Mill."

The city of Providence was founded in 1636 by Roger Williams, and is the second city of New England in wealth and population. It indicates, in its peculiar name, the spirit in which it was founded, and there are few places where the cardinal virtues and higher emotions are signalized in the titles of the thoroughfares as conspicuously as they are here. Thus we have Benevolent Street, Benefit Street, Faith Street, Happy Street, Hope Street, Joy Street, and others of like sort. Amsterdam is perhaps the only city

OLD HOMESTEAD. AN OLD LANDMARK. CITY MONUMENT.

that can go beyond this in the quaintness of the names by which the streets are designated.

What is known as the "Abbott House" is an ancient structure, in which Roger Williams is said to have held his prayer-meetings. It was erected by Samuel Whipple, one of the early settlers of Providence Plantations, and who was the first person buried in the old North Burying-ground. This house must be more than two centuries old, and it is the only structure in the State of which any fragment remains in any way identified with the memory of Williams.

The First Baptist Meeting-House, of which we have a sketch under the title of "An Old Landmark," was erected in 1774-'75, and is eighty feet square, with a spire one hundred and ninety-six feet high. The exterior has remained unaltered from the beginning, and presents a pleasing and picturesque appearance. The steeple is copied from one of Sir Christopher Wren's churches in London, and is singularly symmetrical and beautiful. The edifice stands in an open square on the side of a hill, and is surrounded by trees. The society that worships here was founded by Roger Williams, on his arrival in Providence, and claims to be the oldest church of the Baptist denomination in America. During his time, and for many years after, public services were held in a grove, excepting in stormy weather, when the people assembled in a private house for worship. The first meeting-house was built about the year 1700, at the expense of Pardon Tillinghast, the pastor, who at his death bequeathed the property to the parish. In

1726 a new and larger house was erected, and the record of the great dinner given on this occasion indicates a degree of frugality in striking contrast to the lavish expenditure of our times. The bill-of-fare consists of one sheep, one pound of butter, two loaves of bread, and half a peck of peas—total cost, twenty-seven shillings. The bell which was originally hung in the tower of the present church bore this inscription:

> "For freedom of conscience, the town
> was first planted;
> Persuasion, not force, was used by
> the people.
> This church is the eldest, and has not
> recanted,
> Enjoying and granting bell, temple,
> and steeple."

In the course of a few years this bell was destroyed, and that which was substituted in its place gave such offence to the public that an attempt was made to break it with a sledge-hammer, which, however, succeeded only in knocking off a small chip from the edge. This may have resulted in restoring the right tone to the bell, for it continues to ring the hour of noon and the nine-o'clock vespers down to the present day, and is thought to be pleasant and musical.

The Soldiers and Sailors'

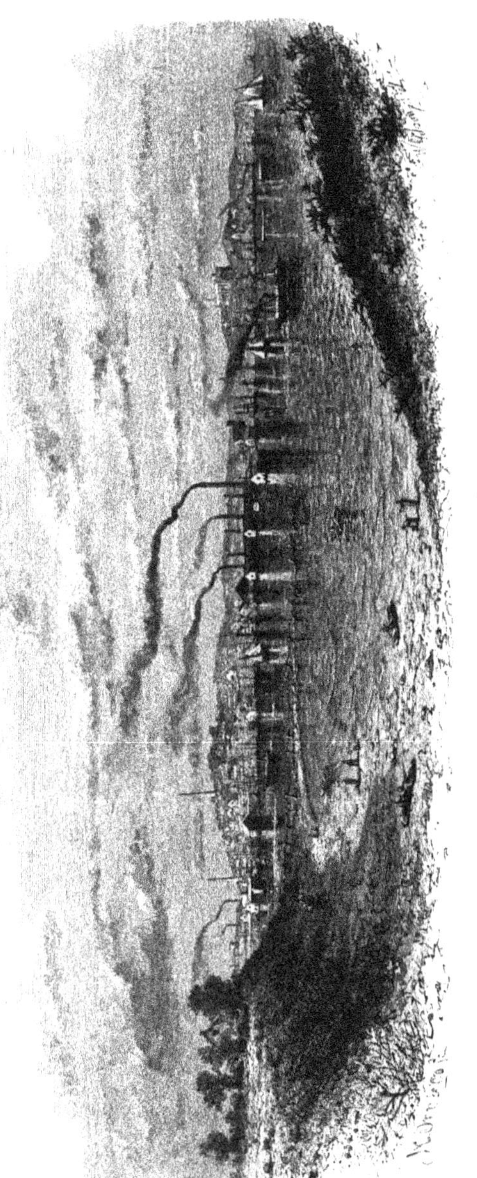

City of Providence, from Southern Suburbs.

Monument stands in an open square adjoining the railroad-station, and was erected in 1871, in accordance with a vote of the State Legislature, at a cost of about sixty thousand dollars. It was modelled by Mr. Randolph Rogers, in Rome, and the castings were made in Munich, under his direction. From the level of the ground to the top of the monument it is forty-six feet, and the general appearance of the structure is very imposing. It is surmounted by a female figure in bronze, eleven feet high, representing America at the close of the late war. The left hand is resting on a sword, and holding a wreath of *immortelles;* and the right hand, extended, holds a wreath of laurels, as if to crown the heroes of the war. The figure is draped in classic robes, hanging easily

The "Abbott House."

and naturally around the form. The face is benign and full of expression. Beneath the plinth upon which the statue stands are stars and wreaths of oak and laurel in bronze. Upon the face of the next section are the arms of the State of Rhode Island, while in the rear are the arms of the United States. On the angles are fasces, indicating that in union there is strength. On the next section, at the front, are the dedication words: "Erected by the people of the State of Rhode Island to the memory of the brave men who died that their country might live." Upon the next section stand four bronze figures, seven feet and three inches in height, representing the infantry, cavalry, artillery, and navy. They are clad in appropriate uniforms, and bear the arms and insignia of their several departments of service. Four bronze *bassi-rilievi*, size of life, appear upon the next section, representing War, Victory, Emancipation, and Peace. War appears with sword and shield, Victory as an angel bearing the sword and palm, Emancipation as a freed-woman with broken chains, and Peace with the olive-branch and horn of plenty. On the projecting abutments are twelve panels, containing bronze tablets on which are engraved the names of the heroic dead—in all, seventeen hundred and forty-one.

We now pass to the outskirts of Providence, in a northwesterly direction, and as we drive through a small manufacturing settlement we come upon a little, double-arched

Whipple's Bridge on Blackstone River.

bridge, where we can get a fair idea of the general style of scenery which is peculiar to the region. It has no startling features, no striking contrasts in the landscape, no mountains, no bold horizon, but there are pleasant walks by the side of running streams, shady nooks and alcoves in the woods, with an occasional glimpse of the distant waters of the bay, which gives a cheerful life to a picture that would otherwise be somewhat tame and monotonous. The territory of this State is so limited, and what there is appears on the map to be so intersected by water, that people sometimes smile when we speak of the *interior* of Rhode Island as if it must be all border; and still it is possible for one to drive a score or two of miles, in a straight line, without getting outside the limits.

BLACKSTONE RIVER.

But, apart from the regions which border upon Narragansett Bay and the ocean, there are few features in the landscape that would arrest the artist's attention. The broad sheet of water which opens directly south of Providence, and stretches for thirty miles down to the Atlantic, constitutes the great attraction of this Commonwealth. The rivers emptying into the bay, whose falling waters are used over and over again to pro-

Mark-Rock Landing.

pel the great wheels of our manufactories, are the main source of the marvellous riches of the State; while hundreds of thousands are drawn every year to the summer resorts which line the shores and adorn the islands of Narragansett Bay.

The sketch of Mark-Rock Landing, with the steamer touching at the wharf, might be repeated almost indefinitely. For miles and miles, on both sides of the bay, places of resort for summer visitors have been established. Let us linger for a while at Rocky

Rocky Point, from Warwick Pier.

Point, on Warwick Neck, about twelve miles south of Providence, and see what it is which attracts such multitudes to this spot. Shady groves, pleasant walks, romantic caverns, a smooth beach, salubrious air, and beautiful views, are among the natural features which attract the weary and the seekers after pleasure and repose. From the high tower which appears in our artist's sketch, the whole bay, from Providence to Newport, with the Atlantic in the distance, comes within the reach of the observer's eye.

But, after all, the great feature of the place is the *clam-bake*, an institution of which Rhode-Islanders are proud, and regard as a connecting link that binds them to the old Narragansetts, with whom it originated. The culinary process may be briefly described: A fire of wood is built in the open air, upon a layer of large stones arranged in a circular form, and when they have become sufficiently heated the embers and ashes are swept off, and a quantity of clams in the shell poured upon the stones, which are immediately covered with a thick layer of fresh sea-weed, and this is also protected from the cooling effects of the atmosphere by an old sail-cloth.

Turning down the west passage of the bay and reaching the open sea, we come upon Narragansett Pier, where the broad ocean rolls in full force, and there is no land that can be reached in an easterly line until we touch the shores of Spain. Not many years ago this place was a waste, and occupied only by a few fishermen's houses, but a strange change has now come over the scene. A thousand bathers may be seen, on a warm summer day, crowding the beach that was once so still and solitary. People from all parts of the Union flock to this spot, for the sake of breathing the cool ocean-air, and plunging in the waves, and watching the breakers as they dash upon the high, precipitous rocks that line the shore, at a little distance south of the smooth, hard beach where the bathing is done. Artists say there are no rocks on our coasts so rich and varied in their coloring as these.

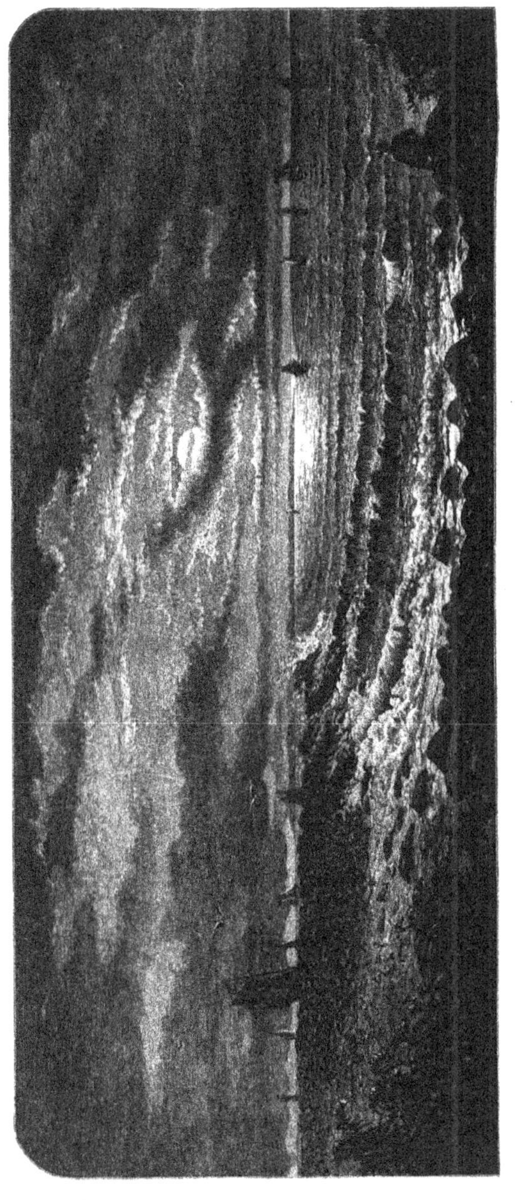

Breakwater, Narragansett Pier.

IN WEST VIRGINIA.

ILLUSTRATIONS DRAWN BY W. L. SHEPPARD, FROM SKETCHES BY DAVID H. STROTHER.

Arched Strata.

IN looking at the map of West Virginia, we may observe that its central regions are so hatched and corrugated with the shadows of mountains, so scribbled over with twisted and meandering lines representing the water-courses, that it is difficult to trace, amid these topographical entanglements, the lighter lines and dots which should indicate the highways and centres of population, or to collect together into words even the bold capitals which tell us the names of the counties.

Yet the adventurous traveller who undertakes to explore this shadowy realm in person will be amazed to find how far the geographical picture has fallen short of the savage and tremendous reality. In its untrodden wilds he will find himself bewildered with difficulties he never dreamed of, and sometimes confronted with dangers he had not provided against. Far beyond the range of pleasure-seeking tourists, he will be often surprised with scenes whose beauty would charm an artist into ecstasies, whose sublimity might awe a poet into silence.

For the sake of convenience and a pleasant starting-point, we will rendezvous at the Berkeley Springs, a famous summer resort on a branch of the Baltimore and Ohio Railroad. Thence, by a good graded road, on wheels or horseback, we can in two days' easy travel reach Moorefield, seventy-five miles distant.

In passing thus lightly over our preliminary journey, it must not be assumed that we have seen nothing worthy of remark by the way. On the contrary, the entire route abounds in objects of interest and beauty. We have seen the imposing cliffs of Candy's Castle, at the crossing of the North Fork of Cacapon River. A few miles distant, on the same stream, is the famous natural ice-house called the Ice Mountain. Then, at Romney, we have the Hanging Rock and the view from the yellow banks; and farther on we pass through Mill-Spring Gap, and wonder at the long, regularly-scalloped ridge of the Trough Mountain, resembling a row of potato-hills; then under the impending cliffs at the Northern Gate, and finally the first glimpse of the great South-Branch Valley, stretching around Moorefield. We cannot conscientiously turn away from the scene immediately around us without something more than a passing word; for, while we may meet with many objects whose rugged and startling features bring them more readily within the power of the graphic arts, we shall see nothing in our travels more softly and magnetically beautiful, to soul and eye, than this same valley of Moorefield.

The South Branch of the Potomac has its sources in the county of Highland, and, after a comparative course of about one hundred miles, running from southwest to northeast, and parallel with the great mountain-ranges, it joins the North Branch in Hampshire County, some fifty miles below Moorefield. Its upper waters flow in three principal streams, called respectively the South, Middle, and North Forks, the channels of which, like that of the main river, are bordered by extensive alluvial levels of extraordinary fertility, alternating with narrow, sharp-cut gorges domineered by bare, perpendicular cliffs of sublime height and picturesque forms.

After the junction of its chief tributaries, and about midway of its course, the river leaves the shadow of the mountains, and winds majestically with its double-fringed borders through an unbroken stretch of bottom-lands, eleven miles in length by three in breadth, lying, like a magnificent billiard-table, cushioned with mountain-ranges of graceful outlines and exquisite coloring, and rising to the imposing height of fifteen and eighteen hundred feet. This rich and verdant plain is mapped into fields and farms of manorial proportions, and dotted over with double-brick, tin-roofed houses and herds of stately cattle, betokening a land of easy wealth and old-fashioned abundance. Like the queen of this fat realm, the pretty village of Moorefield sits sleepily on the river-bank, half embowered in shade, awaiting the homage of her subjects.

Continuing our route southward by a pleasant, graded road, we soon arrive at Baker's, seven miles beyond Moorefield. Just before reaching the house we catch a glimpse of a pretty cove on our left, overlooked by a secondary range of rounded hills faced with some curious rock-work. The view is interrupted by trees, and sufficiently

Hills near Moorefield.

imperfect to stimulate the imagination. So we open the bars, and, riding across cultivated fields for half a mile, find ourselves in the meadow immediately opposite the objects of our curiosity. The closer and more satisfactory view brings no disappointment, but, on the contrary, increases our astonishment. Here are five conically-rounded hills, rising to a height of several hundred feet above the plain, singularly regular in shape and size, each adorned with a half-detached façade of rock-work of the most peculiar and fantastic character. Geologically, these rocks are of stratified sandstone, upheaved perpendicularly; cracked, splintered, and abraded by the elements; their exposed edges wrought into the most strange and startling shapes — images which might be worshipped without breaking the second commandment. So far overtopped by their loftier neighbors, these hills scarcely suggest emotions of sublimity; yet they hold us by the fascination of a curiosity not unmingled with awe.

We will now push on toward Petersburg, ever and anon casting a lingering look

PETERSBURG GAP.

behind, over the level perspective of the beautiful valley, and the fading blue of its northern boundaries.

After a short ride of two miles, we suddenly turn into the cool and shadowy gorge of the Southern Gate, through which the river pours its clear-green waters into the val-

Chimney Rocks.

ley. Crossing by an easy ford, we follow the road, which barely finds room to pass between the stream and the overhanging cliffs. Presently the gorge widens, and we call a halt to view that gigantic wall of naked rock, divided from the clouds by a ragged fringe of evergreens, doubled in height by its mirrored counterfeit in the placid river.

From Petersburg to Seneca—a distance of twenty-two miles—we will follow the river-road, practicable only for cavaliers, which, though rugged, miry, and crossing the stream by frequent plunges, is far the most picturesque. About four miles above Petersburg we see the junction of the North and Middle Forks of the South Branch. Near this point we halt to examine a singularly perfect and beautiful exhibition of arched strata, laid bare by the action of the waters. The breakings of the rock are as clean and square cut as if they had been wrought by a master-mason, its colors and sylvan adornments rich enough to please the most exacting artist. The river sweeps its base in a succession of sparkling rapids; and in the middle of the stream, immediately opposite the centre of the arch, lies a huge, black bowlder, looking as if especially introduced to complete the artificial regularity of the scene.

Within the next mile or two we cross the fork again, and come suddenly upon a scene of quite another character. At the butt of a sharp spur rises a towering architectural mass, which any one familiar with the Old World would pronounce a well-preserved feudal ruin, and a purely American imagination would conceive to be the chimneys of a burnt factory. Even upon a closer inspection, it is difficult to divest one's self of the idea that human hands must have played some part in the erection of the pile before us. So regular and square cut is the masonry, so shapely the towers, so artistically true the embattled summits, the supporting buttresses, the jutting turrets, the cold, gray walls, dappled with lichens, moss, and weather-stains—all combined so artfully to mimic the "ruined castle of romance," that the garish light of a summer morning is scarcely strong enough to dispel the illusion. Yet, by turning on a still stronger light —that of a materialistic age and traditionless country—our castle dwindles into a geological vagary, and we resume our journey, filled with vague regrets.

Since leaving the gorge of the Southern Gate, we have seen rising before us, like a mass of dark, rolling thunder-clouds, the cliffs and pinnacled spurs and grinning summits of the great Alleghany Ridge. Between these and a parallel mountain of gigantic height and savage aspect flows the North Fork, whose borders we are now bent on exploring. Following the river-road, we pass by many a wild and disrupted battle-field, where for unnumbered ages the elements have striven for mastery—

"Crags, knolls, and mounds, in ruin hurled—
The fragments of an earlier world."

The artist is rather annoyed with the superabundance of pictorial attractions and the difficulty of selecting Especially pleased is he when we chance upon a subject so peculiar and impressive in its features that it leaves him no discretion. Such a point we

Karr's Pinnacles.

find at Karr's, eighteen miles beyond Petersburg; and as the day is usually far spent when we reach it, we will pass the night at the farm-house.

To approach the Pinnacles, about a mile distant from the house, one must ford the fork and ride up a narrow ravine, densely wooded. Jutting from the point of the opposite hill we see two thin sheets of rock, towering perpendicularly, side by side, far above

THE CLIFFS OF SENECA.

the tops of the loftiest forest-trees, their jagged and grotesque outlines drawn in dark *silhouette* against the clear-blue sky.

From Karr's, an easy ride of five miles brings us to the mouth of the Seneca. On reaching the open ground, all our faculties are at once concentrated on the magnificent object just across the river—a scene in which all the elements of curiosity, beauty, and sublimity seem to have been accumulated and combined. Imagine a thin, laminated sheet of rock, half a mile long by five hundred feet broad, set up on edge, the base covered for one-third of the height by a forest-grown talus; its sides ribbed with narrow terraces, moss-carpeted and festooned with gay, flowering shrubs; the bare surfaces stained with varied colors, white, yellow, red, brown, gray, and purple; its upper edge riven, splintered, and carved with a succession of grotesque forms which the pencil alone can describe. On the left the cliff abuts against a wooded mountain, defended, as it were, by a double line of bastioned and embattled walls. On the right it terminates abruptly in a sharp precipice. From the opposing hill juts another towering pile of rock, which forms the narrow gate-way through which appears a long vista of woods and mountains. When the sun gilds its painted and festooned sides, we glory in its beauty; when a passing cloud veils it in shadows, we are awe-struck by its weird sublimity. Up the steep bank, and across a shaded plateau, we enter the gate-way. Turning from the horse-path, we clamber up the talus at the base of the right-hand abutment, and, when out of breath, sit down to recover, and look up. We are now directly fronting the perpendicular edge or gable-end of the great cliff. The first emotion is one of bewilderment, not unmingled with dread, at the impending proximity of the awful pile. As we begin to note the details, and comprehend the general effect of the mass, we are troubled with a strange sense of incredulity, a distrust of our senses, even a certain flushing of resentment, as if some imposition were practised upon us. Yet there it stands, in motionless and silent majesty, a vast minster of the Gothic ages, growing more and more marvellous as we scrutinize its carven details and estimate its sublime proportions. There is the grand portal, with its pointed arch, from whose shadowy recesses we may presently expect to hear the organ pealing, and the anthem of chanting priests. There is the heaven-piercing spire, with its pinnacles, finials, turrets, traceries, and all the requisite architectural enrichments, from which anon will ring out the sweet and solemn chimes calling the world to prayer. There, too, sharply traced by sunlight and shadow, are the Gothic oriels and double-arched windows, suggestive of stained-glass pictures only visible from the interior.

And now having faithfully, and we hope satisfactorily, done the valley of the North Fork, we take regretful leave of its wonderful picture-gallery, and follow our adventurous artist across the bleak summits of the Alleghany, through miles of swampy laurel-brakes and dim hemlock-forests, to his camp in the mountain-wilds of Randolph County.

We are here upon the broad, wooded summits of the great dividing ridge of Alle-

Cathedral Rock.

ghany, about three thousand feet above the ocean-tides. Just below us flows the famous Blackwater through an awful rift nearly two thousand feet in depth. This is the land of water-falls. The first leap made by the Skillet Fork is forty feet in the clear, and thence without a halt it goes plunging down a break-neck stairway, with a descent of some four or five hundred feet in half a mile's distance, where it joins the main stream of Blackwater. This initial plunge is selected by our artist as one of the best-arranged pictures to be found in the mountains.

FALLS OF THE BLACKWATER.

Cathedral Rock—Side View.

The tender opal of the narrow strip of sky; the soft, bluish-gray border of distant forest appearing above the fall; the sparkling amber of the water mingling and contrasted with the snowy whiteness of the boiling spray; the dark plumage of the stately hemlock; the glistening foliage and delicate-pink bloom of the rhododendron; the gemmy greenness of the moss-carpeted rocks; the luscious splendor of the pool at our feet—all combine to form a natural picture before which the most ambitious art may hang its head.

Sag Harbor.

SCENES IN EASTERN LONG ISLAND.

WITH ILLUSTRATIONS BY HARRY FENN.

THE eastern end of Long Island is penetrated by a wide bay, extending inland a distance of thirty miles. A large island divides the bay into two distinct parts, the outer division being known as Gardiner's Bay, and the inner, which is subdivided by promontories, as Great Peconic and Little Peconic Bays. This large estuary gives to Long Island the shape of a two-pronged fork. The prongs are of unequal length, that upon the southern side exceeding the northern branch full twenty miles. The southern branch is distinguished as Montauk Point; the northern, until recently, as Oyster-Pond Point, but now is generally called Orient Point, deriving this name from the village of Orient, situated within its limits. Although Orient Point is shorter than Montauk Point, yet a succession of islands carries the line of this fork a long distance northeasterly into the sound—all of the islands, it is generally believed, once forming a portion of the northern peninsula. The most noted of them is Plum Island, upon which is a light-house well known to mariners.

Gardiner's Bay is partly sheltered from the sea by a long, narrow, and low stretch of land, extending, on a line southerly with Plum Island, across the open space that lies between the two points. Westerly, the bay is separated from the inner division of this inland sea by what is appropriately known as Shelter Island, which extends from opposite Greenport on the north branch to near Sag Harbor on the south branch. A more beautiful place could scarcely be found. Unlike all this portion of Long Island, it is crowned by noble hills, from the summits of which superb views can be obtained of the entire width of Long Island, the sound, and long stretches of the open sea. From White Hill, opposite Greenport, Orient Point is visible its entire length, charmingly dotted with villages, while beyond lies the sound, always white with many sails. From Prospect Hill, close at hand, Sag Harbor, and, far off, the open ocean, can be discerned.

View from White Hill, Shelter Island.

Greenport, on the northern branch, is the terminus of the Long Island Railroad. It is comparatively a new settlement, dating only from 1827; while East Hampton and Southampton, on the southerly fork, are nearly two centuries older. There were settlers on Oyster Point, however, as far back as 1646, one Mr. Hallock having, in that year, purchased the district from the Indians. But no towns were built up until long after. The settlers on the southern fork, notwithstanding they came from the neighboring shores of New

EASTERN LONG ISLAND SCENES.

England, passed Orient Point, inviting as it must have been with its rich soil and varied greenery, to the pine-barrens and grassy downs of Montauk. Greenport is a very pretty town—as green, neat, and quiet as the ideal New-England village. The cottages that line the well-shaded streets are hid among trees, and nowhere is decay or unwholesome poverty apparent. The drive from Greenport to the extreme of Orient Point is very charming. Near the town are many handsome villas and cottages, while flourishing farms and neat farm-houses enliven the road during the entire journey. The village of Orient, through which we pass, has a prosperous and pleasing aspect; and all along the drive the scene is varied by frequent glimpses of the sound on one side and the bay on the other. There is animation always in the picture presented here. On the sound, steamers and coasting-vessels come and go incessantly; while, in the bay, fleets of fishing-boats ever hover on the horizon, and yachts and smaller pleasure-boats give life and animation to the nearer scene.

Returning to Greenport, the traveller who explores this region will next desire to reach Sag Harbor, and a pleasant way to make the journey is by sail-boat. The course lies around Shelter Island, and, if winds are fair, the voyage can be accomplished in two hours. Sag Harbor was settled in 1730, nearly one hundred years before Greenport. It is an ancient whaling-place. When Long Island was first settled, whales were common visitors to its shores, and boats were always ready for the pursuit of those welcome strangers. The whales, when caught, were drawn upon the shore, cut in pieces, and sent to primitive boiling-establishments near at hand. From the pursuit of whales on the coast there naturally arose expeditions of a more ambitious character, and in the early part of the century we find the people of this town largely interested in the Pacific and Indian Ocean whale-fishing. The fisheries of the bay are the chief dependence of its citizens, although there are several factories there.

But Sag Harbor has a measure of newness by the side of East Hampton, on the southern branch, and the most easterly town of Long Island. This township was settled in 1649, by thirty families from Lynn and adjacent towns of Massachusetts. The land was purchased of the famous Montauk tribe, remnants of which are still found about Montauk Point.

East Hampton consists simply of one single street, three hundred feet in width. There are no hotels, no shops, no manufactories. The residences are principally farmers' houses, congregated in a village after the French method, with their farms stretching to the ocean-shore on one side, and to the pine-plains that lie between the town and the bay on the other. Its wide street is lined with old trees, and a narrow roadway wanders through a sea of green grass on either side. Few towns in America retain so nearly the primitive habits, tastes, and ideas of our forefathers as East Hampton.

Our illustrations include a view of this primitive village from the belfry of its old church, which the people, since Mr. Fenn made his sketch, have inexcusably destroyed—

EAST HAMPTON, FROM THE CHURCH BELFRY.

the only instance in the town's history of a disregard for its time-honored memorials. The antiquity of this building gave it interest, but it possessed special antiquarian value to the visitor on account of its identification with one of the most famous divines in our history. Here the Rev. Lyman Beecher officiated as minister during a period of twelve years, from 1798 to 1810; and during his residence in the town two of his distinguished children, Catharine and Edward, were born. The view from the belfry of the church is pleasing, the distant glimpse of the sea contrasting charmingly with the embowered cottages in the foreground. The old wind-mill gives quaintness to the picture. Two of these queer piles stand at the east of the village. They are very picturesque,

Home of John Howard Payne.

reminding one forcibly of the quaint old mills in Holland which artists have always delighted to paint. They form a distinctive feature of this part of the island, inasmuch as there are few similar structures existing anywhere in our country.

But East Hampton is not only renowned as the residence of Lyman Beecher, but of one peculiarly associated with our best impulses and feelings. It was here that John Howard Payne, author of "Home, Sweet Home," passed his boyhood. It is commonly asserted that he was born in the very old, shingled cottage pointed out as his residence; but of this there is some doubt. That his father resided here during the tender infancy of the lad is the better-supported story; but here, at least, the precocious lad spent several years of his early boyhood. His father was principal of Clinton Academy, one of the first institutions of the kind established in Long Island. The old house is held very sacred by the villagers, and the ancient kitchen, with its antique fireplace, stands to-day just as it did when Payne left it for his homeless wanderings over the world.

From East Hampton to the easterly extremity of Montauk Point the peninsula possesses a peculiar charm. The road follows the sea-shore over a succession of undulating, grass-covered hills. There is at all times and in all places a fascination in the

Interior of Payne's "Home, Sweet Home."

sea-shore, whether we explore the rocky precipices of Mount Desert, or follow the sandy cliffs of Long Island. But a summer jaunt along the cliffs of Montauk Point has a charm difficult to match. The hills are like the open downs of England, and their rich grasses afford such excellent grazing that great numbers of cattle and sheep are every year driven there for pasturage. The peaceful herds upon the grassy slopes of the hills; the broken, sea-washed cliffs; the beach, with the ever-tumbling surf; the wrecks that strew the shore in pitiful reminder of terrible tragedies passed; the crisp, delicious air from the sea; the long, superb stretch of blue waters—all these make up a picture that

GRIST WIND-MILLS AT EAST HAMPTON.

is full both of exhilaration and of repose. The heart expands and the blood glows under the sweet, subtile stimulant of the scene, even while delicious calm and contentment fill the chambers of the mind. The interest of the scene continually varies, even while its general features are almost monotonously the same. A boat on the beach, half buried

The Downs.

in encroaching sand; a mass of remains of wrecked vessels, such as Mr. Fenn graphically calls "The Graveyard;" a gnarled, wind-beaten tree on the hills; changing groups of cattle, among which occasionally appear drovers or herdsmen on horseback; vessels appearing and disappearing in the horizon of the sea—these make up the changes of the picture, and, simple as they are, give abundant pleasure to the wayfarer.

At last Montauk Point is reached. This is a bold, solitary point of land, composed of sand, bowlders, and pebbles, with far stretches of sea on three of its sides.

The Sand-drift

The storms here are grand, the wide Atlantic rolling in with unbroken force upon the shores. On the extreme point stands a tall, white light-house, erected in 1795, and one of the best-known lights of the coast.

Eastern Long Island is undergoing many physical changes. In reports made to the State Legislature by W. W. Mather, more than fifty years ago, we find a full and interesting description of the action of the sea on this peninsula, and also upon Orient Point. "The coast of Long Island," he says, "on the south side, from Montauk Point to Napeague Beach, a distance of three miles, is constantly washing away by the action of the heavy surf beyond the base of the cliffs, protected only by narrow shingle beaches of a few yards or rods in width. The pebbles and bowlders of the beaches serve as a partial protection to the cliffs during ordinary tides in calm weather; but even then, by the action of the surf as it tumbles upon the shore, they are continually grinding into sand and finer materials, and swept far away by the tidal currents. During storms and high tides the surf breaks directly against the base of the cliffs; and as they are formed only of loose materials, as sand and clay, with a substratum of bowlders, pebbles, gravel, and loam, we can easily appreciate the destructive agency of the heavy waves, rolling in unbroken from the broad Atlantic. The road from Napeague Beach to Montauk Point, which originally was some distance from

The Graveyard.

the shore, has disappeared in several places by the falling of the cliffs; thus is the ocean ever encroaching on the land.

"From Napeague Beach to two miles west of Southampton the coast is protected by a broad and slightly-inclined sand-beach, which breaks the force of the surf as it rolls in from the ocean. From Southampton westward the coast of the island is pro-

Montauk Point.

tected by long, narrow islands, each one of which is from one to five or six miles distant from the main island.

"The eastern parts of Gardiner's and Plum Islands, which are composed of loose materials, are washing away in consequence of the very strong tidal currents, and the heavy sea rolling in upon their shores from the open ocean. Little Gull Island (to the east of Plum Island), on which a light-house is located, was disappearing so rapidly, a few years since, that it became necessary to protect it from the further inroads of the ocean by encircling it with a strong sea-wall.

"Oyster-Pond Point is wearing away rapidly, by the combined action of the waves during heavy northeast storms, and the strong tidal current which flows with great velocity through Plum Gut. During a heavy storm, in 1836, the sea made a clean break over about one-quarter of a mile of the eastern part of the Point, washed away all the lighter materials, and cut a shallow channel, through which the tide now flows.

"Another effect of the sea is the formation of marine alluvion. Northeast storms bring in a heavy sea from the ocean, which, rolling obliquely along the shore, aided by powerful tidal currents, sweep the alluvia along in a westerly direction. Northwest winds do not bring in an ocean-swell, and the waves which they raise fall upon the shore in a line nearly perpendicular to the trend of the coast; so that their effect is to grind the pebbles and sand to gravel by the action of the surf, rather than to transport them coastwise. In this way outlets of small bays are frequently obstructed by bars, shoals, and spits, formed by the tidal currents sweeping past their mouths, and depositing the materials in the eddy formed by the meeting of the currents. Almost every bay and inlet, when not protected from the sea by sandy islands, have their outlets blocked up entirely by the materials deposited, or so nearly as to leave only narrow entrances."

Moonlight on Shore.

THE YOSEMITE.

WITH ILLUSTRATIONS BY JAMES D. SMILLIE.

Half-Dome, from the Merced River.

THE Yosemite Valley lies among the Sierra Nevadas of California, nearly in the centre of the State, north and south, and midway between the east and west bases of the mountains, at this point a little over seventy miles wide. In a direct line it is one hundred and fifty miles almost due east from San Francisco, but at present it can hardly be reached by less than two hundred and fifty miles of travel. The name is an Anglicised or corrupted form of the Indian A-hom-e-tae, which means Great Grizzly Bear, supposed to be the title of a chief, and applied generally to a tribe that held possession of the region from the valley to the plains on the west.

In 1864 Congress passed an act fixing the boundaries, and setting apart, "for public use, resort, and recreation," the Yosemite Valley and the Mariposa Grove of Big Trees. The State of California was to

BIG TREES—MARIPOSA GROVE.

appoint commissioners and assume the trust, which at once she did, and the people of the United States rejoiced in their grand park.

The grant of the Mariposa Grove covers four sections, or two miles square, and is under the charge of the Yosemite commissioners. The first that was known of the big trees was in the spring of 1852, when a hunter discovered what is now called the Calaveras Grove. No one would believe his story, and he had to resort to a trick to get any of his companions to go with him to the trees, so as to verify his statements. Once verified, descriptions were widely published. In 1854 an eminent French botanist, M. Decaisne, at a meeting of the "Société Botanique de France," presented specimens of the big trees and redwood that he had received from the consular agent of France at San Francisco. He explained at length his reasons for considering the big tree and redwood as belonging to the same species, *Sequoia;* so, in accordance with the rules of botanical nomenclature, the new species was called *Sequoia gigantea.* So far as is yet known, there are but eight distinct patches or groves of the big trees. They are very limited in range, and seem to belong exclusively to California. They form groves, largely intermixed with other trees, very little below five thousand and never over seven thousand feet above sea-level. They have been, without difficulty, largely propagated from the seed, and fine specimens are now growing in many parts of America and Europe. A

Fallen Sequoia.

few miles south of the Mariposa Grove, the *Sequoias* seem to find a more congenial home, and may be found of all ages and sizes, from the seedling up. Professor Whitney says: "The big tree is not that wonderfully exceptional thing which popular writers have almost always described it as being. It is not so restricted in its range as some other coniferæ of California." It occurs in great abundance, of all ages and sizes, and there is no reason to suppose that it is now dying out, or that it belongs to a past geological era, any more than the redwood.

The age of the big trees is not so great as that assigned by the highest authorities to some of the English yews. Neither is its height as great, by far, as that of an Australian species, the *Eucalyptus amygdalina*, many of which have been found to measure over four hundred feet. The tallest *Sequoia* that has been measured is in the Calaveras Grove, being three hundred and twenty-five feet high, overtopping Trinity Church spire (a standard of height familiar to most New-Yorkers) by forty feet. The greatest in diameter is the "Grizzly Giant" in the Mariposa Grove, which measures thirty-one feet through at the ground, and twenty feet at eleven feet above the ground. Clarence King described one that he saw in the forest some miles south of Mariposa, "a slowly-tapering, regularly round column, of about forty

Valley Floor, with View of Cathedral Spires.

feet in diameter at the base, and rising two hundred and seventy-four feet." A very large tree in the Calaveras Grove, twenty-four feet in diameter, was, after much labor, cut down, and the base, at six feet from the ground, was smoothed and prepared as a dancing-floor. Thirty feet farther up the trunk was again cut through, and the rings, marking the growth of each year, were carefully counted.

The ride from Wawona to the grove is about six miles. The trail was well worn and easy, the air gloriously pure, and the forest delightful. It would be useless to attempt to describe the confusion of sentiment and impatience that possessed me as I rode

Bridal-Veil Fall.

along, peering anxiously through the labyrinth of the wood for the first glimpse into the vast portals of that grand old grove.

The guide shouts, "There is a big tree!" What! are we so near the sacred precincts? I had built an ideal grove, and at first sight it was demolished, but that was no fault of the Mariposa big trees. There was no gloomily grand grove, there were no profound recesses; the great trees stood widely apart, with many pines and firs interspersed, and sunlight streamed down through all and over all. I wandered about, sorely disappointed that they did not look bigger, and yet every sense told me that they were vast beyond any thing that I had ever seen; and it was not until after I had been among them for hours, and had sketched two or three, that their true proportions loomed upon my understanding. In form they are often savagely gaunt, their respira-

CATHEDRAL SPIRES.

Sentinel Rock, from the North.

tory apparatus of foliage being in remarkably small proportion to their tower-like trunks. The bark is very light and fibrous, like the outer sheath of a cocoa-nut, of a singular cinnamon-color, and running in great ridges that vary from ten inches to three feet in thickness. Some trunks appear quite smooth, but others are warted and gnarled as though wearing the wrinkles of great age. On an area of thirty-seven hundred by twenty-three hundred feet there are just three hundred and sixty-five *Sequoias* of a diameter of one foot or over, but not more than twenty are over twenty feet in diameter. Two or three, greater than any that stand, now lie prone and broken; the trail lies through the hollow section of one that has fallen and been burned out. An ordinary-sized man, sitting upon a horse, can but just touch with his knuckles the blackened arch overhead.

Wawona is the present end of the stage-road, and the beginning of the path into the Yosemite.

It is not necessary to go all the way to the Yosemite to enjoy the picturesque effects of a party of pleasure-seekers, *en route*. The gay colors that inevitably find place, the grouping, action, light and shade in constantly-changing combination with the surrounding landscape, are a never-failing source of pleasure. Now, in bright sunlight, every spot of color tells with intensest power against a mass of sombre green; again, in the deep shadow of a wood, they form yet deeper shadows, and their richer color darkens against the light beyond. Now and then a broad waste of rock had to be passed, and several times, from heights, we had views of the high Sierra peaks. Meadows, covered with natural grasses, following the course of running streams, stretched for miles in narrow belts, where great numbers of horses and cattle roamed and found pasture. But we were impatient to get over the road that intervened between us and Inspiration Point. If, the day before, we rode in the excitement of expectation, it was intensified now; every step brought us nearer to a place that hitherto had been to me like some crater in the moon or spot on the sun. There was no doubt as to its existence, but it belonged to the realm of fancy, now to be transferred to the real—a change almost dreaded. At last, through the trees, there gleamed a pale, mist-like whiteness—it must be a wall of rock—could that be the first sight into the valley? The pulse quickened, the hard saddle and the shabby shamble of the offending beast underneath were forgotten as he forced himself into quicker gait in answer to impatient drubbings; a few moments more, and we rode out to a clear space under pine-trees, where every evidence was presented of the many feet that had halted there before us; so, following their indications, and the unmistakable sug-

Rock Slide.

gestions of our prosaic beasts, we alighted, and fastened them to well-worn branches of pine or manzanita. A few yards only of *chaparral* intervened between us and the cliff—a rush and a bound—in a moment our feet were upon Inspiration Point.

But we must hasten on. Every change of position presented some new charm—trees grouped into picturesque foregrounds, finding bold relief in light and shade against the opal and amethyst tints of distant granite cliffs; flowers nodding in the breeze that

Sentinel Rock and Fall.

brought refreshment to the brow and music to the ear; and little streams dimpling and gurgling across the trail, as if unconscious of the terrible leaps that must be taken before reaching the river below. In strong contrast to this living, moving beauty, beyond all, the walls, towers, and domes of the Yosemite rose grand, serene, impassive, broadly divided into tenderest shadow and sweetest sunlight, giving no impression of cold, implacable, unyielding granite, but of majesty, to which our hearts went out as readily as to the flowers and brooks at our feet. As we approached the level of the valley and the open meadows, the groves of trees and the winding river were more distinctly seen—the glorious, park-like character of the place presented itself. At last the foot of the descent was reached. Trees, bending in graceful framework, enclosed various pictures, one of the most charming being a view of the Bridal-veil Fall as it sprung over the wall nine hundred feet high. Its upper part sparkled a moment in the sunlight, a solid body; then, as though wrestling with invisible spirits, it swept into a wild swirl of spray that came eddying down in soft mists and formless showers. Emerging from the wood, a broad meadow lay before us; and high over all projected, far up against the eastern sky, the Cathedral Rocks, with buttresses cool and spires aglow. At their foot the river crowds so close that the trail is forced to find its way through a wilderness of great granite blocks, that lie embowered in a forest which has grown since they were hurled from their places on the cliffs above. Then followed a long level, and groves of pine and cedar. There was no sentiment of gloom, but rather of deep, slumberous repose; the thick carpeting of sienna-colored pine-spindles that covered the ground hushed each foot-fall; the pillared tree-trunks formed vistas that stretched, like "long-drawn aisles," to profoundest forest-depths; the branches, "intricately crossed," did not obscure the luminous sky above, or hide the tall cathedral-spires that burned ruddy in the last gleam of day; refreshment and invigoration were in

Foot of Sentinel Fall.

the very atmosphere; with thankfulness, my whole being drank deeply; and when, in the gray of evening, the hotel was reached, I was cool, calm, and—very hungry.

The first week after our arrival was spent making acquaintance with the more common points of interest and attraction. At first we rode in beaten paths, and wondered and admired according to regulation; but after a day or two such bonds became irksome, and we ranged at will, there being really no need of a guide in an enclosure six miles long and at most but a mile and a half wide—no need of any one to direct attention to what the eyes could hardly fail to see or the senses discover for themselves; and, then, it was so much more delightful to wander undirected and unattended, on horseback or on foot, regardless of conventional ways, and yielding unreservedly to each new enjoyment. We soon knew each meadow and the separating groves of trees, every stream and every ford across the river. Within the limits we ranged there are but eleven hundred and forty-one acres of level bottom, according to Government reports, and of this seven hundred and forty-five acres are meadow, the rest being covered with trees and *débris* of rock. From Tenaya Cañon, at the upper end of the valley, to Bridal-veil Creek, near the lower end, four and a half miles in a direct line, the decline is only thirty-five feet. Naturally enough, a surface so nearly level is very widely overflowed during the high water in the spring, caused by melting snows among the mountains beyond. The meadows are covered with coarse, scant grass; and innumerable flowers, generally of exceeding delicacy, find choicest beds in slight

Yosemite Fall and Merced River.

depressions, where the water lies longest. Through these meadows the Merced River winds from side to side, during the summer an orderly stream, averaging, maybe, seventy or eighty feet in width, the cold snow-water shimmering in beautiful emerald greens as it flows over the granite-sand of the bottom. Its banks are fringed with alder, willow, poplar, cottonwood, and evergreens; upon the meadow-level are grouped, in groves more or less dense, pines, cedars, and oaks, the latter often bearing large growths of mistletoe; upon the rock-talus, mingling with the pines and firs, the live-oak is a distinctive feature; higher, and clinging in crevices and to small patches of soil, the pungent bay and evergreen oak form patches of verdure. From the foot of Sentinel Fall an excellent view may be had of the meadows, the groves, the river, and the slopes at the foot of the walls of rock on either hand. On the right is El Capitan, three thousand three hundred feet high; on the left are the Cathedral Rocks, nearly two thousand seven hundred feet in height—the two forming what may be called the southern gate to the valley. Each of our illustrations, it is intended, shall present some characteristic feature of the valley. The opening cut was selected from many similar views at the upper end of the valley, where the pine-trees come down to the river's edge, and are mirrored in the still pools. Washington Column, more than two thousand feet high, stands out on the left, casting an afternoon shadow well up on the flank of the Half-dome, whose summit is almost five thousand feet above the river, or nine thousand feet above the sea. The spires are forms of splintered granite, about five hundred feet in height, and altogether not less than two thousand feet above the valley. Sentinel Rock combines more of picturesqueness and grandeur, perhaps, than any other rock-mass in the valley, its obelisk-like top reaching a height of over three thousand feet, the face-wall being almost vertical. The view from the north is taken from a point about midway between the foot of Yosemite Fall and Washington Column; the other is from a point as far south of it, presenting an entirely different aspect, its stupendous proportions dwarfing into littleness every thing at its base. The fall at the right, as shown in the illustration, exists only in the spring, as it depends entirely upon the melting snow for its supply. That its force and volume at times must be terrific is evident from the gorge that it has hollowed at its foot. It is rarely that such exhibitions of destructive energy can be found. The climb up this water-torn gully ends all dreams of a well-ordered park below. Torrents pour into the valley as soon as the snow begins to melt, leaping the cliffs with indescribable fury, carrying immense rocks and great quantities of coarse granite-sand, to work destruction as they spread their burden over the level ground. In some places this detritus has been deposited to the depth of several feet in a single spring. The air then is filled with the roaring of water-falls; the greater portion of the valley is overflowed; and the wayward Merced cuts for itself new channels, making wide waste in the change. At such times the Yosemite Fall is described as grand beyond all power of expression. The summit of the upper fall is a little over two thousand six hundred feet above the

YOSEMITE FALL.

valley; for fifteen hundred feet the descent is absolutely vertical, and the rock is like a wall of masonry. Before this the fall of water sways and sweeps, yielding to the force of the fitful wind with a marvellous grace and endless variety of motion. For a moment it descends with continuous roar; in another instant it is caught, and, reversing its flight, rises upward in wreathing, eddying mists, finally fading out like a summer cloud. The full-page illustration is taken from a clump of pine-trees so near that, by the rapid foreshortening, the entire fall appears in very different proportions from those seen from the opposite side of the valley. Such a glimpse is given in the illustration "Yosemite Fall and Merced River."

In the spring water is an element of destruction, in freezing as well as in thawing. The little rills that filter and percolate into every crack and crevice of rock by day, as

Horse-Racing.

they freeze at night enable the frost to ply its giant leverage; and when disaster from water seems to threaten everything, there is added the shock of falling cliffs. The granite-walls are not homogeneous in structure, some portions being far less durable, under the action of time and the elements, than others. The Half-dome and El Capitan are magnificent masses, at whose feet the *débris* are comparatively slight; but that part known as the Union Rocks, between the Cathedral and Sentinel Rocks, has suffered very much from disintegration. Great cliffs have fallen, and avalanches of rock have ploughed their way down the slope to the bottom of the valley. While climbing in such surroundings the wreck of some world is suggested, so vast the ruin and so pygmy the climber. No words can convey other than a feeble impression of the effects of mountains of granite, sharp and fresh in fracture, piled one upon the other, the torn fragments of a forest underneath, or strewed about, as though the greatest had been but

Merced Gorge.

as straws tossed in the wind. A broad track of desolation leads away up to the heights from which these rocks have been thrown.

The attention may be diverted from cliffs and torrents to the human element characteristic of the place, poor though that element be, and in the change find much that is interesting in the few Indians that straggle, vagrant and worthless, through the region. They seem to be without tribal organization, although they still have "pow-wows," where their leading men, conscious of the inevitable decay of the race, strive to reorganize them and arouse their dying spirit; but the red-men are now hopelessly debauched and demoralized. In general appearance they are robust, and even inclined to be fleshy; this latter is accounted for by the fact that acorns, which is one of their staple articles of food, are extremely fattening.

One Sunday morning I strolled to the upper end of the valley; a quiet like that of languor filled the air; the roar of the Yosemite Fall had died out, and now but a slender stream down the face of the cliff marked its place. In the hush I walked under the pine-trees, whose pendulous branches and long, tremulous needles vibrated into an Æolian melody upon the slightest provocation; a scarcely-perceptible breeze brought whispers, to be caught only by the attentive ear, that swelled through

TENAYA CAÑON, FROM GLACIER POINT.

General View of Yosemite, from Summit of Cloud's Rest.

faultless crescendos into volumes of harmony, rich and deep, yet ever sounding strangely far away. From the shadows and music out to the sunlighted meadow was but a step. At the other extremity of the open space, four or five hundred yards away, was a group of men. Drawing nearer, it was plainly to be seen that they were intent upon the preliminaries of a horse-race. There were Indians, Chinamen, Mexicans, negroes, and very dark-colored specimens of white men. There was a confusion of tongues, through which came the clear ring of clinking gold and silver coin, for all were betting—many of them their last dollar. Several horses were getting ready for the race; the favorites were a sorrel and a roan, or "blue horse;" all were very ordinary animals, and without the slightest training. There were no saddles; the riders, stripped of all superfluous clothing, bareheaded and barefooted, rode with only a sheepskin or bit of blanket under them; over the drawn-up knees and around the horse's body a surcingle was tightly drawn, binding horse and rider into one. Judges, starters, and umpires, were selected and positions taken. The word was given; the horses plunged, started, "bucked;" again they started; again the sorrel bucked. An unlimited amount of profanity expressed the impatience of the crowd. The "blue horse" was now largely the favorite.

At last they came—a cloud of dust, rattling hoofs, and frantic riders plying their whips right and left over the struggling brutes under them; on they came; the squatting crowd sprang to their feet, and up went one simultaneous yell; on they came, the crowd capering, screaming, and "hollerin'," like so many madmen—all alike infected, the stoical Indian as well as the mercurial Mexican. The "blue horse" led, and, in a cloud of dust, all dashed by. It was a whirlpool of excitement, the stake being the vortex. Round and round they went; shouts, laughter, and profanity—one wild, incoherent Babel—losers and winners alike indistinguishable. Their hot temperaments found the excitement they craved, and the losers were rewarded in its drunkenness. Yet another very different interest is to be found in the visitors who throng the valley.

The scenic effects of winter are described as wonderfully beautiful, the ice-forms about the falls being particularly interesting. No doubt in time it will be the fashion to make winter excursions into the Yosemite, but for the present it is safe to advise that, if the visit cannot be made in May or June, it be deferred until another season; for later in the year, to the disappointment of losing some of the finest features in the scenery, are added the discomforts of heat, toil, and an all-pervading dust, that penetrates to the innermost recesses of one's baggage and being. The temperature of spring is delightful, but during summer the thermometer frequently stands as high as 96° and 98°, while on the plains it is away above 100°.

There are now no less than five trails over which a horse may get in or out of the valley: the Mariposa trail, passing Inspiration Point, and entering at the southern end; the Coulterville trail, that comes in at the same end, on the opposite side; a third trail, passing near Glacier Point, and entering at the foot of Sentinel Rock, about midway up the valley on its eastern side; a fourth one, passing through the Merced Gorge by the Vernal and Nevada Fall; and the fifth, through Indian Cañon, on the west side, north of Yosemite Fall.

The trail through Merced Gorge, after reaching the top of Nevada Fall, crosses the stream and the southern end of the Little or Upper Yosemite Valley. This valley, more than two thousand feet above its famous neighbor, is one of the many great granite basins peculiar to this section of country. The bottom is a little more than three miles long, and is a pleasant succession of meadows and forests, through which flows the Merced River. The sides are not so much walls as smooth, bare slopes of seamless granite, ribboned with sienna brown bands from running water, and here and there breaking into strange dome-forms.

Among our more extended excursions we planned one to this place, and as we were to camp out for several days our preparations were careful, and, on starting, our cavalcade was imposing. Five riders led; three pack-horses followed laden with hampers

Gorge of the Merced, from Glacier Point Trail.

and blankets, each pack crowned with an inverted kettle or a broad frying-pan. After beginning the ascent, the way led through woods, close grown, and filled with a tangled undergrowth that, with all its rank vigor, was unable to overtop the great fragments of rock that strewed the forest. In places the trail twisted from right to left in sharp zigzags, and was so exceedingly steep that the horse and rider upon the turn above seemed to be almost overhead. Within sight the river roared and tumbled in a series of cataracts. We left our horses under a great overhanging rock, in charge of the guide, to be taken up the trail to meet us farther on, while we climbed by a foot-path around the base of a magnificent cliff, and out, face to face with that beautiful sheet of falling water called the Vernal Fall. It is a curtain unbroken in its plunge of four hundred feet; on either side the narrow gorge, drenched with spray and glimmering with rainbow-tints, is green with exuberant vegetable life. Climbing long ladders, we reached the top, to find a broad, basined rock and a lovely little lakelet sparkling in the sunlight. Farther on we crossed a slender bridge, Wildcat Cataract flying underneath. Before us Nevada Fall came tumbling over a wall exceeding six hundred feet in height;

to the right the Cap of Liberty, a singular form of granite, rose more than two thousand feet; all about were heights and depths, grand to look up to, terrible to look into. We had rejoined our guide and horses, and, passing through a clump of dark-looking firs that clustered at the foot of the Nevada Fall, we came out upon a slide of freshly-fractured glistening granite that seemed impassable, but a way had been made, and up this avalanche of rock our horses betook themselves, climbing with wonderful pluck and sureness of foot. But one beast had shown a spirit of insubordination, so the guide had tied him close to a leader. At each angle of the zigzagging trail he would balk, refusing to follow; the other horse, keeping on regardless, pulled the obstinate creature into predicaments from which he could not extricate himself; then each pulled against the other, utterly indifferent as to consequences. In one of these contests the foothold of the leader gave way, and in an instant a confused mass of horse, an inextricable jumble of heads, legs, and tails, to say nothing of kettles and frying-pans, came bounding toward me; leaving the trail, the horses turned two or three somersaults among the broken rocks below, and then lay still. We clambered quickly down to them. They were not dead, did not even have any bones broken—their packs had saved them. One, lying wedged, with his feet in the air, received our first attention; ropes and straps were cut, and three of us undertook to roll the beast out of his position. While the packs were being adjusted upon other horses—for these could barely hobble along—I made a sketch of the scene, looking down the gorge. In the distance is a glimpse of the western wall of the Yosemite. Nearer, on the left, is Glacier Point, rounding up to Sentinel Dome. The form to the right, in the middle of the picture, is a point called Crinoline, Sugar-Loaf, Verdant, and several other names. It is a spur from the shoulder of the Half-dome. The rock that forms the right of the sketch is a portion of the base of the Cap of Liberty. Resuming our way, we reached the upper valley late in the afternoon, and found an ingeniously-constructed evergreen brush-house ready for us. It was short work to unpack and unsaddle our horses, turn them loose, gather wood, light a fire, and prepare our evening meal. Bright-colored blankets were spread with an eye critical to effect, and the heavy Mexican saddles made capital lounges and pillows. A stroll in twilight, until it deepened into moonlight, completed the day. In spite of all our precautions, the first night was really uncomfortable, owing to the cold. In the morning a gray rime of frost covered every thing. We were camping at an elevation greater than the summit of Mount Washington.

From camp we made an excursion to the top of Cloud's Rest, a point of view that surpasses all others in its comprehensiveness, as it rises at least six thousand feet above the Yosemite, or ten thousand above the sea. Not very far from the summit we entered a remarkable grove of sugar-pines, through which ran a small stream, where grass grew abundantly. We took our horses to within a few hundred yards of the summit, after cantering over a waste of disintegrating granite, upon which stood, at wide in-

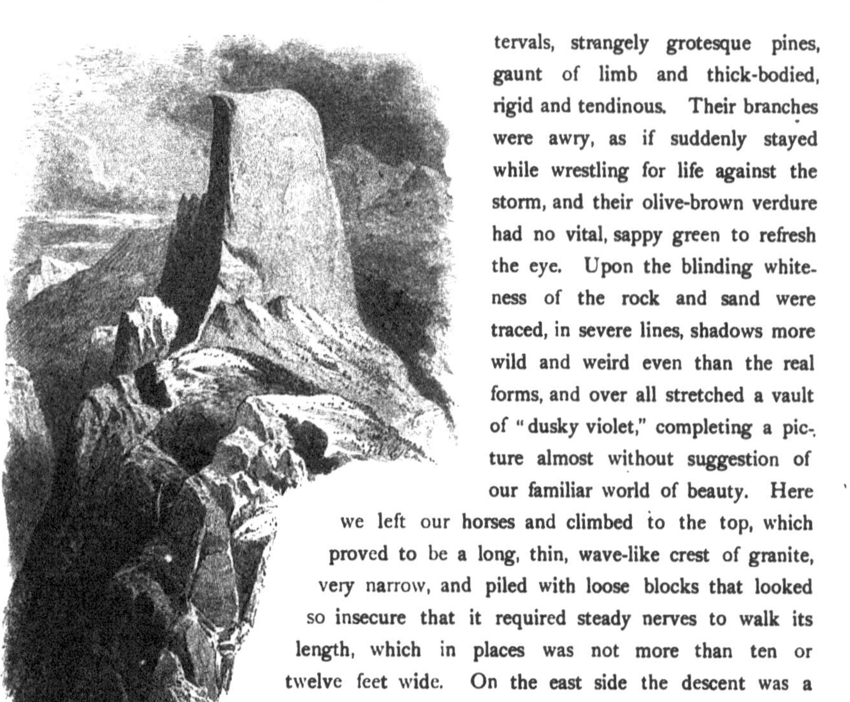

Half-dome.

tervals, strangely grotesque pines, gaunt of limb and thick-bodied, rigid and tendinous. Their branches were awry, as if suddenly stayed while wrestling for life against the storm, and their olive-brown verdure had no vital, sappy green to refresh the eye. Upon the blinding whiteness of the rock and sand were traced, in severe lines, shadows more wild and weird even than the real forms, and over all stretched a vault of "dusky violet," completing a picture almost without suggestion of our familiar world of beauty. Here we left our horses and climbed to the top, which proved to be a long, thin, wave-like crest of granite, very narrow, and piled with loose blocks that looked so insecure that it required steady nerves to walk its length, which in places was not more than ten or twelve feet wide. On the east side the descent was a steep sweep for hundreds of feet; on the west it was thousands. It fell away in one unbroken surface of granite, at an angle of not less than 45°, with no obstacle to stay a falling body until it should reach the depths of Tenaya Cañon, at least a mile and a half distant. This slope is shown in the full-page illustration of Tenaya Cañon, where Cloud's Rest is the point just to the left of the Half-dome. It required some minutes to settle the nerves and look calmly about. To the north, over intervening cañons and gorges, the Sierra peaks rose grandly desolate, pale and delicately tinted with many tones, warm and cool, against the cloudless vacuum of the sky beyond, that, by contrast, wore a strangely sombre hue. Their shoulders were robed with snow and ice, and their flanks were grooved and scarred by glaciers long since extinct. Upon lower levels a sparse growth of evergreens hardly served to cover the naked appearance of the landscape, and bald spots of rock showed almost as white as the snow beyond. This peculiar appearance of sterility, and meagre, patchy forest-growth, characterizes all the surrounding country when seen from such a height. Turning from the Sierras, that were from three to five thousand feet above our level, we looked down six thousand feet into the Yosemite, whose peculiar, trough-like formation was readily recognizable, running almost at right angles to the regular trend of the mountains, and fully four thousand feet below the average level of the surrounding country. The familiar forms of the enclosing walls, and the green groves and meadows of the valley-floor upon which the Merced sparkled, could be plainly seen, but angles of rock hid each water-fall.

After a day or two we broke camp, and, by a new trail, over which we were the first to pass, made a *détour*, keeping along the upper edge of the Merced Gorge, crossing the Too-lulu-wack a few hundred yards above its fall, and thence to Glacier Point. This is one of the most interesting rides about the valley, presenting many grand and even startling views. From one point we could look down into what seemed a bottomless abyss, for it was impossible to see its greatest depth. Out of it came the roaring of distant waters and the lulling song of pine-tree forests. The Too-lulu-wack Fall was almost under us, and could not be seen; but on the opposite side were the Vernal and Nevada Falls, and the many cataracts of the Merced, that, unlike most of the other streams that enter the Yosemite, are very imposing all the year round. The Cap of Liberty rose prominently in the centre; back of that the upper Yosemite opened, and beyond all were the snow-capped High Sierras. In the engraving of this view the peculiar rock-form and character of the upper valley walls or slopes have been quite lost. Passing on, we soon reached Glacier Point, which is a spur of rock or mountain jutting out on the west or right-hand side of the valley, where it divides. From its terraced summit we looked down thirty-two hundred feet to the meadows at our very feet. Directly opposite, on the other side, perhaps a mile and a half away, the Yosemite Fall came down half a mile in three leaps, its truly graceful proportions seen to greater advantage than from any other point. To the right, or north, we looked up Tenaya Cañon, its narrow floor beautiful with tall pines that almost hid its one jewel, Mirror Lake; but with walls grim and vast that swept on the right up five thousand feet to the grand, dominating form of the valley, the Half-dome. The bald slope and crest of Cloud's Rest towered beyond, and back of all the Sierras lifted their peaks, as yet untrodden by the foot of man. There can be but few places where so much of the terrible and the beautiful are at once combined.

From Glacier Point a trail leads to the summit of Sentinel Dome. Upon this height we spent an hour or more, enjoying already familiar features as viewed from a new stand-point. We returned to our old quarters in the valley, and worked hard for two months to bring away some limned shadow, however faint, of the wonders about us. At last our work was done, and our traps were packed for departure. A long, lingering glance as the cold moon tips with silver those giant, sleeping forms of granite, and then by its growing light I cleared my palette, and closed the box upon my last study of the Yosemite and Sierras.

MOUNT MANSFIELD.

WITH ILLUSTRATIONS BY HARRY FENN.

Rock of Terror.

VERMONT is, and perhaps ever will be, the most purely rural of all the older States. Though bordered by Lake Champlain, and pretty well supplied with railways, she seems to be aside from any great thoroughfare, and to hold her greenness nearly unsoiled by the dust of travel and traffic. Between the unyielding granite masses of the White-Mountain range on the one side, and the Adirondack Wilderness on the other, lies this happy valley of simple contentment, with its mellower soil and gentler water-courses, its thriftier farmers and more numerous herds, its marble-ledges, its fertile uplands, and its own mountains of gentler slope and softened outline.

Nearly through the middle runs the Green-Mountain range, giving rise to a thousand murmuring rivulets and modest rivers, that lapse down through green-browed hills and crumbling limestone-cliffs and sunny meadow-lands, now turned quickly by a mossy ledge, and now skirting a bit of native forest, until they lose themselves on the one side in the deep-channelled Connecticut, or on the other in the historic waters of Lake Champlain. Quiet industry, pastoral contentment, out-door luxury, and in-door comfort—these are the characteristics that continually suggest themselves to the visitor, wherever he loiters among the valley-farms or pleasant villages of the Green-Mountain State. It impresses him as a land where wealth will seldom accumulate, and men should never decay—whose dwell-

The Old Woman of the Mountain.

ers may forever praise God for the greenness of the hills, the fertility of the soil, the purity of the streams, the delicious atmosphere, and the mellow sunshine—where the earth extends such a genial invitation to labor that all must be allies, striving together for a living out of the ground, and none need be enemies, scheming to get it out of each other.

When Jacques Cartier, a third of a millennium ago, descried these peaks from

Corduroy-Bridge, Mount-Mansfield Road.

Mount Royal, by the St. Lawrence, he looked upon a land whose history was yet to be, where we look upon one whose history, in the romantic sense of the term, is probably closed. For nicely-worded statutes and accurate surveyors' lines have taken the place of vague royal patents, bounded by unknown rivers; and the contention between New Hampshire and New York, that kept Vermont out of the Union during the Revolution, can have no repetition or parallel. There was one Bennington—there need be no more; there was one Ethan Allen—there can never be another. But, though the days of colonial jealousies and rebellious warfare are over, and this quiet people are counting their cattle and weighing their butter-firkins where their grandsires shouldered their muskets and lighted beacon-fires, the glory of manhood has not departed with the romance of frontier-life. It was the sons of the men who carried Ticonderoga and Crown Point who annihilated Lee's forlorn hope at Gettysburg, turning the battle that turned the civil war. Vermont, too, may have a history of literature and art, which is but just begun. Here lies the marble-quarry of America, and here sprung America's earliest and best-known sculptor. One of her most famous journalists here spent his boyhood, learning the use of pen and type; and here, also, his aptest pupil was reared. And, for the extremes of literature, one of our earliest humorists, and one of our most celebrated philologists, were born in these same verdurous valleys.

If Professor Rogers's theory of mountain-formation be correct—that elevated ranges have been produced by a sort of tidal wave of the earth's once plastic crust—then the Green Mountains must be the softened undulation that followed the greater billow which crested and broke in Mount Washington and Mount Lafayette, leaving its form forever fixed in the abrupt and rugged declivities of the White Hills and the Franconia group. The Green Mountains form the northern portion of what is known as the Appalachian Chain. Their wooded sides obtained for them from the early French settlers the term *Monts Verts*, and from this phrase is derived the name of the State in which they are situated. The continuation of the range through Massachusetts and Connecticut is also known to geographers as the Green Mountains, but by the inhabitants of those States other names are applied to them —as the Hoosac Mountains, in Massachusetts, for that portion lying near the Connecticut River, and constituting the most elevated portion of the State between this river and the Housatonic; and the Taconic Mountains for the western part of the range, which lies along the New-York line. These ranges extend into Vermont near the southwest corner of the State, and join in a continuous line of hills that pass through the western portion of the State nearly to Montpelier. Without attaining very great elevation, these hills form an unbroken water-shed between the affluents of the Connecticut on the east, and the Hudson and Lake Champlain on the west, and about equidistant between them. South from Montpelier two ranges extend—one toward the northeast, nearly parallel with the Connecticut River, dividing the waters flowing east from those flowing west; and the other, which is the higher and more broken, extending nearly north, and near Lake Champlain. Through this range run the Onion, Lamoille, and Winooski Rivers.

View from Mountain-Road.

Mount Mansfield, except near the summit, is very heavily timbered; and the glimpses downward, through entanglements of trees into the deep ravines, are full of superb beauty. Now the road passes over a terraced solid rock, and now it jolts over the

Glimpse of Lake Champlain, from Summit.

crazy scaffolding of a corduroy bridge that spans a chasm in the mountain-side; soon the forest-growths begin to thin out perceptibly; and at last we reach the Summit House, amid masses of bare rocks, at the foot of the huge cliff known as the Nose.

Cave under Lower Lip.

The path up the Nose, on its western side, is quite as rugged as the ordinary climber will wish; but, with the help of the cable, its ascent may be accomplished. The view from the top is one of the finest in our country. To the eastward are the White Mountains, dwindled by distance. The isolated and symmetrical form of Ascutney rises on the southeast. Southward are Camel's Hump and Killington Peak, and innumerable smaller elevations of the Green-Mountain range—respectable and respected in their own townships, doubtless, but here losing much of their individual importance, like monstrosities at a fair. Westward lies a considerable expanse of lowland, with many sparkling streams winding about among the farms and forests and villages, the city of Burlington in the distance, and beyond them the beautiful expanse of Lake Champlain, with the blue ridges of the Adirondacks serrating the farthest horizon. On the northwest is the Lamoille Valley, watered by the Lamoille and Winooski Rivers, that tumble through the depressions of the outliers, and dream their way across

Climbing the Nose.

the plain. And far northward are Jay Peak and Owl's Head, the stately St. Lawrence, the spires of Montreal, a score of nameless mountains, and Lake Memphremagog, familiar to many readers by the means of Whittier's pleasing verse. The difficulty, however, with all views from mountain-tops is, to find an occasion when the atmosphere is sufficiently clear to take in the prospect. Mr. Fenn was three days on the summit of Mansfield, during all which time a dense, gray vapor enveloped all the facial features of that grand profile, and veiled the surrounding scene as completely as the curtain at the play shuts from view the splendors behind it. At last, the misty veil lifted a little; and we have as a result, in one of the illustrations, a glimpse, through this parting vapor, of Lake Champlain and the distant Adirondacks. Another view shows us the mountain-cliffs looming through the mist, affording a glimpse of what is known as Smuggler's Notch, one of the most interesting features of this mountain. In the far West this notch would be called a cañon. It differs from the cañons of the Sierras mainly in being more picturesque and beautiful—not so ruggedly grand as those rocky

Smuggler's Notch.

walls, it must be understood, but the abundant moisture has filled it with superb forest-growths, has covered all the rocks with ferns and lichens, has painted the stone with delicious tints. The sides of the Notch rise to an altitude of about a thousand feet, the

Rocks in Smuggler's Notch.

upper verge of the cliffs rising above the fringe of mountain-trees that cling to their sides. The floor of the Notch is covered with immense bowlders and fallen masses of rocks, which in this half-lighted vault have partly crumbled, and given foothold for vege-

LOOKING TOWARD SMUGGLER'S NOTCH, FROM THE NOSE.

tation. Mosses and ferns cover them, and in many instances great trees have found nourishment in the crevices, sometimes huge, gnarled roots encircling the rocks like immense anacondas. The painter could find no more delightful studies in color than this scene affords. At the time visited by the artist and the writer, there had been a three days' rain. The stream that flowed through the gorge was swollen into a torrent. Over the top of every cliff came pouring extemporized water-falls and cascades, while the foliage, of fairly tropical abundance, shone with a brilliant intensity of green. Smuggler's Notch has a hundred poetical charms that deserve for it a better name. It is so called because once used as a hiding-place for goods smuggled over the Canada border.

Moss-Glen Cascade.

Another very charming picture in this Mansfield gallery is Moss-Glen Cascade, a water-fall that comes tumbling down, in successive leaps, through a narrow gorge. The pipe, or flume, supported by the rude ladders on the right, conveys a portion of the water to the wheel of a saw-mill. It seems like an impertinence to introduce any mechanical contrivance into so exquisitely wild a bit of scenery as this; for the brook is emphatically "a gushing child of Nature," not intended for homely usefulness.

THE JUNIATA.

WITH ILLUSTRATIONS BY GRANVILLE PERKINS.

Duncannon, Mouth of the Juniata.

THE Juniata is a tributary of the far-famed Susquehanna; and though its short life begins at a point beyond Clearfield, and ends at Duncannon—a distance of one hundred and fifty miles—yet it presents many scenes of entrancing beauty. It falls into the Susquehanna, about a mile from the last-named place, at a spot well worthy the inspiration of an artist's pencil, as is shown in the illustration.

The mouth of the Juniata is not very broad, and seems quite narrow when compared with the flood of her big sister; but her stream is much deeper, and her waters of a deep blue. The poets of the locality love to write about the blue Juniata, and speak of it as the gently-gliding stream. In summer-time, no doubt, this name is appropriate; but from the hill of observation above Duncannon one can see the remains of four stone piers—all that is left of the bridge that spanned the Juniata at this point. Regularly every spring, when the snows melt and the ice piles up in masses, the Juniata sweeps away her bridges as if they were feathers, and comes rushing into the broad Susquehanna with a wealth of blue water that materially changes the color of the big, brown stream.

Following the bank of the blue Juniata, side by side with the canal, one is for a

Night Scene on the Juniata, near Perryville.

few miles, at first, in a level country. The stream is not broad, but tolerably deep, and abounding in fish, which rise every moment at the flies that hover over the placid surface. Between here and Perryville the river is full of beautiful islands, covered with trees whose branches sweep down to the ground and often hide the bank. With the branches are interlaced wild vines with huge leaves, and between them the golden-rod, and the big yellow daisy, and the large-leaved fern, make their appearance. In the low parts of these islands there are beautiful mosses, and a species of water-grass which becomes a deep orange in circular patches.

Approaching Perryville, the foot-hills disappear, and the bright glimpse of champaign country vanishes. The mountains are once more upon us, looming up into the clear sky like giants. They are on both sides, and in front likewise. On the right there is one huge, solid wall, with hardly an irregularity or a break along the crest, which is straight as a piece of masonry. On the left the mountains are strung along like a chain of gigantic agates. Each seems to be triangular, and between each is a ravine, where there are not only tall trees, but also fine slopes of high grass. At night-time, when there is a full moon, the river near Perryville is exceedingly grand: the solemn stillness of the hour; the lapping sound of the gentle water; the whisper of the wind among the trees, that seems more like the falling of a distant cascade than the rustling of leaf on leaf and the chafing of bough against bough.

But if the approach to Perryville be most beautiful by night, it is not so beyond. On the left bank the mountains still show their bold fronts, and the stream, forced

Windings of the Juniata, near Perryville.

Moss Islands in the Juniata.

around the capes on the one side, has worn similar indentations on the other, presenting a most beautiful appearance. The most picturesque part of this lovely region is after we pass the little village of Mexico. It is difficult to say whether the river is finer looking forward or looking back, for there are so many phases of beauty about the

Narrows near Lewistown.

scene. How shall we decide? Perhaps looking forward is the best, if one can leave out of the perspective a wretched mountain called Slip Hill, which, having been deprived by the wood-cutters of its forest mantle, has ever since taken to rolling stones down its great slope, and presents a hideously forlorn appearance. It is covered from apex to base

TYRONE GAP, VIEW FROM THE BRIDGE

with a mass of small, flat stones, like scales, and about every half-hour there is a movement, and a miniature land-slip goes gliding into the river.

The next point along the line of the Juniata is one where the river sinks into a very subordinate position indeed. The hills on both sides, that have hitherto been so amiable, suddenly break off, and the great wall comes into view on the right hand, while on the left we get the side of a mountain instead of its front. On both banks the hills are remarkably steep, and they approach so closely together as to confine the little river

The Forks of the Juniata, near Huntingdon.

within extremely narrow bounds. For seven miles and a half this imprisonment lasts; and here, perhaps, the mountains show their grandest forms. As the entrance into the Narrows was sudden, so the exit is abrupt. One wanders along the tow-path of the canal looking up at the mountains, and wondering how much nearer they intend to come, and whether they are going to close in and crush us utterly, when suddenly the Juniata makes a bold fling to the right, and we find ourselves in Lewistown, with the mountains behind us, and a pleasant valley smiling welcome in our front.

SINKING RUN, ABOVE TYRONE.

Between Lewistown and Huntingdon the scenery is extremely beautiful; but to describe it would be simply a repetition of the phrases applied to Perryville, where the curves of the river are so lovely. But the mountains are decidedly bolder, and the river becomes wilder, and curves in such a multitudinous fashion as to make frequent bridging absolutely necessary. At Huntingdon the hills retire and leave a pleasant level. Here the Juniata forks, the larger but less picturesque fork striking southward toward Hollidaysburg, and the smaller branch, known as the Little Juniata, going west in the direction of Tyrone. But henceforth the Juniata ceases to be a river, both branches being just trout-streams, and nothing more. And, what is still more cruel, the Little Juniata loses its beautiful blue color, because it flows through a mining region, and the miners will persist in washing their ore in its clear wave.

After we leave Huntingdon we are in the mountains altogether. Various creeks join the Little Juniata, which winds so that it has to be bridged every three or four miles. At the junction of Spruce Creek the mountains on the left, which have been shouldering us for some time back, suddenly hurl a huge barrier over our path in the shape of Tussey's Mountain—a great turtle-backed monster, several thousand feet high. We are now seven miles from Tyrone, the centre of the mountains, and the pines are quite thick. The mountains show us now their fronts and now their bases, but are never out of sight, and at intervals come right up to us. At Tyrone they look as if they had been cleft asunder, for there is a great gap cut between two mountains. This in times past was doubtless the work of the Juniata, and was not so difficult as it looks; for the shaly mountains are very different from the firm limestone through which the Kanata cuts its way at Trenton Falls. The scenery around is decidedly Alpine in character; and some of the roads made for the lumber business traverse regions of savage beauty. The clouds whirl about the mountains so furiously that one is sure to be caught several times in the daily thunder-storms in these heights; and the writer was wetted to the skin three distinct times when descending Sinking-Run Hill, a mountain about six miles from Tyrone. The view presented by the artist is taken from an old road now discontinued for lumber travel, which starts from the side of the mountain, about half-way up, and descends circuitously to the base of the opposite mountain. Wild cherries and whortleberries grow in abundance, and the route is shaded by pines and hickories, while an occasional spruce-tree adds variety to the foliage.

THE MAMMOTH CAVE.

WITH ILLUSTRATIONS BY ALFRED R. WAUD.

THE Mammoth Cave of Kentucky is the largest known cave in the world. It is near Green River, on the road from Louisville to Nashville. The railway takes you directly to an old-fashioned hotel. Each person is provided with a lamp; and then you are led, by a negro guide, down a path that enters a wooded ravine, and, slanting aside, terminates suddenly at the almost obscure entrance of the cave.

The conductor lights the lamps, and in a severe voice calls "Forward!" You are ushered into a primitive chaos of wild limestone forms, moist with the water oozing from above. For nearly half a mile on your way you see, in the dim light, the ruins of the saltpetre works, that were built in 1808 by persons in the employ of the United States Government. The huge vats and tools still remain undecayed. The print of an ox's hoof is embedded in the hard floor, and the ruts of cart-wheels are also traceable.

Advancing farther, you enter the Rotunda, which is illuminated for a moment by the guide. It is over seventy-five feet high, one hundred and sixty feet across, directly

SCENES IN MAMMOTH CAVE

SCENES IN MAMMOTH CAVE.

under the dining-room of the hotel, and the beginning of the main cave. The lamps throw a feeble light on the dark, irregular walls, broken in places by the mysterious entrances to several avenues which lead from the main cave, and are said to extend a distance of over one hundred miles! As you tramp onward, your companions ahead are rimmed with light, and the supernatural aspect of the scene is from time to time heightened by the fluttering of a bat that spins out of a dark crevice for an instant and disappears again in the all-enveloping darkness. One chamber, entered from the Rotunda, bears the unattractive name of the Great Bat-Room; and here thousands of the little creatures are found snarling and curling their delicate lips at all intruders.

From the Rotunda you pass beneath the beetling Kentucky Cliffs, and enter the Gothic Chapel, a low-roofed chamber of considerable extent. Several twisted pillars ascend from the ground into arches formed of jagged rock, and in the distance there are two which form an altar of glittering splendor as the light falls on their brilliant stalactites. Near here is the Bridal Chamber, and the guide will tell you how a certain maiden, having promised at the death-bed of her mother that she would not marry any man on the face of the earth, came down to this dark place and was married. He will also tell you that these great stalactites that are so massive take fifty years to grow to the thickness of a sheet of paper.

There are rivers and lakes in the Mammoth Cave, and you are floated in a boat on the dark waters, among columns and walls, arches and spires, leaden-hued rock and jewelled stalactites, lighted up by a torch in the guide's hand.

In the centre of another wide room rests an immense rock, called the Giant's Coffin, and the guide, leaving you for a minute, reappears on its lid, his form, shadowed on the wall, imitating all his movements. Above the shadow you will notice the figure of an ant-eater, one of the many shapes with which the ceilings of the caverns are adorned by iron oxide. After resting a while under the Mammoth Dome, which appears much over a hundred feet high, with its magnificent walls of sheer rock, and at Napoleon's Dome, which is smaller than the former, the guide will conduct you to the edge of a projecting rock overlooking a hollow, the surface of which consists of masses of rock, called the Lover's Leap. In the Star-Chamber the stalactites assume curious and beautiful forms; and in Shelby's Dome you are ushered into a scene of indescribable grandeur. Under the dome is the Bottomless Pit, and a wooden Bridge of Sighs, which leads from this chasm to another, called the Side-Saddle Pit. A railing surrounds the principal pit, and as you stand holding to it and peering into the depths, the guide illuminates the dome above, affording one of the grandest sights in the cave.

At a point called the Acute Angle there is a rude pile of unhewn stone, called McPherson's Monument, which was built by the surviving staff-officers of that general. A stone is occasionally added to the pile by those of McPherson's soldiers or friends who visit the cave.

THE ADIRONDACK REGION.

WITH ILLUSTRATIONS BY HARRY FENN.

Ascent of Whiteface.

IT is a common notion among Europeans — even those who have travelled extensively in this country—that there is very little grand scenery in the United States east of the Mississippi River. The cause of this delusion is obvious enough. The great routes of travel run through the fertile plains, where the mass of the population is naturally found, and where the great cities have consequently arisen. The grand and picturesque scenery of the country lies far aloof from the great lines of railroad; and the traveller whirls on for hundreds of miles through the level region, and decides that the aspect of America is very tame and monotonous, and that it has no scenery to show except the Highlands of the Hudson, Lake George, and the Falls of Niagara.

In the State of New York alone, however—to say nothing of the mountains and the sea-coast of New England, or the mountains of Pennsylvania, Virginia, North Carolina, and Tennessee—there are vast regions of the most beautiful and picturesque scenery to which the traveller seldom penetrates, and of which, until recently, no glimpses could be obtained from the great lines of railroads, for they were established for purposes of trade, and not for sight-seeing. West of the Hudson lies a mountainous region, half as large as Wales, abounding in grand scenery, formerly known only to the wandering artist or the adventurous hunter; and beyond that, in the centre of the State, a lower and still larger region, studded with the loveliest lakes in the world, and adorned with beautiful villages romantically situated amid rocky glens, exhibiting some of the strangest freaks of Nature anywhere to be seen, and water-falls of prodigious height and of the wildest beauty.

The Ausable Chasm.

But the grandeur of the Catskills and the loveliness of the lake-region of Central New York are

Birmingham Falls, Ausable Chasm.

both surpassed in the great Wilderness of Northern New York, the Adirondack, where the mountains tower far above the loftiest of the Catskills, and where the lakes are to be counted by the hundreds, and are not surpassed in beauty even by Lakes George, Otsego, or Seneca. This remarkable tract, which fifty years ago was known, even by name, only to a few hunters, trappers, and lumbermen, lies between Lakes George and Champlain on the east and the St. Lawrence on the northwest. It extends, on the north, to Canada, and on the south nearly to the Mohawk. In area it is considerably larger than Connecticut, and, in fact, nearly approaches Wales in size, and resembles that country also in its mountainous character, though many of the mountains are a thousand or two thousand feet higher than the highest of the Welsh.

Five ranges of mountains, running nearly parallel, traverse the Adirondack from southwest to northeast, where they terminate on the shores of Lake Champlain. The fifth and most westerly range begins at Little Falls, and terminates at Trembleau Point, on Lake Champlain. It bears the name Clinton Range, though it is also sometimes called the Adirondack Range. It contains the highest peaks of the whole region, the loftiest being Mount Marcy, or Tahawus, five thousand three hundred and thirty-four feet high. Though none of these peaks attain to the height of the loftiest summits of the White Mountains of New Hampshire, or the Black Mountains of North Carolina, their general elevation surpasses that of any range east of the Rocky Mountains. The entire number of mountains in this region is supposed to exceed five hundred, of which only a few have received separate names. The highest peaks, besides Tahawus, are Whiteface, Dix Peak, Seward, Colden, McIntyre, Santanoni, Snowy Mountain, and Pharaoh, all of which are not far from five thousand feet in height above the

The Stairway, Ausable Chasm.

CLEARING A JAM, GREAT FALLS OF THE AUSABLE.

Climbing Tahawus.

sea. They are all wild and savage and covered with the "forest primeval," except the stony summits of the highest, which rise above all vegetation but that of mosses, grasses, and dwarf Alpine plants. These high summits are thought by geologists to be the oldest land on the globe, or the first which showed itself above the waters.

In the valleys between the mountains lie many beautiful lakes and ponds, to the number, perhaps, of more than a thousand. The general level of these lakes is about fifteen hundred feet above the sea; but Avalanche Lake, the highest of them, is at nearly twice that elevation above tide-water. Some of them are twenty miles in length, while others cover only a few acres. The largest of these lakes are Long Lake, the Saranacs, Tupper, the Fulton Lakes, and Lakes Colden, Henderson, Sanford, Eckford, Raquette, Forked, Newcomb, and Pleasant. Steep, densely-wooded mountains rise from their margins; beautiful bays indent their borders, and leafy points jut out; spring-brooks tinkle in; while the shallows are fringed with water-grasses and flowering plants, and covered sometimes with acres of white and yellow water-lilies. The lakes are all lovely and romantic in everything except their names, and the scenery they offer, in combination with the towering mountains and the old and savage forest, is not surpassed on earth. In natural features it greatly resembles Switzerland and the Scottish Highlands as they must have been before those regions were settled and cultivated. An excellent writer says that an American artist, travelling in Switzerland, wrote home, a year or two ago, that, "having trav-

Whiteface, from Lake Placid.

elled over all Switzerland and the Rhine and Rhone regions, he had not met with scenery which, judged from a purely artistic point of view, combined so many beauties in connection with such grandeur as the lakes, mountains, and forests of the Adirondack region presented to the gazer's eye."

This labyrinth of lakes is intertwined and connected by a very intricate system of rivers, brooks, and rills. The Saranac, the Ausable, the Boquet, and the Raquette, rise in and flow through this wilderness; and in its loftiest and most dismal recesses are found the springs of the Hudson and its earliest branches.

Mr. Alfred B. Street, the poet of these woods, describes them as follows: "Let the eye become a little accustomed to the scene, and how the picturesque beauties, the delicate, minute charms, the small, overlooked things, steal out, like lurking tints in an old picture! See that wreath of fern, graceful as the garland of a Greek victor at the games; how it hides the dark, crooked root writhing, snake-like, from yon beech! Look at the beech's instep steeped in moss green as emerald, with other moss twining round the silver-spotted trunk in garlands, or in broad, thick, velvety spots! Behold yonder stump, charred with the hunter's camp-fire, and glistening, black, and satin-like in its cracked ebony! Mark yon mass of creeping pine, mantling the black mould with furzy softness! View those polished cohosh-berries, white as drops of pearl! See the purple barberries and crimson clusters of the hopple contrasting their vivid hues! and the massive logs, peeled by decay—what gray, downy smoothness! and the grasses in which they are weltering—how full of beautiful motions and outlines!"

Lower Saranac Lake.

Among the favorite spots are Lake Paradox, whose outlet in high water flows back on the lake; the

Indian Carry, Upper Saranac.

pond on the summit of Mount Joseph, whose rim is close upon the edge; the mingling of the fountains of the Hudson and Ausable, in freshets, in the Indian Pass; the torrent-dashes or lace-work from the greater or lesser rain down the grooved side of Mount Colden toward Lake Avalanche; the three lakes on the top of Wallface, sending streams into the St. Lawrence by Cold River and the Raquette, into Lake Champlain by the Ausable, and the Atlantic by the Hudson; the enormous rocks of the Indian Pass standing upon sharp edges on steep slopes, and looking as if the deer, breaking off against them his yearly antlers, would topple them headlong, yet defying unmoved the mighty agencies of frost, and plumed

Round Lake, from Bartlett's.

with towering trees; with all the cavern intricacy between and underneath the fallen masses, where the ice gleams unmelted throughout the year; and the same rock intricacy in the Panther Gorge of Mount Marcy, or Tahawus.

The Wilmington Notch and the Indian Pass are great curiosities. The former is thus described by Mr. Street in his "Woods and Waters:"

"At North Elba we crossed a bridge where the Ausable came winding down, and then followed its bank toward the northeast, over a good, hard wheel-track, generally descending, with the thick woods almost continually around us, and the little river shooting darts of light at us through the leaves.

"At length a broad summit, rising to a taller one, broke above the foliage at our right, and at the same time a gigantic mass of rock and forest saluted us upon our left—the giant portals of the notch. We entered. The pass suddenly shrank, pressing the rocky river and rough road close together. It was a chasm cloven boldly through the flank of Whiteface. On each side towered the mountains, but at our left the range rose in still

St. Regis Lake.

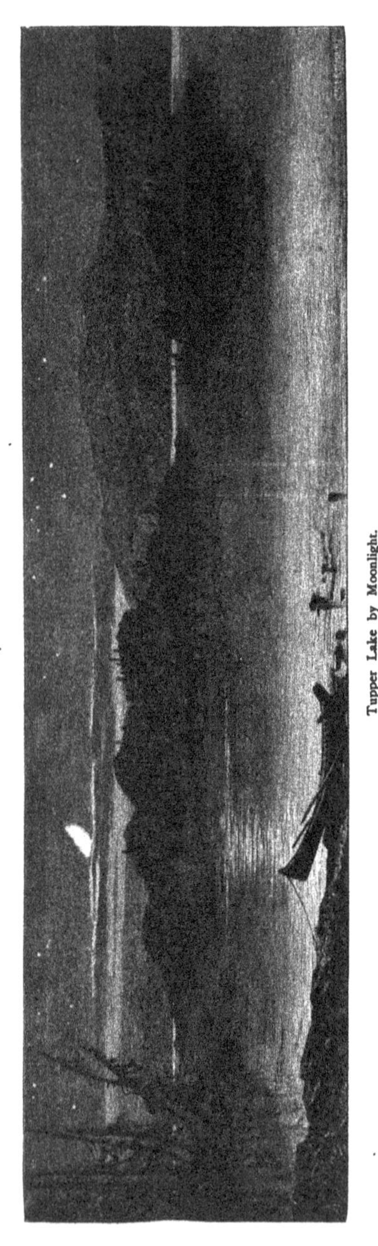

Tupper Lake by Moonlight.

sublimer altitude, with grand precipices like a majestic wall, or a line of palisades climbing sheer from the half-way forests upward. The crowded row of pines along the broken and wavy crest was diminished to a fringe. The whole prospect, except the rocks, was dark with thickest, wildest woods. As we rode slowly through the still narrowing gorge the mountains soared higher and higher, as if to scale the clouds, presenting truly a terrific majesty. I shrank within myself; I seemed to dwindle beneath it. Something alike to dread pervaded the scene. The mountains appeared knitting their stern brows into one threatening frown at our daring intrusion into their stately solitudes. Nothing seemed native to the awful landscape but the plunge of the torrent and the scream of the eagle. Even the shy, wild deer, drinking at the stream, would have been out of keeping. Below, at our left, the dark Ausable dashed onward with hoarse, foreboding murmurs, in harmony with the loneliness and wildness of the spot.

"We passed two miles through this sublime avenue, which at mid-day was only partially lighted from the narrow roof of sky.

"At length the peak of Whiteface itself appeared above the acclivity at our left, and, once emerging, kept in view in misty azure. There it stood, its crest—whence I had gazed a few days before—rising like some pedestal built up by Jove or Pan to overlook his realm. The pinnacles piled about it seemed but vast steps reared for its ascent. One dark, wooded summit, a mere bulwark of the mighty mass above, showed athwart its heart a broad, pale streak, either the channel of a vanished torrent, or another but far less formidable slide. The notch now broadened, and, in a rapid descent of the road, the Au-sable came again in view, plunging and twisting down a gorge of rocks, with the foam flung at intervals through the skirting trees. At last the pass opened into cultivated fields; the acclivities at our right wheeled away sharply east, but Whiteface yet waved along the western horizon."

Tahawus has often been ascended, though the task is by no means an easy one.

On Tupper Lake.

Its summit commands a magnificent prospect, which is thus described by Mr. Street in his "Indian Pass:"

"What a multitude of peaks! The whole horizon is full to repletion. As a guide said, 'Where there wasn't a big peak, a little one was stuck up.' Really true—and how savage! how wild! Close on my right rises Haystack, a truncated cone, the top shaved apparently to a smooth level. To the west soars the sublime slope of Mount Colden, with McIntyre looking over its shoulder; a little above, point the purple peaks of Mount Seward—a grand mountain-cathedral—with the tops of Mount Henderson and

Bog-River Falls, Tupper Lake.

Santanoni in misty sapphire. At the southwest shimmers a dreamy summit—Blue Mountain; while to the south stands the near and lesser top of Skylight. Beyond, at the southeast, wave the stern crests of the Boreas Mountain. Thence ascends the Dial, with its leaning cone, like the Tower of Pisa; and close to it swells the majesty of Dix's Peak, shaped like a slumbering lion. Thence stagger the wild, savage, splintered

Sand Point, Little Tupper Lake.

tops of the Gothic Mountains at the Lower Ausable Pond—a ragged thunder-cloud—linking themselves, on the east, with the Noon-Mark and Rogers's Mountain, that watch over the valley of Keene. To the northeast rise the Edmunds's Pond summits—the mountain-picture closed by the sharp crest of old Whiteface on the north—stately outpost of the Adirondacks. Scattered through this picture are manifold expanses of water—those almost indispensable eyes of a landscape. That glitter at the north, by old Whiteface, is Lake Placid; and the spangle, Bennett's Pond. Yon streak running south from Mount Seward, as if a silver vein had been opened in the stern mountain, is Long Lake; and between it and our vision shine Lakes Henderson and Sanford, with the sparkles of Lake Harkness, and the twin-lakes Jamie and Sallie. At the southwest glances beautiful Blue-Mountain Lake—name most suggestive and poetic. South, lies Boreas Pond, with its green beaver-meadow and a mass of rock at the edge. To the southeast glisten the Upper and Lower Ausable Ponds; and farther off, in the same direction, Mud and Clear Ponds, by the Dial and Dix's Peak. But what is that long, long gleam at the east? Lake Champlain! And that glittering line north? The St. Lawrence, above the dark sea of the Canadian woods!"

The Indian Pass is a stupendous gorge in the wildest part of the Adirondack Mountains, in that lonely and savage region which the aborigines named Conyacraga, or the Dismal Wilderness, the

A Carry near Little Tupper Lake.

larger portions of which have only recently been visited by white men, and which still remains the secure haunt of the wolf, the panther, the great black bear, and the rarer lynx, wolverine, and moose. The springs which form the source are found at an elevation of more than four thousand feet above the sea, in rocky recesses, in whose cold depths the ice of winter never melts entirely away, but remains in some measure even in the hottest months of the year. Here, in the centre of the pass, rise also the springs of the Ausable, which flows into Lake Champlain, and whose waters reach the Atlantic through the mouth of the St. Lawrence several hundred miles from the mouth of the Hudson; and yet, so close are the springs of the two rivers, that the wild-cat, lapping the water of the one, may bathe his hind-feet in the other, and a rock rolling from the precipices above could scatter spray from both in the same concussion. In freshets the waters of the two streams actually mingle. The main stream of the Ausable, however, flows from the northeast portal of the pass; and the main stream of the

Hudson from the southwest. It is locally known as the Adirondack River, and, after leaving the pass, flows into Lakes Henderson and Sanford. On issuing from them it receives the name of Hudson, and passes into Warren County, receiving the Boreas and the Schroon, which, with their branches, bring to it the waters of a score or more of mountain lakes, and of tarns innumerable.

Fifty years ago the Adirondack region was almost as unknown as the interior of Africa. There were few huts or houses there, and very few visitors. But of late the number of sportsmen and tourists has greatly increased, and taverns have been established in some of the wildest spots. In summer the lakes swarm with the boats of travellers in search of game, or health, or mere contemplation of beautiful scenery, and the strange sights and sounds of primitive Nature. Formerly all travelling was done by means of boats of small size and slight build, rowed by a single guide, and made so

Long Lake, from the Lower Island.

light that the craft could be lifted from the water and carried on the guide's shoulders from pond to pond or from stream to stream. Each traveller provided himself with a guide and a boat, and the cost of maintenance in the woods did not exceed a dollar a week for each man of the party. The fare was trout and venison, of which there was generally an abundance to be procured by gun and rod. A good-sized valise or carpet-bag would hold all the clothes that one person required for a two months' trip in the woods. Nothing was needed, besides those he wore, but woollen and flannel.

In those days the following list comprised the essentials of what constituted an outfit: A complete undersuit of woollen or flannel, with a "change;" stout pantaloons, vest, and coat; a felt hat; two pairs of stockings; a pair of common winter-boots and camp-shoes; a rubber blanket or coat; a hunting-knife, belt, and pint tin cup; a pair of warm blankets, towel, soap, etc.

There are several routes by which the Adirondacks can be reached, but that followed by our artist over twenty years ago was that by Lake Champlain. The steamer from Whitehall will land the traveller at Port Kent, on the west side of the lake, nearly opposite Burlington, Vermont, where in those days coaches conveyed the tourist six miles, to Keeseville. Here conveyances for the Wilderness can be had, that will carry the traveller to the Lower Saranac, a distance of about fifty miles, in a day. It is a long drive, but a very pleasant and interesting one. The tourist may then move about alto-

Mount Seward, from Long Lake.

gether in boats, or may, as he pleases, camp out in his tent, or so time his day's voyage as to pass each night in some one of the comfortable taverns which are now to be found in almost all of the easily-accessible parts of the Wilderness.

It was from this quarter that our artist entered the Adirondack, and we will follow him in the old route, which has now largely been abandoned for the more direct methods of transportation. At Keeseville he paused to sketch the falls and walled rocks of the Ausable chasm, which afford some of the wildest and most impressive scenes to be found

Round Island, Long Lake.

on this side of the Rocky Mountains. At the distance of a mile or so from Keeseville is Birmingham Falls, where the Ausable descends about thirty feet into a semicircular basin of great beauty. A mile farther down are the Great Falls, one hundred and fifty feet high, surrounded by the wildest scenery. Below this the stream grows narrower and deeper, and rushes rapidly through the chasm, where, at the narrowest point, a wedged bowlder cramps the channel to the width of five or six feet. From the main stream branches run at right angles through fissures, down one of which, between almost perpendicular rocks a hundred feet high, hangs an equally steep stairway of over two hundred steps, at the bottom of which is a narrow platform of rock forming the floor of the fissure.

From Keeseville the traveller rides westward on a road leading to Martin's, on Lower Saranac Lake. He will pass most of the way in sight of Whiteface Mountain, the great outpost of the Adirondacks. At the village of Ausable Forks, about twelve miles from Keeseville, he can turn off into a road which leads through the famous Whiteface or Wilmington Notch, and can regain the main road about a dozen miles before it reaches Saranac Lake. The distance by this route is not much longer than by the main road, and the scenery is incomparably finer. The view of Whiteface from Wilmington was pronounced by Professor Agassiz to be one of the finest mountain-views he had ever seen, and few men were better acquainted with mountain-scenery than Agassiz. Through the notch flows the Ausable River, with a succession of rapids and cataracts, down which is floated much of the timber cut in the Adirondack forests by the hardy and adventurous lumberers, some idea of whose toils and dangers may be formed from the sketch of "Clearing a Jam," the scene of which is at the head of one of the falls of the Ausable, in the Wilmington Notch. From the village of Wilmington our artist ascended Whiteface, which is second only to Tahawus among the mountains, its height being nearly five thousand feet. At its foot, on the southwest side, lies Lake Placid, one of the loveliest lakes of the Wilderness. From this lake, which is a favorite summer resort, one of the best views of Whiteface can be obtained.

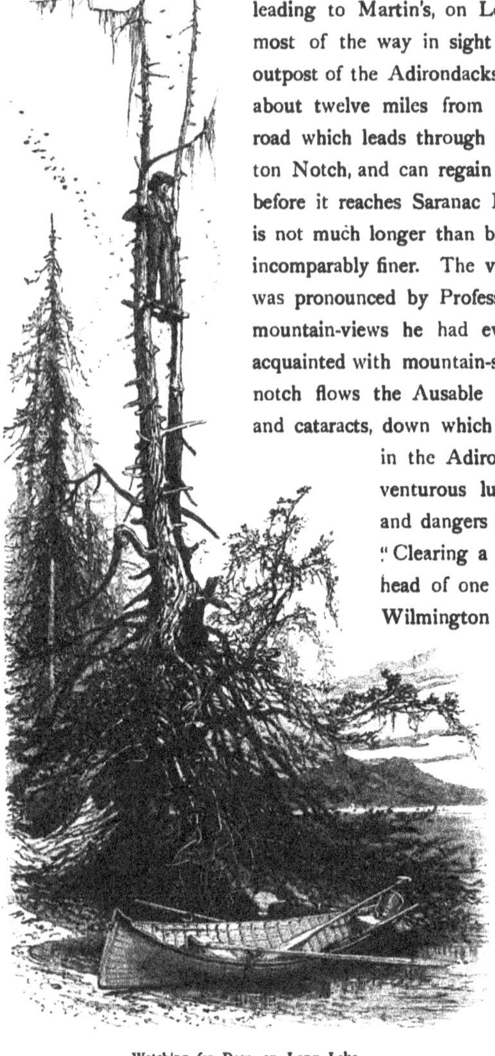

Watching for Deer, on Long Lake.

From Lake Placid to Martin's is a few hours' drive over a rough but picturesque road. Martin's is a large and comfortable hotel on the very edge of the Lower Saranac, a beautiful lake six or seven miles long and two miles wide, studded with romantic islands fifty-two in number. The Saranac River connects it with Round Lake, three miles to the westward. Round Lake is about two miles in diameter, and is famous for its storms. It is in its turn connected with the Upper Saranac Lake by another stretch of the Saranac River, on which are now several of the best and most frequented of the Adirondack hotels. From a point at no great distance from the house a fine view can be obtained of Round Lake and the surrounding mountains. A short "carry," of a mile or so in length, conducts from Bartlett's to the Upper Saranac, whence it is easy to pass in boats to St. Regis Lake, our view of which gives a singularly good and accurate idea of the general characteristics of Adirondack scenery. A short voyage in the opposite direction across the Upper Saranac will take the traveller's boat to the Indian carry, or Carey's carry, as it is sometimes called, to distinguish it from another carry, Sweeny's, established more recently. Both lead to the Raquette River, the great artery of the Wilderness.

The Indian Pass.

Source of the Hudson.

A few hours' row down the Raquette brings you to the outlet of Lake Tupper, so named not from the author of "Proverbial Philosophy," but from the hunter or guide who discovered it. It is several miles in length, and contains many picturesque, rocky islands, covered with evergreens. At its head the wild and little-explored Bog River flows into the lake over a romantic cascade, which forms one of the great attractions of the Adirondacks, being a famous place for trout, and having near by one of the most popular taverns of the Wilderness. Since the completion of the railroad through the centre of the Adirondack region Tupper Lake has become a favorite resort, and has now several large hotels on its shores. It may be approached either from the north or from the south, according as the fancy of the tourist may dictate.

From Tupper Lake the route of the traveller is up Bog River, through a series of ponds and an occasional "carry"—where the guides take the boats on their backs, as represented in our engraving—to Little Tupper Lake. Thence a series of ponds and carries leads to Long Lake, which for more than twenty miles resembles a great river. It is the longest of the Adirondack lakes, though there are many broader ones. From this lake a fine view can be had of Mount Seward, four thousand three hundred and forty-eight feet high. We give also an illustration of the way in which the guides of this region station themselves in trees to watch for deer. The deer are hunted by powerful hounds, which are put on their trail in the woods, and pursue them with such tenacity that the frightened animal at last takes to the water. The hunters, with their boats stationed at intervals along the shore, watch patiently till the deer breaks from the woods and plunges into the water. The nearest hunter immediately enters his boat, gives chase, and generally succeeds in overtaking and killing the game.

From Long Lake to the Indian Pass is a very rough journey through the wildest part of the Wilderness. We give an illustration which conveys some idea of the kind of road the explorer who ventures thither may expect to encounter. He will find in it the source of the Hudson at an elevation of four thousand three hundred feet above the sea. From this lofty pool the water flows through Feldspar Brook into the Opalescent River, on which there is one of the most picturesque cascades of the Adirondacks.

Opalescent Falls.

Of the scenery of the source of the Hudson, Mr. Lossing, in "The Hudson from the Wilderness to the Sea," writes as follows: "We entered the rocky gorge between the steep slopes of Mount McIntyre and the cliffs of Wallface Mountain. There we encountered enormous masses of rocks, some worn by the abrasion of the elements, some angular, some bare, and some covered with moss, and many of them bearing large trees,

The Hudson, Twenty Miles from its Source.

whose roots, clasping them on all sides, strike into the earth for sustenance. One of the masses presented a singular appearance; it is of cubic form, its summit full thirty feet from its base, and upon it was quite a grove of hemlock and cedar trees. Around and partly under this and others lying loosely, apparently kept from rolling by roots and vines, we were compelled to clamber a long distance, when we reached a point more than one hundred feet above the bottom of the gorge, where we could see the famous Indian Pass in all its wild grandeur. Before us arose a perpendicular cliff, nearly twelve hundred feet from base to summit, as raw in appearance as if cleft only yesterday. Above us sloped McIntyre, still more lofty than the cliff of Wallface, and in the gorge lay huge piles of rock, chaotic in position, grand in dimensions, and awful in general aspect. They appear to have been cast in there by some terrible convulsion not very remote. Through these the waters of this branch of the Hudson, bubbling from a spring not far distant (close by a fountain of the Ausable), find their way. Here the head-waters of these rivers commingle in the spring season, and when they separate they find their way to the Atlantic Ocean at points a thousand miles apart."

THE LOWER MISSISSIPPI.

WITH ILLUSTRATIONS BY ALFRED R. WAUD.

A Bayou of the Mississippi.

JUST fifty years after Columbus discovered the islands of the Bahamas, De Soto, accompanied by a broken-down and dispirited remnant of a once powerful expedition, reached the banks of the Mississippi a thousand miles or more from its mouth. The discovery gave him a lasting fame and furnished him a fitting grave. This river, ever changing and yet ever the same, after more than three centuries still answers to the original description of the adventurous Spaniards, for their chief chronicler wrote that "the river was so broad that if a man stood still on the other side it could not be told whether he was a man or no. The channel," he continues, "was very deep, the current strong, the water muddy, and filled with floating trees."

More than a century elapsed after the discovery of the river before its solitude was again disturbed by the presence of the white man. During this time its mouth became involved in popular mystery. Tales were circulated that the flood of water, where the great outlet should be, was precipitated into the earth; that great dragons and sullen

Southwest Pass.

mists guarded the vicinity from man's approach; and these tales, so harmonious with the spirit of the age, found confirmation in the traditions of the Indians, who lived thousands of miles away, on the banks of the Fox and the Illinois.

In the year 1673 Marquette, a French monk, left Quebec, traversed the great northern lakes, and reached the "Upper Mississippi" by way of the Fox and Wisconsin

Rivers. Having accomplished what was then supposed to be a heroic task, he returned to Quebec, and announced that, from what he saw, he was certain the Gulf of Mexico could be reached by uninterrupted navigation. Great rejoicings ensued; the *Te Deum* was sung in all the churches; the military fired salutes, and the great "Western Valley," by the right of discovery, was declared to belong to France. La Salle followed, and, from the Falls of St. Anthony, made the first continuous voyage of the whole length of the river. He entered the Gulf of Mexico, April 9, 1682, founded the fort of St. Louis, and gave to the adjacent lands the name of Louisiana.

The mouth of the river was discovered by Iberville eighteen years later. Instead of one vast current pouring into the Gulf, it was found to consist of numerous arms, or passes, through low swamps and islands formed by the sediment brought down by the river. This net-work of creeks, bayous, and passes is known as the Delta of the

A Bayou of the Mississippi.

Mississippi. It covers an area estimated at fourteen thousand square miles, and is slowly advancing into the Gulf by the shoaling caused by the deposition of fresh sediment brought down by the river. Three of the main passes bear the practical names of Southwest, South, Northeast, and the fourth is called à l'Outre.

The ragged and unformed arms of the "passes" are involved in what appears, even after careful examination, to be an interminable marsh. For miles before you reach the passes you observe the muddy Mississippi water in great masses, rolling and tumbling unmingled with the briny blue sea.

As you ascend the channels of the passes, if on board of a swiftly-moving steamer, you perceive that coarse grass finally appears in consecutive lines, and then crop out here and there great lumps of mud, around which seethes and boils what now has become a rushing current. It is apparent that the sediment of the river has obtained a foothold.

Steadily moving onward, the shore at last becomes defined, and water-soaked shrubs are noticeable, ever moving and fretting from the lashings of the deflecting waves. If the day is bright and the sun well toward the horizon, as the swelling tide moves grandly onward, its surface glistens with the hues of brass and bronze. Vegetation now rapidly asserts its supremacy; the low banks are covered with ferns, and here and there is an ill-shapen tree; while, landward, a dark line indicates the perfectly-developed forest.

Naught but the sameness and monotony of the river now impresses you, save the consciousness that you are borne upon a mighty, sweeping flood. Mile after mile, and still the same. The bittern screams, the wild-fowl start in upward flight; and, if night

Sunset in the Mississippi Swamp.

sets in, you seem to be moving through an unvaried waste. The low and scarcely perceptible walls of Forts Jackson and St. Philip are just discernible, when lights dancing ahead give the first signs of intelligible settlement. The "quarantine" is reached, the official visit is made, and again you begin your monotonous upward trip.

If the morning sun greets you within fifty miles of New Orleans, you find the banks of the river above the flood-tide, and evidences of permanent cultivation and happy homesteads attract the eye. Along the "coast," as the river-banks are denominated, are the "gardens," upon which the city depends for vegetable food. Then come large sugar-plantations, the dwelling-houses made imposing by their verandas, and picturesque by

Cypress-Swamp.

being half hidden in an untold variety oι magnificent trees. Thus is displayed, in the upward trip to within twoscore miles of New Orleans, the gradual development of the banks of the Mississippi. The constant creation goes on seemingly under your own eye. From water to ooze, to mud, to soil; from grass to shrubs, to ferns, to forest-trees.

The first grand tree-development of the "swamps" is the tall and ghostly cypress.

It flourishes in our semi-tropical climate of the South, being nourished by warmth, water, and the richest possible soil, often reaching the height of one hundred and thirty feet. The base of the trunk, generally covered with ooze and mud, conceals the formidable "spikes," called "knees," which spring up from the roots. These excrescences, when young, are sharp and formidable weapons, and, young or old, are nearly as hard as steel. To travel in safety through a flooded cypress-swamp on horseback, the greatest care must be taken to avoid the concealed cypress-knees; for if your generous steed, while floundering in the soft mud, settles down upon one of them, he may never recover from the injury. The bark of the tree is spongy and fibrous, and the trunk of the tree often attains fifty or sixty feet without a branch. The foliage, as seen from below, is as soft as green silken fringe, and strangely beautiful and delicate when contrasted with the tree itself and the gloomy, repulsive place of its nativity. The wood, though light and soft, is of extraordinary durability. It has been asserted that cypress-trees which have been buried a thousand years under the solid but always damp earth retain every quality of the most perfect wood. At the root of the cypress the palmetto flourishes in vigor; and its intensely green, spear-like foliage adds to the variety of the vegetable productions in the forest solitudes.

Coming to the unsubmerged lands, which, like islands, are everywhere interspersed in this immense swamp, you meet with broad expanses on which grow the renowned "cane-brakes;" and, leaving them, you possibly come upon vistas of prairie, which, open to the constant influence of sunshine and sea-air, are dotted over with the magnificent "live-oak," the most picturesque tree of our continent.

In contrast to the oak is the wonderful magnolia, a flowering giant, often reaching an altitude of ninety feet. Its form is attractive, and each particular bough has characteristics of its own. Its leaves are large and crisp; the surface, exposed to the sun, is of a polished, dark green, while underneath it is almost as gray and velvety as the mullein. When the ever-green foliage of the live-oaks is trembling and whispering in the slightest breeze, or waving in great swaths in the rushing wind, the magnolia stands firm and unmoved. Its large imperial blossoms of pure white look like great ivory eggs, enveloped in green and brown. When the petals finally open, you have that bridal gift, the orange-blossom, enlarged to a span in diameter, and so fragrant that it oppresses the senses. The magnolia-tree in full blossom, with the Spanish moss enshrouding it in a gray, neutral haze, makes a superb picture.

The scenery of the undisturbed forests of the Lower Mississippi is of a mysterious interest. Destitute though it be of the charms of mountains and water-falls, with no distant views, no great comprehensive exhibitions, it nevertheless inspires a sort of awe which it is difficult to define or account for. All objects are upon a water-level; and when you look aloft through the gloom of the towering trees, you feel as if you were in a well, and that the place is born of the overflowing waters.

The grape-vines which festoon the trees curl round their supports with the force of cordage, and their trunks, slimy and grim, spring from the ground, and, writhing upward, like great pythons, grasp a supporting limb sixty feet in the air. The shimmer of remote lagoons greets you in the distance, and there are water-marks on the trees

Magnolia-Swamp.

twenty feet above your head. If some passing storm has made a "window" and let in the sunshine, the undergrowth, heretofore stunted, now starts into life, and seems to rejoice in new-born luxuriance. The bright colors are metallic in intensity. The flower of the scarlet lobelia trembles and flashes as if a living coal of fire. The hydrangea, a modest shrub in the North, becomes a tree, a very mound of delicate blue flowers.

THE MOSS-GATHERERS

A deep and lasting impression was made upon the early discoverers of the Mississippi by the drapery which festooned the trees, and which is generally known as Spanish moss. This moss is a parasite, that lives by inserting its delicate suckers under the bark and draws its sustenance from the flowing sap. It is repelled by trees in perfect vigor, but in one enfeebled by age or accident the moss gains foothold, and goes on with its quiet work of destruction until it consumes the heart's-blood of its helpless victim, and then enwraps it in a weird winding-sheet. Many trees apparently stricken with age, when artificially relieved of this burden, have revived and assumed almost their natural vigor. In the great order of Nature the moss has its purposes. It consumes the hard and iron-like woods which would otherwise for long years—a century perhaps—be a vegetable wreck, and thus quietly and surely makes way for a new growth.

Bienville, the first governor of Louisiana, is represented as laying the foundation of New Orleans on the first available high land he met with in ascending the river. Below the city there are now, along the banks, nearly fifty miles of continuous cultivation, and this arable land is the result of the accretions of the hundred and fifty years which have passed since the city was founded. As you ascend the river, evidences multiply that you are approaching the great Southern metropolis. Large fleets of sailing-vessels in tow pass on their way to the ocean. Nondescript craft of all kinds line the shores; at last the "Crescent City" appears, stretching miles away along the coast, and opening wide its enfolding arms as a welcome to the arriving stranger.

The river opposite the city is more than a mile and a half in width, and, notwithstanding the velocity of its movement and the distance from the sea (one hundred and eight miles), the tide regularly ebbs and flows, modifying somewhat the sweep of the downward current. Here we have a magnificent bay, grand in dimensions as any arm of the sea. The city extends along the eastern bank as far as the eye can reach; the western side is dotted over with villages, highly-cultivated farms, and great workshops. A consecutive mile or more of steamers is in sight, including the magnificent "floating palaces," which "carry" between the "great cities of the West," down through every conceivable representative graduation to the absurd "stern-wheeler," which works its way up the shallower streams and "damp places" tributary to the Arkansas and Red Rivers. Ships of stately proportions from every land lie side by side, their masts and cordage revealing in rich confusion a leafless forest. The ferry-boats are constantly in motion, while the great steam-tugs, bringing up with ease a fleet of sailing-vessels from the mouth of the river, make the lowlands echo with their high-pressure puffing, and send great clouds of bituminous smoke from their chimneys.

Reaching the shore, you find that the "levee," which below was a narrow embankment, is now a wide, artificial plateau, extending miles each way, and crowded with the teeming productions of the counties and States which lie on the tributary stream of the great river.

THE MISSISSIPPI AT NEW ORLEANS.

To float down the Western rivers was as easy as healthy respiration; to stem the swift current on the upward trip was a task of almost superhuman labor. If artificial means had not come to the rescue, much of the great West, which to-day is enriched by cities and towns, and teeming with intelligent populations, might have remained a primitive wilderness. Before the application of steam for the propulsion of water-craft, commerce was carried on by means of "broad-horns" and "keel-boats." The broad-horn accomplished its purpose when, floating down the current, it arrived at its place of destination and delivered its cargo. The keel-boat not only brought down a cargo, but loaded with foreign products, was "cordelled" by months of hard work up the river to its original starting-point.

A favorably-situated series of plantations, with land more than ordinarily high, and therefore comparatively free from overflow, in the course of long years of cultivation becomes the centre of charming landscape scenery, which combines the novelty of many exotics growing side by side with the best-preserved specimens of the original forest.

On these old plantations, modified by climate, are developed in the greatest perfection some of the choicest tropical plants. Orange-trees may be met with which are three-quarters of a century old, with great, gnarled trunks and strong arms, still bearing in perfection their delicious fruit. The sugar-cane, usually a tender, sensitive plant, has become acclimated, and repays most liberally for its cultivation. The magnificent banana, with its great, sweeping leaves of emerald green waving in the breeze with the dignity of a banner, has within a comparatively few years almost overcome its susceptibility to cold, and is now successfully cultivated.

In the rear of the garden you find the elm-shaped pecan, of immense height and beautiful proportions, bearing abundantly an oval-shaped, thin-shelled fruit, possessing all the sweetness of the hickory-nut and almond combined. As you go farther south, below the Louisiana coast, these trees form forests, and yield to their possessors princely incomes. Hedges of jasmine lead up to the door-ways of the planters' residences, and vie in fragrance with the flowing pomegranate and night-blooming cereus, and an endless variety of the queenly family of the rose. And just where the cultivated line disappears, and the natural swamp begins, will often be found the yellow jasmine climbing up some blasted tree and usurping its dead branches for its own uses, and covering it over with a canopy of blossoms which shed a fragrance that, in descending, is palpable to the touch and oppressive to the nostrils. Here the honey-bee revels, and the humming-bird, glancing in the sunlight as if made of living sapphires, dashes to and fro with lightning rapidity, shaking from its tiny, quivering wings the golden pollen.

At nightfall, when the warm spring day has disappeared, to be followed by the cool sea-breeze, and the atmosphere predisposes to lassitude and dreamy repose, the minstrel of the Southern landscape, the wonderful mocking-bird, will find a commanding perch near the house, where he can enjoy the fragrance of flowers in the sea-cooled air, and know that his human admirers are listening, and he will then carol forth songs of praise and admiration, of joy and humor, of sweet strains and discords, like a very "Puck of the woods," a marvel of music and song.

The settlers who first gained foothold were of French origin, and the original impress is still maintained. Up to within a few years communities existed in Louisiana of the most charming rural population: the little chapel, with its social French priest; the men temperate and of good bearing, because the genial climate called for moderate labor; the women bright, fond of home, and inheriting a natural taste for dress worthy of the mother-country. Unprovided with the theatre and opera, these rural populations

Market-Garden on the Coast.

were content in matters of display with the imposing ceremonies of their church, and for amusement with the weekly enjoyment of their extemporized balls. Among this population originally were many scions of the best families of France, whose historic names are still preserved, who shed over their simple settlements in the far-off wilds of the Mississippi something of the style pertaining to the villa and *château*. In course of time many of these old mansions along the river have disappeared, or, falling into the possession of the irreverent Anglo-Saxon, have had their outward faces buried under broadly-constructed verandas and galleries.

The Mississippi, left to itself for hundreds of miles above its mouth in the spring floods, would overflow its banks from two to three feet. To obviate such a catastrophe, there has been built by the enterprising planters a continuous line of levee, or earth intrenchments, upon which slight barrier depends the material wealth of the people. The alluvium, or sediment, of the river, which is deposited most abundantly upon its banks, makes the frontage the highest surface; and as you go inland, you unconsciously but steadily descend at least four feet to the mile, until you often find the water-level marked on the trees at times of overflow far above your head.

For nine months of the year the Louisiana planter pays but little attention to the levee, but when the spring comes, and the melted snows, which fall even as far off as the foot of the Rocky Mountains, find their way past his residence to the sea, he is

Planter's House on the Mississippi.

suddenly awakened to the most intense anxiety; and when, at last, the great flood of water—the drainage, in fact, of two-thirds of the lands of the centre of the continent—dashes over the frail embankment of the levee, he realizes what a slender hold he has upon his young crop and the earthy improvements of a large estate. The rains at these times assist in making the water-soaked barrier unstable; rats, mice, and beetles have their burrows, and thousands of crawfish, with their claws as hard and sharp as a chisel of iron, riddle the levee with holes. Under these critical conditions even a light wind may invite the impending catastrophe. In an unexpected moment the alarm is given that a "crevasse" is threatened. All is confusion and consternation. The plantation-bells are rung, the news is carried to out-of-the-way places by fleet horsemen, the laboring population assemble, and, armed with such implements as are at command, the attempt is made to stay the threatening waves. The levee at the point of assault, in spite of all action to the contrary, moves from its foundation and crumbles away, and the river, raised to an artificial height, now finds relief in a current that roars like a cataract. If the break is of formidable proportions, the passing flat-boat is drawn into the vortex and sent like a chip high and dry into the distant fields.

Added to the danger of overflow is that of caving banks. By a natural law in

A "Crevasse" on the Mississippi.

the formation of the banks of the Mississippi, the alluvium is rapidly deposited upon the "points," and dissolves away from the "bends." It is not an extraordinary sight to see a grandly-constructed and ancient house hanging outside the levee and over the edge of the river-bank, destined sooner or later to drop into the river. You will find these things occur where the mighty current, sweeping round a bend, has worn away the soft earth, often dissolving it by acres. If this occurs in front of a plantation, the house and improvements, perhaps originally a mile from the river, will be gradually brought to the edge of the bank, to be finally engulfed. The point directly opposite the bend, however, makes, in accretions, exactly what is taken away from the opposite side of the river.

THE SOUTH SHORE OF LAKE ERIE.

WITH ILLUSTRATIONS BY J. DOUGLAS WOODWARD.

Erie Canal Basin and Elevator, Buffalo.

AMONG the five great lakes of the Western chain, Erie occupies the fourth place as regards size, the last place in point of beauty, and no place at all in romance, but it excels them all in historic interest.

Lake Erie is two hundred and fifty miles long, sixty miles broad, and two hundred and four feet at its greatest depth, although, on an average, it is not more than ninety feet deep. Compared with the other lakes, it is shallow; and the difference has been

Ship-Canal, Buffalo.

described as follows: "The surplus waters poured from the vast *basins* of Superior, Michigan, and Huron, flow across the *plate* of Erie into the deep *bowl* of Ontario." Lake Erie is the only member of the chain which is reputed to have any current; but it has another reputation, which is founded on certainty — it is the most dangerous of the fresh-water seas. Its waves are short and chopping, its harbors insecure, especially along the northern shore, and it has little sea-room. The mouths of its streams are clogged with sand-bars; and in the early days, before improvements were made, the lake-captains kept out in the offing, and landed their cargoes in small boats, rather than risk the perils of the so-called harbors.

The rivers are docked, and rows of canal-boats usurp their sides—canal-boats decked with lines of drying clothes, for it always seems to be washing-day on a canal-boat. A giant elevator is sucking grain from the hold of one vessel; red iron from Lake Superior is being unloaded from another; wood from Lake Huron, and limestone

Grain Propeller, Buffalo.

from the western islands, are coming in; coal and petroleum are going out; and the lines of slow-moving lumber barges, the schooners and barks, the canal-boats, propellers, and side-wheel steamers, have only the narrow, crooked river for a roadway. The incoming tug catches a sight of all this confusion from the light-house at the end of the pier, and whistles defiantly. Hers it is to take the vessel safely to its berth, a mile up the river, and she does it.

The shores of Lake Erie are wooded, rising, on an average, sixty feet above the water. Through this plateau the streams come down in gorges and ravines, and the banks are full of springs and quicksands. In a north wind the water is dark, and the waves dash on the beach with a loud roar; in an east wind it is sea-green, the white-caps curl toward the west, and it has a treacherous aspect; but when the west wind blows, it is a blue summer sea, over which the ships sail gayly under a cloud of canvas. Only when the south wind comes off the land, bringing a gray rain-storm, does the lake lose all its beauty; then it sullenly sinks into lethargy; the woods on its shores stand desolate; and the little villages, each with its long, dripping dock and warehouse, look so miserable that the lake-traveller hastily disappears into the cabin.

Lake Erie derived its name from the Eries, or tribe of the Cat, who lived upon its shores when the Jesuit missionaries first visited the country, two centuries ago. Every thing connected with the Eries—who have left only a name behind them—is involved in obscurity. The city of Buffalo, taking its name from the American bison who roamed in herds along the shore as late as 1720, lies at the eastern end of Lake Erie. It was

MAIN STREET, BUFFALO, FROM ST. PAUL'S CHURCH

first settled in 1801. Notwithstanding attacks by the Indians, the dangers of starvation and cold, the little settlement kept itself alive, immigrants came, and in 1810 Tushuway, or Buffalo, including all that part of the State which lies west of the west transit line, was set off from the neighboring settlement of Clarence. Thus, at its first organization, Buffalo contained an area of about three hundred thousand acres; which was an ambitious beginning, even for the "Queen City of the Lakes," as it is called. Shortly after this the progress of Buffalo was checked by the War of 1812; and near the close of 1813, Fort Niagara was taken by the British, and the surrounding villages, including Buffalo, burned to the ground. When peace was declared it was rebuilt, and in 1832 it took its place as a city, ranking now the third in size in the State of New York. The Buffalo of to-day is a large, bright, busy town, with broad streets of well-built residences and business blocks.

The most noticeable feature of Buffalo is its mode of handling grain in bulk by means of its numerous elevators. These wooden monsters, with long trunks and high heads, stand on the bank of the river waiting for their prey. In from the lake come the vessels and propellers laden with grain from Milwaukee and Chicago, and the tugs carry them up within reach and leave them to their fate; then down out of the long neck comes the trunk, and, plunging itself deep into the hold of the craft, begins to suck up the grain, nor pauses until the last atom is gone. It may be that this grain is to go eastward by the Erie Canal; in that case the canal-boat is waiting on the other side, a man opens another door, the grain runs down another trunk into its hold, and behold it ready for its journey to New York City.

Light-House, Buffalo.

Buffalo is attractive by force of its situation at the eastern end of Lake Erie. It does not lie on a side-bank, as Cleveland lies; it does not stand back on a bay, as Toledo and Sandusky stand; it does not retreat up a river, like Detroit; it takes its place boldly at the foot of the lake, and catches every breeze and every gale in their full strength. Buffalo harbor is the largest on the lake, but, owing to its situation, it is often the last gathering-place for the weakened ice, so that when the other coast cities are sending out their vessels in the early spring, when Detroit River is open, and the iron fleets of Cleveland are starting for Lake Superior, the harbor of Buffalo is still reported by telegraph as "closed."

But at length the ice goes, and as the steamer leaves Buffalo Light behind, the

Ship-Canal and Coal-Docks, Buffalo.

lake broadens, and after passing Sturgeon Point the breeze is almost sure to freshen into a strong wind. Along this portion of the coast, in winter, the snow sweeps with fierce fury, and even in summer it is a bleak coast, with little to attract the eye. Occasionally a village is passed, where the smoke of a furnace or a mill and the masts of vessels show that a city is growing up; but even should the steamer turn into the wharf, there is nothing to be seen save the never-ending loading and unloading of the lake schooners, the dock-hands with their wheel-barrows, and on shore the newness and the rawness of a Western town in its awkward, growing youth.

The situation of Erie is picturesque, owing to the beauty of its bay and outlying island. As early as 1753 the French landed at this point and erected a little fort.

ERIE, FROM FEDERAL HILL

naming it Presque Isle; it was one of a chain which was to connect the St. Lawrence and *la belle rivière*, as they called the Ohio. In 1760 Presque Isle was surrendered to the British, and soon after it was destroyed by the Indians. From that time the beautiful bay was solitary until the arrival of the surveyors. The present town of Erie was incorporated as a borough in 1805, is now a thriving city—the outlet of the iron

Lake-Shore, above Erie.

and coal district of Western Pennsylvania; it is the principal market for bituminous coal on the lakes. In its bay Commodore Perry built most of the vessels of his famous little fleet, having for material only the trees of the forest, and for plans only his own iron determination.

Dotted along the coast stand the light-houses, picturesque towers finding a footing on lonely islets and rocky ledges, wherever they can command a wide sweep of the horizon. The farm-buildings cluster inland; but the light-house, with the keeper's little cabin at the base, stands alone on its point, where its tower gleams white by day and red by night far out at sea. To the traveller over the Western waters the light-houses seem both picturesque and friendly, for the steamers keep within sight of the shore, and, a pillar of cloud by day and fire by night, they greet the voyager as he journeys, one fading astern as the next shines out ahead.

Cleveland, the city of the Western Reserve, is considered the most beautiful town on the Great Lakes. It was named after General Moses Cleaveland, the agent of the Connecticut Land Company, and was first settled in 1796. The town lies on both sides

Main Light, at Erie.

of the Cuyahoga River, a narrow, crooked stream, which flows through a deep valley into the lake, leaving on either side the bluffs whose shaded streets have gained the name of "Forest City." Not long ago it was a marshy meadow, where the river meandered in peace, with nothing to disturb its sedgy margin save the cows and water-birds. Now it is a dense mass of iron-mills, lumber-yards, and oil-refineries—a seething basin of life, movement, noise, and smoke. Above the lake on either side of the river stretch the long avenues, with miles of pleasant residences, gardens, velvet lawns, vines, and flowers. Each house is isolated in green, and one of the avenues is lined with rows of country-seats, with extensive grounds, such as are seldom seen within the limits of a city. But Cleveland on the hill is not like a city; it is like a suburban village multiplied by ten, and miraculously endowed with gas and pavements. Even in its central square, with its post-office, court-house, business-blocks, and street-cars, it has an air of leisure; and the statue of Commodore Perry, the flag-staff, and the little seats scattered over the grass, seem quite appropriate to its elegant ease. But step to the verge of the hill, and everything is different. Down on the Flat we see Cleveland at work, Cleveland grimy, Cleveland toiling in the sweat of her brow. Slowly through the oily river, whose name fitly signifies "crooked," wind the heavily-laden boats, bringing work for all these puffing engines, and taking away the product in its new shape as fast as the en-

Mouth of Cuyahoga River, Cleveland.

gines let it go. Here are seen all varieties of the lake-craft, from the scow to "The Last of the Mohicans" among boats, the two large side-wheel steamers which once ran between Cleveland and Detroit.

As Buffalo has its elevators, so Cleveland has its oil-refineries, which line the river-valley for miles. Hither, from the petroleum district, comes that fiery fluid which, hidden through all these centuries, has crowned the nineteenth with its luminous splendor. Here it is purified and sent forth into the wide world to fulfil its mission.

The sunsets are the glory of Cleveland. The sun, throughout the summer, sinks directly into the bosom of the water, lighting up the floating clouds with gorgeous tints, which cannot be surpassed the world over. Crimson mountains lie on the hori-

CLEVELAND, FROM SCRANTON'S HILL.

zon, their soft peaks fading into rose; then comes faint pink, tipped with gold, which lies against a deep-violet background, shading away higher and higher, until it mingles with the quiet blue of the zenith. The evening is the Western sailor's favorite starting-hour; and one by one, against the glowing sky, the ships steal out of the harbor, and, setting their white sails, glide away over the hazy water and vanish into night. The gazer stands enchanted; he has no words; a silence falls upon him; and, motionless, he

Euclid Avenue, Cleveland.

watches until the last vessel is lost in the twilight haze, and the last tint has faded into the usual blue of the summer night; then, over the lake, shines out the evening star and he turns homeward with a sigh.

West of Cleveland the coast grows more picturesque; the shore is high and precipitous, and the streams come rushing down in falls and rapids. Seven miles from the city is Rocky River, which flows through a deep gorge between perpendicular cliffs, that

SUPERIOR STREET, CLEVELAND, FROM PRESBYTERIAN CHURCH.

jut boldly into the lake and command a wide prospect. Here is the most extensive unbroken view of Lake Erie. Black-River Point is seen on the west, and the spires of Cleveland shine out against the green curve of the eastern shore; but far away toward the north stretches the unbroken expanse of water, and one can see on the horizon-line distant sails, which are still only in mid-lake, with miles of blue waves beyond.

West of Rocky River the Black, Vermilion, and Huron Rivers flow into the lake through ravines of wild beauty. The Black River is a beautiful stream. On a peninsula formed by its forks stands the town of Elyria—a name which is unique, having been derived from the surname of the first proprietor, "Ely," and the last syllable of his wife's Christian name, "ria," from Maria. The river falls over a rocky ledge forty-five feet in height, in two streams; and its whole course is full of picturesque beauties, making it remarkable among the Lake-Erie tributaries, which for the most part are

Mouth of Rocky River.

decorous, uninteresting creeks, coursing along slowly between tame shores, and making an undignified entrance into the lake by oozing through the sand-bars which clog up their passage.

Sandusky, the "Bay City," has spread out before it a lovely view. The town itself is not busy and breezy, like Buffalo, nor adorned with costly residences, like Cleveland, neither does it command, like Rocky River, a broad, landless ocean, whose waves roll in unbroken and dash against steep cliffs. But lovely is the bay with its gently-sloping shores and island—its river coming from the south and sweeping past the town, the peninsula opposite with its vineyard, and beyond, in the broad lake outside, the wine-islands, near and far, stretching one after the other, green, purple, a cloud, a speck, a mist, toward the Canada shore. It is a peaceful view also; one is not here called upon to calculate the statistics of grain, oil, or iron, and count the profits. The artist and the poet, who are out of place where traffic and dollars rule, might take to themselves homes on these lovely shores, nor ask a more beautiful prospect than this. It was first settled in 1817. During the civil war Johnson's Island, lying opposite the city, was used as a depot for Confederate prisoners, principally officers.

Sandusky has a mysterious name, whose derivation is said to be from a Wyandot word signifying "wells of cold water." The searcher for the picturesque, whether for

Black River, near Elyria, Ohio.

eye or ear, will certainly accept this derivation, and all along shore he will do his best to fix these half-forgotten titles on the bays and cliffs, where they belong, so that they who come after may at least catch the echo of the lingering names which belonged to the vanished races of the lake. The beautiful country around Sandusky was a favorite resort of the Indians; they hunted on the slopes and fished in the bay, whose upper

Red-Mill Falls, Black River, Elyria.

waters are an archipelago of green islets, abounding in ducks and other water-birds. The poor red-men have never been credited with a taste for the beautiful; indeed, the pioneers, who have fixed their place in the world's estimation, considered them little better than the bears. Yet all along the lake-shore, if we discover a peculiarly lovely island or bay, like this of Sandusky, we are sure to find also the tradition that it was dear to the Indians.

Sailing out through the bay, passing the unwieldy lumber-boats coming in heavily laden from the lumber-country of Lake Huron, the little fishing-smacks, and the light-house on its point, the steamer enters the lake, and turns toward Kelley's Island. This group of islands, fifteen or more in number, lying in the southwestern corner of Lake Erie, has come into notice at a comparatively recent date. The first pioneers preferred the solid main-land; they found enough to do in forcing their forest fields to yield them sustenance without encountering in addition the dangers of this inland sea. Even the grasping land companies did not stretch their hands as far as this vaguely-known group, which was, therefore, left to the adventurers who hover in front of civilization and disappear before its advance. But at length United States surveys were made, the land was entered and purchased, farm-houses were built, and fishermen, attracted by the number of bass, made their homes upon the shores, and called certain of the group Bass Islands.

Kelley's Island is the largest of the American group, containing about two thousand eight hundred acres. There is here an Indian-writing upon the rock, which has been pronounced the best-sculptured and best-preserved inscription in the West; it probably owes its distinctness to its remote situation, at the end of an island which has remained uninhabited until within a few years. The almost mythical tribe of Eries had here a fortified retreat, whose outlines can still be traced, and, according to interpretation, the inscription refers to them and their final destruction by the Iroquois.

Put-in-Bay Island received its name from Commodore Perry, who put in there with his fleet before and after the battle of Lake Erie, during the War of 1812. After leaving the harbor of Presque Isle, where he had built his war-vessels from the growing

Lumber-Boats, Sandusky, Ohio.

LAKE ERIE, FROM BLUFF, MOUTH OF ROCKY RIVER.

forest, Perry made sail for the head of the lake, and anchored in Put-in-Bay, opposite the British fleet, which lay under the guns of Malden, on the Canadian shore. Here he remained for some days watching the movements of the enemy, in order, if possible,

Glimpse of Sandusky, from St. Paul's Church.

to bring on an engagement. Every one knows the story of Perry's victory; and his famous dispatch, "We have met the enemy, and they are ours," is immortal.

At Erie they have the old flag-ship, the Lawrence; at Cleveland they have the commemorative statue; the islands are clustered over with associations of the engage-

South Coast, Kelley's Island.

ment; and one county, four towns, and twenty-six townships, in Ohio alone, recall the young commodore's name.

Put-in-Bay is a lovely sheet of water, with Little Gibraltar islet nestled in its crescent. Put-in-Bay Island has large summer hotels standing among its vineyards. Roses bloom in its gardens in December. Some of the islands are still wild and uninhabited, and several have only a single family. They abound in caves and rocky formations, to which, in many instances, Perry's name is attached. Little Gibraltar is crowned with the towers of a picturesque villa; it has also its Sphinx Head, which may be called a fresh-water imitation of the Egyptian queen. The Rattlesnake Island and its rattles alone preserve the memory of the real aboriginal inhabitants of the group, who, according to the geographies of the last century, "lay in acres upon the lily-leaves basking in the sun, and hissing out a breath which struck death to the incautious mariner who ventured near these isles of terror."

Along the Sandusky peninsula and over the islands stretch the vineyards, whose grapes and wine form the feature of this portion of the shore. Here in the sunny autumn, when the long aisles are full of gatherers and the trellises are heavy with purple bunches, when the little steamers go away loaded with grapes, and the presses in the wine-houses crush out their juice by day and by night, the islands are like an enchanted land, watching the autumn out and the winter in with light-hearted joyousness. The water is still and blue, the colored trees are reflected in its mirror, a golden haze shines over the near islands, and a purple shadow lies on those afar.

Kelley's Island.

West of the Fire-Lands lies the country called the Black Swamp, well known in the early settlement of the lake-shore, and long retained enough of its primitive character to justify the name. It is a singular region, and not without its charm; its level surface and the uniformity of its soil gave to the forest a remarkable regularity—the trees being of the same height, extending in straight ranks mile after mile, resembling from a distance an even, blue wall against the sky. The foliage was so dense that when the first roads were built through to the West the immigrants travelled for days along the shadowed aisles, nor saw the sun from border to border. This long twilight and silence impressed them strangely, and, bold frontiersmen as they were, they drew a long breath when they emerged into daylight and the open country beyond; and ever afterward they spoke of the journey in terms which seem almost poetical when compared

Put-in-Bay.

with the practical prose of their ordinary language. But it was not the poetry of admiration; it was a vague fear, a vague wonder over the mystery of the dark labyrinth and what it might contain. Yet it was not a land of desolation. Vines and blossoms were everywhere, and birds sang among the branches. It was the mystery that impressed them—"a land of the shadow of death," they called it. The soil of the Black Swamp is very fertile, and as soon as drained it becomes a garden; fruit, grain, and vegetables spring up with wonderful rapidity.

As the lake-shore is divided into districts whose boundaries, although not to be found on any map of the day, are yet better known than the carefully-marked lines of the counties, and as these districts have names of their own, often spoken, although not set down in the geographies; so each has its one city, and one only. Thus the Holland Purchase has Buffalo; the Triangle has Erie; the Western Reserve has Cleveland; the Fire-Lands have Sandusky; and the Black Swamp has Toledo.

This city, with its Spanish name, stands on the Maumee, a river which once bore the melodious title of Miami of the Lakes. Ohio having already two Miamis, the name of the northern river was changed. Toledo is four miles from the mouth of the river, and ten miles from the lake, Maumee Bay lying between.

A few miles beyond Maumee Bay the coast turns sharply to the north, the Black Swamp is left in the southwest, and the boundary-line of Michigan is passed. The eastern end of Lake Erie slopes to a point at Buffalo, both shores coming toward each

Perry's Cave, Put-in-Bay Island.

other and making a natural gate-way for the Niagara River. But the western end is blunt and unyielding. The Detroit River has no gate-way; it comes unexpectedly into the lake from a broad shore; its mouth is clogged with islands; and there is nothing to indicate the entrance of a grand strait, which in its peculiar beauty has no peer throughout a chain that holds the Saut Ste. Marie, the St. Clair, the Niagara, and the St. Lawrence. The northward-sloping coast of Michigan—sixty miles in length, between the Ohio boundary and the city of Detroit—is a green, fertile shore, with numerous little rivers flowing through it, and a more gentle aspect than the north and south coast-lines of Ohio and Canada. This territory has had two distinct settlements—the French, which is ancient, and the American which is comparatively modern. It was not until 1830 that American emigration flowed freely into Michigan Territory; and Ohio had a settled population of New-England colonists, with their schools and churches, and had sent her pioneers into Indiana and Illinois, while the Detroit shore remained wholly

French. The unextinguished Indian titles, the foreign ideas of the inhabitants, and the barrier of the Black Swamp lying in the way, kept the emigrants from this lovely land. In the mean time the French settlers remained undisturbed in their little houses along the shore. From the river Raisin, which flows into Lake Erie near the present town of Monroe, as far north as Lake St. Clair, this line of French cabins extended. The people were a gay, contented race, raising the same little crops in the same little fields year after year, and grinding their Indian-corn and wheat in rude windmills, some of which are still to be seen on the shore. Alone of all the colonists of the New World, these Frenchmen readily assimilated themselves with the Indians; and, by adopting some of the forest customs, they lived in peaceful friendship with the very tribes whom the English and Americans regarded as treacherous and cruel. French traders established posts along the frontier, French *coureurs des bois* penetrated beyond the Mississippi, French *voyageurs* paddled their canoes from Lake Superior to the St. Lawrence, unmolested and successful; while the hunters and traders of other nations lived in constant danger of massacre.

At a later date, when the compact, white houses of New-England settlers began to appear among the French cabins, and when courts of the United States were established, much difficulty was experienced from these feudal customs. The French witnesses could speak no English; and, accustomed as they were to the plain "yes" and "no" of military rule, they could not understand the

Perry's Lookout, Gibraltar Island.

law's delays and finely-drawn lines, and in not a few instances they took to the law of steel and cudgel to defend their rights against the lawyers.

There are fifteen islands within the first twelve miles of the Detroit River. Father Hennepin, who passed up the strait in 1679, enthusiastically writes: "The islands are the finest in the world; the strait is finer than Niagara; the banks are vast meadows; and the prospect is terminated with some hills crowned with vineyards, trees bearing good fruit, groves and forests so well disposed that one would think that Nature alone could not have made, without the help of art, so charming a prospect." The good father spoke but the truth. The river has neither foam, rapids, nor mountains; it has not that sweep to the sea, that incoming of the salt tide, which give to the ocean-rivers their majesty; yet it is a grand strait, full to the very brim of its green shores, calm, deep, and beautiful.

The city of Detroit, with the exception of Mackinac, the first white settlement in the Northwest, was visited by the French in 1610 A permanent settlement was made there in 1701 by La Motte Cadillac, where a fort was built and named Pontchartrain, after the French colonial minister. Some years later a colony of French emigrants was sent out from France, who, mingling with the Indians, began that race of half-breeds whose history is indissolubly connected with the history of the fur-trade. A French military and trading post, Detroit was unlike the other lake-cities, and many of its original characteristics still appear—French names and customs, a deference to military rule, and a certain *insouciance.* In 1805 the old town was burned, and the new town which arose on the site was laid out with more regularity, but at the expense of its picturesque quaintness. The flag flying over it has been changed five times in the following order: French, British,

Sphinx Head, Gibraltar.

Toledo, Ohio.

American, British, American; and it has been the scene of one surrender, twelve massacres, and fifty battles. It is a veteran town compared with Cleveland and Buffalo. It was a century old when they were born.

The central figure of Detroit history is Pontiac, the great Ottawa chieftain. He was the king of the river—the only Indian who, in the history of America, proved himself a match for the white man in far-reaching sagacity—the only Indian who succeeded in forming and maintaining powerful combinations among the discordant tribes. The masterpiece of Pontiac's life was a conspiracy to capture simultaneously on a fixed day all the British posts in the West, twelve garrisoned forts, extending from Niagara to Pittsburg, along the lake-shore, and on as far as the Mississippi. In such a wide field many tribes must act, and many clashing interests must be reconciled; and yet such was the personal influence of Pontiac that the plan was carried out: nine of the posts were taken upon the same day, and their garrisons massacred. Detroit made a successful resistance, owing to the warning of an Indian girl—the Pocahontas of the West.

Above the city the Detroit River curves to the eastward and enters Lake St. Clair. Here are long lines of lumber-barges with their tugs, schooners with their raking masts leaning far over under a cloud of canvas, brigs with their high-lifted, aggressive sails, scows with their yellow wings spread widely to the breeze, and steamers coming up and passing them all in the evening race to the Flats, through whose narrow canal or tortuous channel one and all must pass before darkness comes, or lie at anchor until morning. On they sail through the golden afternoon, the red sunset and dusky twilight, and as they pass Fort Gratiot and enter the broad Lake Huron, night closes down on the dark water, lights are run up to the mast-heads of the steamers, the vessels twinkle in red and green, and Lake Erie, its scenery, history, and associations, vanish in dreams.

Windmill, opposite Detroit.

GLANCE AT DETROIT FROM THE CITY HALL

ON THE OHIO.

WITH ILLUSTRATIONS BY ALLRED R. WAUD.

The Ohio, below Pittsburg.

O-HE-YO is a Wyandot word, signifying "Fair to look upon." The early French explorers, floating down the river's gentle tide, adopted the name, translating it into their own tongue as *la Belle Rivière;* and the English, who here, as elsewhere throughout the West, stepped into the possessions of the French, took the word and its spelling, but gave it their own pronunciation, so that, instead of O-he-yo, we now have the Ohio. It is a lovely, gentle stream, flowing on between the North and South. In and out it meanders for one thousand and seven miles; it is never in a hurry; it never seems to be going anywhere in particular, but has time to loiter about among the coal and iron mines of Pennsylvania; to ripple around the mountains of West Virginia; to make deep bends in order to take in the Southern rivers, knowing well that thrifty Ohio, with her cornfields and villages, will fill up all the angles; then it curves up northward toward Cincinnati, as if to leave a broad land-sweep for the beautiful blue-grass meadows of Kentucky; and at North Bend away it glides again on a long south-western stretch, down, down, along the southern borders of Indiana and Illinois, and

Pittsburg, from Soldiers' Monument.

after making a last curve to receive the twin-rivers—the Cumberland and the long, mountain-born Tennessee—it mixes its waters with the Mississippi, one thousand miles above the ocean.

The Ohio is formed from the junction of two rivers as unlike as two rivers can be. The northern parent, named Alleghany, which signifies "Clear water," is a quick, transparent stream, coming down directly from the north; while the southern parent, named Monongahela, which signifies "Falling-in banks," comes even more directly from the south. These two rivers, so unlike in their sources, their natures, and the people along their banks, unite at Pittsburg, forming the Ohio, which from that point to its mouth receives into itself seventy-five tributaries, crosses seven States, and holds in its embrace one hundred islands. On the northern shore of the Upper Ohio the railroad to Pittsburg is seen; the long trains of yellow cars rush by, their shrill whistles coming from the steep hill-side over the water, as if remonstrating with the boats for their lazy progress. In truth, the boats

Pittsburg, from Reservoir.

do their work in a leisurely way. A man appears on the bank and signals, but even he is not in a hurry, finding a comfortable seat before he begins his waving; then the captain confers with the mate, the deck-hands gather on the side to inspect the man, and all so slowly that you feel sure the boat will not stop, and you look forward toward the next bend. But the engine pauses, the steamer veers slowly round, runs its head into the bank, out comes the plank, and out come the motley crew, who proceed to bring on board whatever the waving man has ready for them, while he, still seated, watches the work, and fans himself with his straw hat.

The present city of Pittsburg for many years was described as having the picturesque aspect of a volcano, owing to its numerous manufactories; a cloud of smoke hangs over it, and at night it is illuminated by the glow and flash of the iron-mills filling its valley and stretching up its hill-sides, resting not day or night, but ever ceaselessly gleaming, smoking, and roaring. Looking down on Pittsburg at night from the summit of its surrounding hills, the city, with its red fires and smoke, seems Satanic. Quiet streets there are, and pleasant residences; the two rivers winding down on either side, and uniting at the point of the peninsula, the graceful bridges, the water-craft of all kinds lying at the levee, some coming from far New Orleans, and others bound up the slack-

SOUTH PITTSBURG AND ALLEGHANY CITY.

water into the interior, are all picturesque. But it is the smoke and the fires of Pittsburg that give it its character.

Anthony Trollope wrote, "It is the blackest place I ever saw, but its very blackness is picturesque." Parton said, "It is all hell with the lid taken off." With the discovery of natural gas came its use as a fuel, and Pittsburg, long known as the "Smoky City," was christened again, becoming the "Iron City."

The river starts away in a northwestern direction. On its banks, nineteen miles from Pittsburg, is the quaint German town of Economy, founded by Father Rapp, a German pietist, who emigrated with a colony from Würtemberg in 1804. The little band of believers, in what seems to us a dreary creed, made one or two changes of loca-

The Ohio, from Marietta.

tion; but, after selling their possessions in Indiana to Robert Owen, a man of kindred enthusiasm but opposite belief, they came to the Ohio River, where their village, with its Old-World houses, tiled roofs, grass-grown streets, and quiet air, seems hardly to belong to this practical, busy, American world.

The State of Ohio reaches the river at Columbiana County. This was a fancy name, formed from Columbus and Anna. Opposite, as the river turns abruptly down toward the south, is the queer little strip of land which Virginia thrusts up toward the north, the ownership of which is probably due to some of the fierce quarrels and compromises over land-titles which came after the Revolution, and made almost as much trouble as the great struggle itself. This northern arm is called the Pan-Handle—Virginia, undivided, being the pan.

Baltimore and Ohio Railroad Bridge, Parkersburg, Va.

Three miles below Steubenville was an old Mingo town, the residence of Logan, the Mingo chief. This celebrated Indian was the son of a Cayuga chieftain of Pennsylvania, who was converted to Christianity by the Moravian missionaries, the only rivals of the Jesuit fathers in the West. The Cayuga chief, greatly admiring James Logan, the secretary of the province, named his son after him. Logan took no part in the old French War, and remained a firm friend of the whites until the causeless murder of all his family on the Ohio River, above Steubenville. From that time his hand was against the white man, although, from the curt records of the day, we learn that he was singularly magnanimous to all white prisoners. The last years of Logan were lonely. He wandered from tribe to tribe, and was finally murdered by one of his own race on the banks of the Detroit River, as he sat before a camp-fire, with his blanket over his head, buried in thought.

The river, as it stretches southward, is here fair enough to justify its name. The

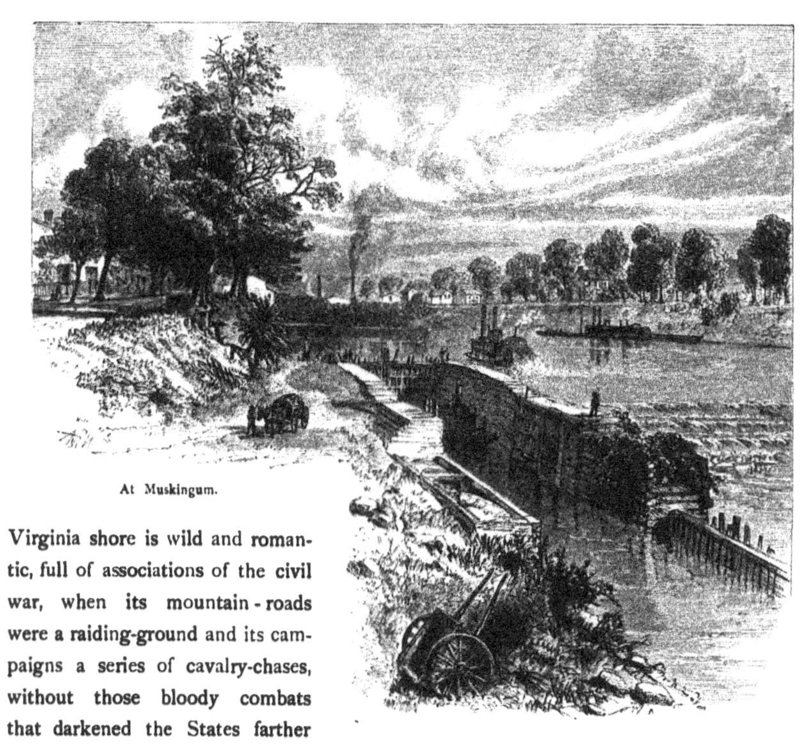
At Muskingum.

Virginia shore is wild and romantic, full of associations of the civil war, when its mountain-roads were a raiding-ground and its campaigns a series of cavalry-chases, without those bloody combats that darkened the States farther south. At Wheeling, the national road, a relic of stage-coach days, crosses the river on its westward way. This turnpike was constructed by the national government, beginning at Cumberland, in Maryland, crossing the mountains, and intended to run indefinitely on westward as the country became settled. But railroads took away its glory, and the occasional traveller now finds it difficult to get an explanation of this neglected work, its laborious construction and solid stone bridges striking him as he passes through Central Ohio, although the careless inhabitants neither know nor care about its origin. In the Old World it would pass as a Roman road.

Marietta, in Washington County, Ohio, is the oldest town in the State. It has a picturesque position, lying in a deep bend where the Muskingum flows into the Ohio, with a slender, curved island opposite, like a green crescent, and, beyond, the high, rolling hills of Virginia on the southern shore. It was settled by New-Englanders, in November, 1787, who floated down the Ohio in a flat-boat called the Mayflower. Their first act on landing was to write a set of laws and nail them to a tree. Washington said of them, "No colony in America was settled under such favorable auspices as that on the Muskingum." At Marietta were found the remains of an ancient fortification—a square,

SCENES ON THE OHIO, ABOVE AND BELOW CINCINNATI.

enclosed by a wall of earth ten feet high, with twelve entrances, containing a covert way, bulwarks to defend the gate-ways, and various works of elaborate construction, including a moat fifteen feet wide, defended by a parapet. These are supposed to belong to the era of the mound-builders. Thirteen miles below Marietta is Parkersburg, in West Virginia; the old Belpré, or Beautiful Meadow, in Ohio, opposite; and near by, in the river, Blennerhassett's Island, which has gone into history with Aaron Burr.

At Parkersburg the Little Kanawha flows into the Ohio, which is here crossed by the massive iron bridge of the Baltimore and Ohio Railroad. Farther on is Gallipolis, where, in 1790, a French colony laid out a village of eighty cabins, protected by a

"The Rhine"

stockade, and even in the face of starvation took time to build a ballroom, and danced there twice a week.

At the mouth of the Great Kanawha, on the Virginia side, is Point Pleasant. This stream is the principal river of West Virginia, rising in the mountains and winding through a picturesque country northward to the Ohio. Point Pleasant was the site of the bloodiest Indian battle of the river-valley, when, in 1774, one thousand Americans were attacked by the flower of the Western tribes under the chieftain Cornstalk. The battle raged all day, but the Indians were finally overpowered, and retreated to their towns on the Chillicothe plains.

Fourth Street, Cincinnati.

Kentucky, which comes up to the Ohio at the mouth of the big Sandy River, is one of the most beautiful States in the country. It is wild without being rugged, luxuriant but not closely cultivated. Once seen, its rolling meadows are never forgotten. Stretching back from the river are vast parks; there is no underbrush, few fences, and few grain-fields; the trees are majestic, each one by itself, and here and there stands a bold hill, or a river comes sweeping over a limestone-bed. It is the grazing-country of America; the wealth of its people is in their flocks and herds; and there is a tradition that they love their horses better than their sweethearts. Some miles back from the river lies the famous Blue-Grass Country, so called from the blue tinge of the grass when in blossom. This district embraces five counties, the loveliest in Ken-

A VIEW OF CINCINNATI.

tucky, where you may ride for miles through a park dotted with herds, single trees, and here and there a grove shadowing the rolling, green turf.

Cincinnati, the Queen of the West, was first settled in 1778. It lies in Symmes's Purchase—land stretching between the Great and Little Miami, called in early descriptions the Miami Country. It received its high-sounding name from General Arthur St. Clair, in honor of a military society to which he belonged.

Cincinnati was founded in romance. There were two other rival settlements on the river, and all three were striving for the possession of the United States fort. North Bend was selected, the work begun, when one of the settlers, observing that the bright

View on the Rhine, Cincinnati.

eyes of his wife had attracted the attention of the commanding officer, moved to Cincinnati. But immediately Cincinnati was discovered to be the better site, and materials and men were moved up the river without delay. North Bend was left to its fate, and Cincinnati obtained an advantage over her rivals from that time, steadily progressing toward her present population, which, including her suburbs, is nearly four hundred thousand. The city proper is closely built in solid blocks, rising in several plateaus back from the river; it is surrounded by a circle of hills, through which flow the Little Miami and Mill Creek. There are many fine buildings in Cincinnati; but the beauty of the city is in its suburbs, where, upon the Clifton Hills, are the most picturesque residences of the entire West—beautiful homes, with sweeping grounds and a wide outlook over the valley. In the centre of the city is the Tyler-Davidson Fountain—one of the most beautiful fountains in the world. The figures are of bronze, and were cast in Munich, Bavaria, at a cost of one hundred thousand dollars. The fountain is a memorial, presented to the city by one of its millionaires in memory of a relative. It

The Tyler-Davidson Fountain.

bears the inscription, "To the People of Cincinnati;" and the Genius of Water, that surmounts the fountain, from her beneficent, outstretched hands seems to be sending down rain upon a thirsty land. The people of Cincinnati do not live in their city; they attend to their business affairs there, and retire out to the hills when work is over. They have an air of calm contentment and indifference to the rest of the world; they know they are masters of the river. Pittsburg is lurid and busy; Louisville is fair and

LOUISVILLE, FROM THE BLIND ASYLUM.

indolent; but Cincinnati is the queen. She has no specialty, like Buffalo with her elevators, Louisville with her bourbon-warehouses, Cleveland with her oil-refineries, and Pittsburg with her iron-mills; or, rather, she has them all, and therefore any one is not noticeable. Within the city is one picturesque locality—the German quarter—known as "Over the Rhine," the Miami Canal representing the Rhine. Here the German signs, the flaxen-haired children, the old women in kerchiefs knitting at the doors, the lager-beer, the window-gardens and climbing vines, the dense population, and, at evening, the street-music of all kinds, are at once foreign and southern.

Below Cincinnati are the vineyards, stretching up the hills along the northern shore. Floating down the river in the spring and seeing the green ranks of the vines, one is moved to exclaim, "*This* is the most beautiful of all!" forgetting that the mountains of Virginia and the parks of Kentucky have already called forth the same words. The

New Albany, Indiana.

native Catawba wine of the West was first made in Cincinnati, and the juices of the vineyards of the Beautiful River have gained an honorable name among wines.

Bellevue, in Kentucky, and Patriot, in Indiana, are charming specimens of river scenery, the latter showing the hill-side vineyards, and may be seen in the illustration.

The navigation of the Ohio is obstructed by tow-heads and sand-bars, and by the remarkable changes in its depth, there being a variation of fifty feet between high and low water mark. In the early days a broad river was the safest highway, as the forests on shore concealed a treacherous foe who coveted the goods of the immigrant; hence once over the mountains, families purchased a flat-boat and floated down-stream, hugging the Kentucky shore. These Kentucky flats were made of green oak-plank, fastened by wooden pins to a frame of timber and calked with tow, and upon reaching their destination the immigrants used the material in building their cabins.

The boatmen of the Ohio were a hardy, merry race, poling their unwieldy craft slowly along, or gliding on under sail, sounding a bugle as they approached a village, and shouting out their compliments to the girls, who, attracted by the music, came down to the shore to see them pass. They wore red handkerchiefs on their heads, turban fashion, and talked in a jargon of their own, half French, half Indian; a violin formed part of their equipment; and at night, drawn up at some village, they danced on the flat tops of their boats—the original minstrels. In this way, as the old song has it, "They glided down the river, the O-hi-o." The majority of the Ohio-River craft are tow-boats, black, puffing monsters, mere grimy shells to cover a powerful engine. If tow means to pull, then the name of tow-boat is a misnomer; for these boats never pull, but always push. Their tows go in front, and then comes the steamer pushing them along, her stern-wheel churning up the water, and her smoke-stacks belching forth black streams.

Jeffersonville, Indiana, is a thriving town nearly opposite Louisville. Here is the only fall in the Ohio River—a descent of twenty-three feet in two miles—a very mild cataract, hardly more than a rapid. Such as it is, however, it obstructs navigation at low stages of water, and a canal has been cut around it through the solid rock. New Albany, Indiana, a few miles below, is an important and handsomely-situated town.

Louisville is a large, bright city, the pride of Kentucky. It was first settled by Virginians in 1773, and remained for some time under the protection of the mother State; even now, to have been born in Virginia is a Louisville patent of nobility. The city is built on a sloping plane seventy feet above low-water mark, with broad streets lined with stately stone warehouses on and near the river, and beautiful residences farther back. Louisville has a more Southern aspect than Pittsburg and Cincinnati. Here you meet great wains piled with cotton-bales; the windows are shaded with awnings; and the residences swarm with servants—turbaned negro cooks, who are artists in their line; waiting-maids with the stately manners of their old mistresses; and innumerable children—eight or ten pairs of hands to do the work for one family. In the Court-House is a life-like statue of Henry Clay, a man whose memory Kentucky delights to honor. His grave is at Lexington—the most stately tomb in the West.

Jeffersonville, Indiana.

THE ST. LAWRENCE AND THE SAGUENAY.

WITH ILLUSTRATIONS BY JAMES D. SMILLIE.

Entrance to the Thousand Islands.

IT is a June morning on the St. Lawrence; the little city of Kingston is as fast asleep as its founder, the old Frenchman De Courcelles; the moon is ebbing before the breaking day; a phantom-like sloop is creeping slowly across the smooth stream. At the steamboat-wharf there is a little blaze of light and a rush of noisy life, which breaks but does not penetrate the surrounding silence. "All aboard!" The bell rings out its farewell notes, the whistle pipes its shrill warning into the night, and the steamer slips her moorings, to the pleasure of the sleepy travellers who crowd her decks and

Light-Houses among the Thousand Islands.

cabins. By this time the east is tinted purple, amber, and roseate. Night is fast retreating. Soon after leaving Kingston we bestir ourselves and choose eligible seats in the forward part of the boat. We chat without restraint, and expectation is rife as we near the famed Thousand Islands. The descriptions we have read and the stories we have heard of the panorama before us flock vividly into our memories. We are soon

Among the Thousand Islands.

in the midst of the one thousand six hundred and ninety-two dots of land which, according to the Treaty of Ghent, constitute the Thousand Islands.

Are we disappointed? That is the question which most of us propound before we proceed many miles. There is little variety in their form and covering. So much alike are they in these respects that our steamer might be almost at a stand-still for all the change we notice as she threads her way through the thirty-nine miles which they thickly intersperse. In size they differ much, however, some being only a few yards in extent, and others several miles. The verdure on most of them is limited to a sturdy growth of fir and pine, with occasionally some scrubby undergrowth, which sprouts with northern vigor from crevices in the rocky bed. The light-houses which mark out our channel are a picturesque feature, and are nearly as frequent as the islands themselves; but all are drearily alike—fragile wooden structures, about twenty feet high, uniformly whitewashed. As the sun mounts yet higher, and the mist and haze disperse, we run between Welles-

Between Wellesley Island and the Canadian Shore.

ley Island and the Canadian shore, and obtain one of the most charming views of the passage. The verdure is more plentiful and the forms are more graceful than we have previously seen. Tall reeds and water-grasses crop out of the shoals. The banks of the island and the main-land slope with easy gradations, inclining into several bays; and afar a barrier seems to arise where the river turns and is lost in the distance. Thence we steam on in an enthusiastic mood toward Prescott, satisfied with the beauties we have seen, and arrive there at breakfast-time, a little over four hours after leaving Kingston.

Below Prescott we pass an old windmill on a low cape, where the insurrectionists established themselves in 1837; and, two miles farther, we catch a glimpse of a gray old

French fortification on Chimney Island. Here, too, we descend the first rapids of the river—the Gallope and the Deplau Rapids—with full steam on. No excitement, no breathlessness, attends us so far in our journey. Illustrations usually represent the water as seething white, with a preposterous steamer reeling through it at a fearful rate, but the Gallopes and Deplaus are passed almost without a sensation. Soon we are nearing

Entering the Rapids.

the famous Long-Sault Rapids. An Indian pilot comes on board to guide us through, and as he enters the wheel-house on the upper deck he is an absorbing object of interest. A stout, sailorly fellow he appears, without an aboriginal trait about him, or a single feather, or a dab of paint. There are some bustling preparations among the crew for what is coming. Four men stand by the double wheel in the house overhead, and two others man the tiller astern, as a precaution against the breaking of a rudder-rope. The captain stands on the upper deck, with one hand calmly folded in his breast and the other grasping the signal-bell. In a few seconds more we shall be in the rapids. The attention of the passengers is engrossed by the movements of the captain's hand. As he is seen to raise it, and the bell is heard in the engine-room, the vibrations of the huge vessel die away; the water leaps tempestuously around her, and she pauses an instant like a thing of life, bracing herself for a crisis, before she plunges into the boiling current and rides defiantly down it. It is a grand, thrilling moment; but it is only a moment. The next instant she is speeding on as quietly as ever, without other percep-

River Front, Montreal.

tible motion than a slight roll. The rapids are nine miles long, and are divided in the centre by a picturesque island, the southern course usually being chosen by the steamers. The distance is made in half an hour, without steam, and then the steamer emerges into the waters of Lake St. Francis, whose banks are deserted, and the only human habitations seen are in the little village of Lancaster. The drear monotony of our passage through Lake St. Francis is followed by renewed excitement in the descent of the Cedar Rapids, at the foot of which we enter Lake St. Louis. Still more uninteresting is the sheet of water now before us, bordered as it is by flat lands reminding us of the Southern bayous. But it is here we get our first glimpse of the bold outlines of Montreal Island, rising

MONTREAL.

softly in the background; and here, too, the river Ottawa, ending in the rapids of St. Anne's, pours its volume into the greater St. Lawrence.

A queer-looking barge, with a square sail set, lumbering across our course, and throwing a black shadow on the water that is now richly tinted with purple and deep red; a light-house at the extremity of a shoal, yet unlighted; a mass of drift-wood, sluggishly moving with the current; a puff of smoke, hovering about the isolated village of St. Clair—these things are all we meet in our voyage across the broad St. Louis.

The steamer stops at the Indian village of Caughnawaga, and then proceeds toward the Lachine Rapids. In the descent of these we are wrought to a feverish degree of

Mount Murray Bay, St. Lawrence.

excitement, exceeding that produced in the descent of the Long Sault. Once—twice—we seem to be hurrying on to a rock, and are within an ace of total destruction, when the steamer yields to her helm and sweeps into another channel. As we reach calm water again we can faintly distinguish in the growing night the prim form of the Victoria Bridge, and the spires, domes, and towers of Montreal.

A night's rest in a modern hotel prepares us for the following day's tramp through this ancient metropolis of the Indians and modern metropolis of the Canadians. Montreal does not resemble an English city—the streets are too regular; and it does not resemble our own American cities, than which it is more substantially built. Its sub-

MARKET-HALL AND BOAT-LANDING, QUEBEC.

stantiality is particularly impressive—the limestone wharves extending for miles, the finely-paved streets lined with massive edifices of the most enduring materials, imprinted with their constructors' determination that they shall not be swept away in many generations. The site is naturally picturesque. It is on the southern slope of a mountain in the chain which divides the verdant, fertile island of Montreal. There are a high town and a low town, as at Quebec; and on the up-reaching ground, leafy roads winding through, are the villa residences of the fashionable. The prospect from these bosky heights repays with liberal interest the toil of the pedestrian who seeks them from the city, for he may survey, on the fair level beneath him, the humming streets; the long line of wharves, with their clustering argosies; the vast iron tube which binds the opposite sparsely-settled shore to the arterial city; Nun's Island, with its flowery grounds neatly laid out; beautiful Helen's Island, thick with wood; the village of Laprairie, its spire glistening like a spike of silver; the golden thread of the St. Lawrence, stretching beyond the Lachine Rapids into mazes of heavy, green foliage; the pretty villages

Breakneck Stairs, Quebec.

of St. Lambert, Longueuil, and Vercheres; and afar off, bathed in haze and mystery, the purple hills of Vermont. Perchance, while his eye roams over the varied picture with keen delight, there booms over the roofs of the town the great bell of Notre-Dame, and he saunters down the height in answer to its summons—through hilly lanes of pretty cottages on the outskirts into the resonant St. James Street; through Victoria Square, and on until he reaches the Place d'Armes. Here is the cathedral of Notre-Dame, a massive structure capable of holding ten thousand people, with a front on the square of one hundred and forty feet, and two towers soaring two hundred and twenty feet above. Climbing one of these towers, the view of the river and city

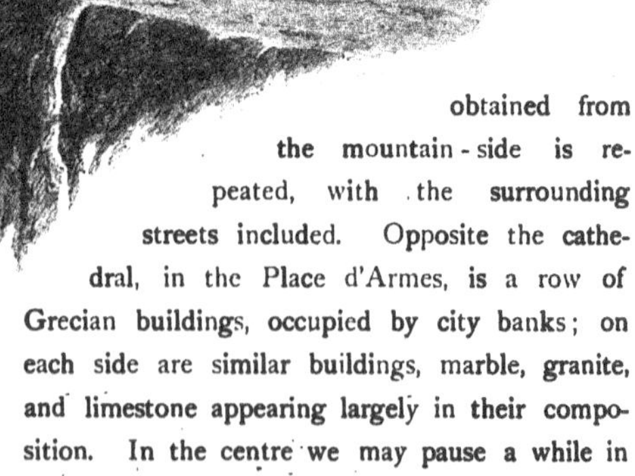

The Citadel, Quebec.

obtained from the mountain-side is repeated, with the surrounding streets included. Opposite the cathedral, in the Place d'Armes, is a row of Grecian buildings, occupied by city banks; on each side are similar buildings, marble, granite, and limestone appearing largely in their composition. In the centre we may pause a while in the refreshing shade of the park, and hear the musical plashing of the handsome fountain as it glints in the bright sunlight. Then we go to see the Bonsecours Market, the nunneries, Mount-Royal Cemetery, the imposing Custom-House, the Nelson Monument, and the water-works; and in the evening we resume our journey to Quebec.

We might be travelling through some broad river of France, so thoroughly French are the names of the villages. On one bank are L'Assomption, St. Sulpice, La Vittre,

From the Top of Montmorency Falls, looking toward Quebec.

Berthier, Fond du Lac, and Batiscon; on the other, Becancour, Gentilly, St. Pierre, Dechellons, and Lothinier. The people of these villages are descended from the old French settlers, and, crossed with the Indian and American, they retain some of the traits of each. Their high cheek-bones, aquiline nose, and thin, compressed lips, refer us to the aboriginal; but they are below the average height, while stouter and stronger and less graceful than the natives of France.

Quebec! The historic city of Canada; the city of conquests, of military glory, of bewildering contrasts! It is yet early morning when we arrive there; a veil of mist obscures the more distant objects. Who that loves the ancient, the gray, the quaint, is not touched with emotion on finding himself at the portals of the noble old fortress looking down upon the ample water-path to the heart of the continent? Our minds are fraught with memories of the early explorers, of battles and their heroes, of strange social conditions that have existed and exist in the shadow of yon looming rock, whither our steamer's bow is directed. Illustrious names are woven in its pages—Richelieu, Condé, Beauharnais, Montmorency, Laval, and Montcalm. Two nations struggled for its possession. We see old Jacques Cartier ascending the river in 1534, and holding a conference with the Indians. Half a century later Champlain enters the scene at the head of a vigorous colony. He is not fairly settled before an English fleet speeds up the St. Lawrence, captures Quebec, and carries him off a prisoner to England. Then a treaty of peace is signed, and the city is restored to France, Champlain resuming his place as governor of the colony. Thereafter, for a hundred and fifty years, France rules unmolested, and the lily-flag waves from the heights of the citadel. But a storm impends, and soon England shall add New France to her colonial empire. Two armies contend for the prize: Wolfe, on the land below, at the head of the English; Montcalm, on the heights above, at the head of the French. With the armies thus arrayed, Wolfe is at a disadvantage, which he determines to overcome by strategy. A narrow path twisting up the precipice is discovered, and on a starlight night the valiant young general leads his men through the defile. On the following day the battle is fought: Montcalm advances, and covers the English with an incessant fire. Wolfe is wounded in the wrist, nevertheless he hastens from rank to rank exhorting his men to be steady and to reserve their shots. Twice again Wolfe is wounded, the last time mortally, but his army is victorious; and he sinks from his horse as the French are retreating, and Montcalm, too, is mortally wounded.

The Montreal steamer, after passing Wolfe's Cove and Cape Diamond, lands us at an old wharf a few yards above the Champlain Market, where we get our first glimpse of Quebec. At our back is the placid river, with a crowd of row-boats and sloops and schooners drifting easily in the stilly morning air; to the right is the Market-Hall, a pleasing building of important size, with several rows of broad stairs running from its portals to the water's edge; behind it are the dormer-windowed, slated and tinned roofs of the lower town; behind these, again, on the heights, the gray ramparts, the famous Terrace, resting on the buttress arches of the old castle of St. Louis, the foliage of the Government Garden, and the obelisk erected to Wolfe and Montcalm.

In Breakneck Stairs, which every tourist religiously visits, we have one of those alleys that are often seen in the old towns of England and France—a passage, scarcely fifteen feet wide, between two rows of leaning houses, the road-bed consisting of several successive flights of stairs. In the evening Dufferin Terrace offers a telling contrast to the more sombre quarters of the city. It is one of the finest promenades in the world;

FALLS OF MONTMORENCY.

Under Trinity Rock, Saguenay.

from the railing that surrounds it the view down the river is enchanting. Seen from the elevation of the Terrace, the lower town, with its tinned roofs, seems to be under a veil of gold. It is here, on this lofty esplanade, that Quebec airs itself; and, at twilight, throngs of people lounge on benches near the mouths of beetling cannon, and roam among the fountains and shrubbery of the Place d'Armes.

On the next day we go to Montmorency. We jolt in a calash across the St. Charles River by the Dorchester Bridge, and then enter a macadamized road leading through a very pretty country filled with attractive residences. Farther away, we pass the Canadian village of Beauport, and get an insight of old colonial life. The houses are all alike in size, form, and feature. Thence we follow an English lane through sweet-scented meadows until we arrive at the falls, and are admitted to some grounds where, from a perch at the very edge of the rock, we can look upon the fleecy cataract as it pours its volume into the river. Hereabout the banks are precipitous—two hundred and fifty feet high—and covered with luxuriant verdure. The falls are deep-set in a small bay or chasm, and descend in a sheet twenty-five yards wide, broken midway by an immense rock hidden beneath the seething foam. The surrounding forms are picturesque in the extreme. Half a mile above the falls we visit the Natural Steps, where the limestone-rock bordering on the river has been hewn by Nature into several successive flights of steps, all remarkably regular in form; and in the evening we are returning to Quebec, which, as it is seen from the Beauport road, strikes one as the most beautiful city on the continent.

In the morning we are on board the Saguenay boat bound for Murray Bay and Cacouna, where the Canadians go for their sea-bathing, which they cannot have at Quebec, as the water there is fresh. For nearly an hour we retrace by water the trip we made yesterday by land, and are soon abreast of the Montmorency Falls, which are seen to still better advantage than on the day before. Afar off, the stately range of the Laurentian Hills roll upward in a delicate haze; and through the trees on the summit of the bank the river Montmorency shimmers in perfect calm, with something like the placid resignation of a brave soul conscious of an approaching death. The stream is divided here by the island of Orleans, a low-lying reach of farm-land, with groves of pine and oak embowering romantic little farm-houses and cottages, such as lovers dream of. But, as we journey on, this exquisite picture passes out of view and the river widens, and the banks are nothing more than indistinct blue lines marking the boundary of the lonely waters. Late in the afternoon Murray Bay is reached. About the little landing-place some evidences of fashionable civilization are apparent, and in the background is a summer hotel of the period. At Cacouna more fashionable people are waiting for the steamer, the arrival of which is the event of the day. The sun has set before our steamer crosses the St. Lawrence toward the mouth of the Saguenay, and black clouds are lowering in the sky as we glide to the landing at Tadousac. An hour or two is spent here, and the opportunity afforded to visit the oldest church in America north of Florida, which Tadousac contains among its other curiosities. It is a frame building, on a high, alluvial bank, and the interior, as we see it lighted by one small taper, appears scarcely more than thirty feet square. A handsome altar is placed in an octagon alcove in the rear, with altar-pieces symbolizing the crucifixion; and the walls

POINT NOIR, TRINITY ROCK, AND CAPE ETERNITY, SAGUENAY RIVER.

are adorned with two pictures, one a scriptural scene, the other a portrait of the first priest who visited Canada.

It is not surprising that the Saguenay, with its massive, desolate scenery, should have inspired early mariners with terror. To them it was a river with marvellous surroundings, with an unnavigable current, immeasurable depths, terrible hurricanes, inaccessible and dangerous rocks, and destructive eddies and whirlpools. It is formed by the junction of two outlets of St. John's Lake, which lies in the wilderness, one hundred and thirty miles northwest of Tadousac, and covers five hundred square miles of surface. At some distance below the lake the river passes over cliffs in several magnificent

St. Louis Island, from the West Bank of the Saguenay.

cascades, rushing between rocky banks from two hundred to one thousand feet high; and for a distance of sixty miles from the mouth it is about one mile wide. In some parts soundings cannot be found with three hundred and thirty fathoms; and at all points the water is exceedingly deep, presenting an inky-black appearance.

During the night the steamer has threaded its way through the hills, and in the morning we arrive at a little village in Ha-ha Bay, the nominal head of navigation. The scenery is less massive and sullen here than at any other point, and the character of the crowd at the landing is diversified in the extreme. We soon resume our journey down the dark river. Ha-ha Bay, with its shrubbery and beaches, fades from sight;

Point Crêpe, near the Mouth of the Saguenay.

we are sailing between two towering walls of rock, so dreary, so desolate, that those of us who are impressionable become dejected and nervous. Nature has formed it in her sternest mood, lavishing scarcely one grace on her monstrous offspring. Wherever a promontory juts out on one side of the river, a corresponding indentation is found upon the opposite shore; and this has been made the basis of a theory that the chasm through which the black waters flow was formed by an earthquake's separation of a solid mountain. Occasionally an island lies in our path, but it is as rugged and barren as the shore, formed out of primitive granite, offering no relief to the terrible monotony that impresses us. And, once in a while, a ravine breaks the precipitous walls, and exposes in its darkling hollow the white foam of a mountain-torrent.

Presently we near Trinity Rock and Cape Eternity. As these two capes are accounted among the grandest sights of the voyage, there is a flutter of anticipation among the passengers, and the decks are crowded again. A slight curve brings us into Trinity Bay, a semicircular estuary, flanked at the entrance by two precipices, each rising, almost perpendicularly, eighteen hundred feet above the river. The steepest is Trinity, so called because of the three distinct peaks on its northern summit; and that on the other side is Cape Eternity. Trinity presents a face of fractured granite, which appears almost white in contrast to the sombre pine-clad front of Eternity. For the rest of the day we are toiling through like wildernesses of bowlders, precipices, and mountains. We bid adieu to Trinity and Eternity at Point Noir, thread the desolate mazes of St. Louis Island, and soon are passing Point Crèpe, where the rocks, the everlasting rocks, look in the distance like the channel of a dried-up cataract. Toward night we are in the St. Lawrence again, and as we speed across the brighter waters the moon is rising over Murray Bay, and the wreck of a canoe reposing on the low beach reminds us of the desert through which we have passed. Here our trip down the St. Lawrence comes to an end. The great river broadens from this point until it loses itself in the gulf of the same name, whence its waters pass to join those of the mighty ocean—which is alike the source and final refuge of all rivers. It is not to be wondered at that the early navigators should have been persuaded, by the wide and spacious entrance of the Gulf of St. Lawrence, into a hope that they had at last found the long-sought northwest passage. They had hoped to gain untold wealth, but they were more successful, for they acquired an everlasting reputation.

Montreal Island.

THE CAÑONS OF THE COLORADO.

WITH ILLUSTRATIONS BY THOMAS MORAN.

Bonita Bend.

NONE of the works of Nature on the American Continent, where many things are done by her upon a scale of grandeur elsewhere unknown, approach in magnificence and wonder the cañons of the Colorado.

The Colorado is formed by the junction of the Grand and Green Rivers in the eastern part of Utah. The cañons begin very soon after leaving the railway, and in the series named are Flaming Gorge, Kingfisher, and Red Cañons, Cañon of Lodore, Whirlpool and Yampa Cañons, Cañon of Desolation, Gray, Labyrinth, Still-water, Cataract, Narrow, Glen, and Marble Cañons. Each has some characteristic, which in most instances is indicated by the name. There is generally no break in the walls between the different cañons, the divisions being marked by changes in their geological structure. The cañons whose names above precede Cataract are on Green River before it joins the Grand.

Labyrinth is one of the lower cañons of the Green River. It is a wide and beautiful cañon, with comparatively low walls, but perpendicular and impassable. Indeed, from Gunnison's Crossing, one hundred and sixteen miles above the junction of the Grand and Green, to the running out of the Grand Cañon, there are only two places, and they are not more than a mile apart, where the river and its chasm can be crossed. At one point in Labyrinth Cañon the river makes a long bend, in the bow of which it sweeps around a huge circular *butte*, whose regular and perpendicular walls look as though they might have been laid by a race of giant craftsmen. At a distance the pile called Bonita Bend resembles a vast, turret-shaped fortress.

The Grand and Green Rivers meet in a gorge more than two thousand feet deep; and here the cañons of the Colorado begin.

The first is called Cataract Cañon. It is about forty miles long. The descent of the river through this cañon is very great, and the velocity acquired by the current is sometimes equal to the speed of the fastest railroad-train. Great buttresses of the walls stand out into the rushing flood at frequent intervals, turning the rapid current into boiling whirlpools, which were encountered by the adventurous boatmen with great peril and

GLEN CAÑON.

labor. At the foot of Cataract Cañon, the walls of the chasm approach each other, and, for a distance of seven miles, the water rushes through Narrow Cañon at the rate of forty miles an hour.

At the end of Narrow Cañon, the character of the gorge changes, and, from that point to the place where the Paria River enters the Colorado, a distance of a hundred and forty and a half miles, it is called Glen Cañon. At the mouth of the Paria, a trail leads down the cliffs to the bottom of the cañon on both sides, and animals and wagons can be taken down and crossed over in boats. The Indians swim across on logs.

A mile above the Paria is the Crossing of the Fathers, where Father Escalante and his hundred priests passed across the cañon. An alcove in this cañon, which the artist has drawn, illustrates the general character of the walls, and the scenery from which the cañon takes its name. The smooth and precipitous character of the walls of Glen Cañon is well shown in the illustration. The chasm is carved in homogeneous red sandstone, and in some places, for a thousand feet on the face of the rock, there is scarce a check or seam.

Buttresses of Marble Cañon.

The most beautiful of all the cañons begins at the mouth of the Paria, and extends to the junction of the Little Colorado, or Chiquito, as it is called by the Indians. This part of the gorge is named Marble Cañon and is sixty-five and a half miles long. The walls are of limestone or marble, beautifully carved and polished, and the forms assumed have the most remarkable resemblances to ruined architecture. The colors of the marble are various—pink, brown, gray, white, slate-color, and vermilion. The beautiful forms, with a suggestion of the grand scale on which they are constructed, are given by the two views in this cañon, which the artist has drawn. But it is only on large canvas, and by the use of the many-tinted brush, that any reproduction can be made, approaching truthfulness, of the combination of the grand and beautiful exhibited in the sculpturing, the colors, and the awful depth, of Marble Cañon.

MARBLE CAÑON.

The Marble Cañon runs out at the junction of the Chiquito and Colorado, at which point the Grand Cañon begins. The head of the Grand Cañon is in the north-eastern part of Arizona, and it runs out in the northwestern part, lying wholly within that Territory. Its general course is westerly, but it makes two great bends to the south. It is two hundred and seventeen and a half miles long, and the walls vary in height from four thousand to six thousand two hundred and thirty-three feet. It is cut through a series of levels of varying altitudes, the chasm being deepest, of course, where it passes through the highest. There are in the cañon no perpendicular cliffs more than three thousand feet in height. At that elevation from the river, the sides slope back, and rise by a series of perpendicular cliffs and benches to the level of the surrounding country. In many places it is possible to find gorges or side-cañons, cutting down through the upper cliffs, by which it is possible, and in some instances easy, to approach to the edge of the wall which rises perpendicularly from the river. At three thousand feet above the river, the chasm is often but a few hundred feet wide. At the highest elevation mentioned, the distance across is generally from five to ten miles.

At various places the chasm is cleft through the primal granite rock to the depth of twenty-eight hundred feet. In those parts of the cañon, which are many miles of its whole extent, the chasm is narrow, the walls rugged, broken, and precipitous, and the navigation of the river dangerous. The daring voyagers gave profound thanks, as though they had escaped from death, whenever they passed out from between the walls of granite into waters confined by lime or sandstone. Mr. Moran has drawn a section of these granite walls, showing some of the pinnacles and buttresses which are met at every turn of the river. The waters rush through the granite cañons at terrific speed. Great waves, formed by the irregular sides and bottom, threatened every moment to engulf the boats. Spray dashes upon the rocks fifty feet above the edge of the river, and the gorge is filled with a roar as of thunder, which is heard many miles away.

Fortunately, the wonders of the Grand Cañon can now be seen without incurring any of the peril, and but little of the hardship, endured by Major Powell and his companions. The writer of this, and Mr. Moran, the artist, visited two of the most interesting points in the cañon in July and August, 1873. We travelled by stage in hired vehicles—they could not be called carriages—and on horseback from Salt-Lake City to Toquerville, in Southwestern Utah, and thence about sixty miles to Kanab, just north of the Arizona line. Quite passable roads have been constructed by the Mormons this whole distance of about four hundred miles. At Kanab we met Professor A. H. Thompson, in charge of the topographical work of Major Powell's survey, and, with guides and companions from his camp, we visited the cañon.

Our first journey was to the Toroweap Valley, about seventy miles. By following down this valley we passed through the upper line of cliffs to the edge of a chasm cut

WALLS OF THE GRAND CAÑON.

in red sandstone and vermilion-colored limestone, or marble, twenty-eight hundred feet deep, and about one thousand feet wide. Creeping out carefully on the edge of the precipice, we could look down directly upon the river, fifteen times as far away as the waters of the Niagara are below the bridge. Mr. Hillers, who has passed through the cañon with Major Powell, was with us, and he informed us that the river below was a raging torrent; and yet it looked, from the top of the cliff, like a small, smooth, and sluggish river. The view looking up the cañon is magnificent and beautiful beyond the most extravagant conception of the imagination. In the foreground lies the profound gorge, with a mile or two of the river seen in its deep bed. The eye looks twenty miles or more through what appears like a narrow valley, formed by the upper line of cliffs. The many-colored rocks in which this valley is carved, project into it in vast headlands, two thousand feet high, wrought into beautiful but gigantic architectural forms. Within an hour of the time of sunset the effect is strangely awful, weird, and dazzling. Every moment until light is gone the scene shifts, as one monumental pile passes into shade, and another, before unobserved, into light. But no power of description can aid the imagination to picture it, and only the most gifted artist, with all the materials that artists can command, is able to suggest any thing like it.

Our next visit was to the Kai-bal Plateau, the highest plateau through which the cañon cuts. It was only after much hard labor, and possibly a little danger, that we reached a point where we could see the river, which we did from the edge of Powell Plateau, a small plain severed from the main-land by a precipitous gorge, two thousand feet deep, across which we succeeded in making a passage. Here we beheld one of the most awful scenes upon our globe. While upon the highest point of the plateau, a terrific thunder-storm burst over the cañon. The lighting flashed from crag to crag. A thousand streams gathered on the surrounding plains, and dashed down into the depths of the cañon in water-falls many times the height of Niagara. The vast chasm which we saw before us, stretching away forty miles in one direction and twenty miles in another, was nearly seven thousand feet deep. Into it all the domes of the Yosemite, if plucked up from the level of that valley, might be cast, together with all the mass of the White Mountains in New Hampshire, and still the chasm would not be filled.

Kanab Cañon is about sixty miles long, and, by following its bed, one can descend to the bottom of the Grand Cañon. It is a very difficult task, requiring several days' severe labor. We were forced, by lack of time, which other engagements absorbed, to abandon the undertaking The picture drawn by the artist of a pinnacle in one of the angles of the Kanab is from a photograph taken by Mr. Hillers. The pinnacle itself is about eight hundred, and the wall in the background of the illustration more than four thousand feet in altitude. A railroad is projected from Salt-Lake City to the southern settlements, and, when it is constructed, some of the most remarkable portions of the Grand Cañon of the Colorado will be as accessible as the valley of the Yosemite.

KANAB CAÑON.

WASHINGTON AND ITS VICINITY.

WITH ILLUSTRATIONS BY W. L. SHEPPARD.

The Capitol, from the Botanic Gardens.

THE site chosen by the first Congress for the capital of the United States, and christened by the name of the first President, is a broad plateau, which, on the eastern side, rises to a graceful elevation, and is bounded on two sides by the river Potomac and its tributary called the "Eastern Branch." The main portion of the city, including its business quarter, its public buildings, its main thoroughfares, and its aristocratic residences, stands upon a rather level plain, terminated at the rear by a series of wooded and irregular hills; while the Capitol rears itself upon a sloping elevation, and overlooks a wide extent of country.

Washington was not until recent years celebrated for its beauty. Formerly it was an unattractive place, composed in large part of low and mostly wooden buildings, with streets ill-paved and little cared for. But now the national metropolis—thanks to liberal expenditures and a newly-born pride in the government that its seat should be worthy of its distinction—presents an aspect not only of prosperity, but of sights agreeable to the eye and mostly in good taste. Its adornment has demonstrated that its natural advantages were greater than had been supposed; and the seeker after the picturesque may find ample opportunity to gratify his quest while observing, at "magnificent distances," the official palaces which have been erected at the service of the republic.

The most striking object at Washington is undoubtedly the magnificent white-marble Capitol, a glimpse of which is caught as the city is approached by rail from Baltimore. It rises majestically far above all surrounding objects, amid a nest of thick and darkly verdant foliage, on the brow of the hill to which it gives its name; its very lofty dome, with its tiers of columns, its rich ornamentation, and its summit surmounted by the colossal statue of Liberty, presents a noble appearance, and may be seen for many miles around; while its broad, white wings, low in proportion to the dome, give an idea of spaciousness which no palace of European potentate surpasses. There are few more beautiful—though there are many larger—parks in the United States than that which surrounds the Capitol. The edifice is approached through an avenue entered by high iron gates; on either side of this are beautiful flower-plots, paths shaded by arching branches, fountains, and copses. A double tier of green terraces is ascended before the base of the

Capitol, Western Terrace.

Capitol is reached; then you find yourself on a broad marble terrace, semicircular in form, with a large fountain beside you, whence you may see the silvery windings of the Potomac miles away, disappearing at last amid the abundant foliage where the Maryland and Virginia coasts seem to blend in the far distance. From this lookout you may discern every part of the metropolis. In the midst of the mass of houses rise the white-marble Post-Office Department and the yet handsomer Patent-Office just beside it. Some distance farther on is to be descried the long colonnade of the Treasury, and the top of the White House, just beyond, peeps from among the crests of flourishing groups of trees. More to the left are seen the picturesque, castle-like, red-sandstone towers and turrets of the Smithsonian Institution, standing solitary on a broad plain

In the White-House Grounds.

already sprouting with young foliage. Between the Smithsonian and the creek the conspicuous shaft of the Washington Monument, a noble marble pyramid towering skyward, meets the view; while the eye, spanning the Potomac, may catch sight, in the distance, of that lordly old manor-house of Arlington, identified, in very different ways, with the earlier and later history of the country. Georgetown Heights form the far background in the west; more to the north the picturesque hills, with their wild, straggling growths, which, from the main suburbs and sites of suburban residences of the city, form a striking framework to the scene. A small park also stretches out at the rear of the Capitol, on the east. This presents, however, nothing notable in scenery, its chief adornment being the sitting statue of Washington, in Roman costume, which has been so sharply criticised and so warmly defended. Just outside the limits of this park stands the "Old Capitol," a quaint brick building used by Congress when the Capitol was burned by the British in 1814, in which Calhoun died, and which was used as a prison during the civil war.

At the opposite end of the city from the Capitol is the group of departments surrounding the presidential mansion, and enclosing with it pleasant, umbrageous parks and grounds. On one side are the Treasury and State Departments; on the other, the rather plain, old-fashioned, cosey-looking War and Navy Departments—oddly enough the most

Smithsonian Institution, near White-House Grounds.

placid and modest of the Washington purlieus. The White House is situated midway between these two groups of edifices, and is completely surrounded by open and ornamental spaces. In front of its high, glaringly white portico, with its *porte cochère*, is a lawn that reaches to the thoroughfare. Beyond is Lafayette Square, thickly planted with trees, among which stands Clark Mills's equestrian statue of Jackson and the newer statue of Lafayette surrounded by his compatriots, while on the sides are elegant residences occupied by senators, diplomats, cabinet ministers, and wealthy bankers. The most picturesque view of the White House, however, is from its rear. The front is not imposing. At the back, a small but beautiful park, profusely adorned with plants and flowers,

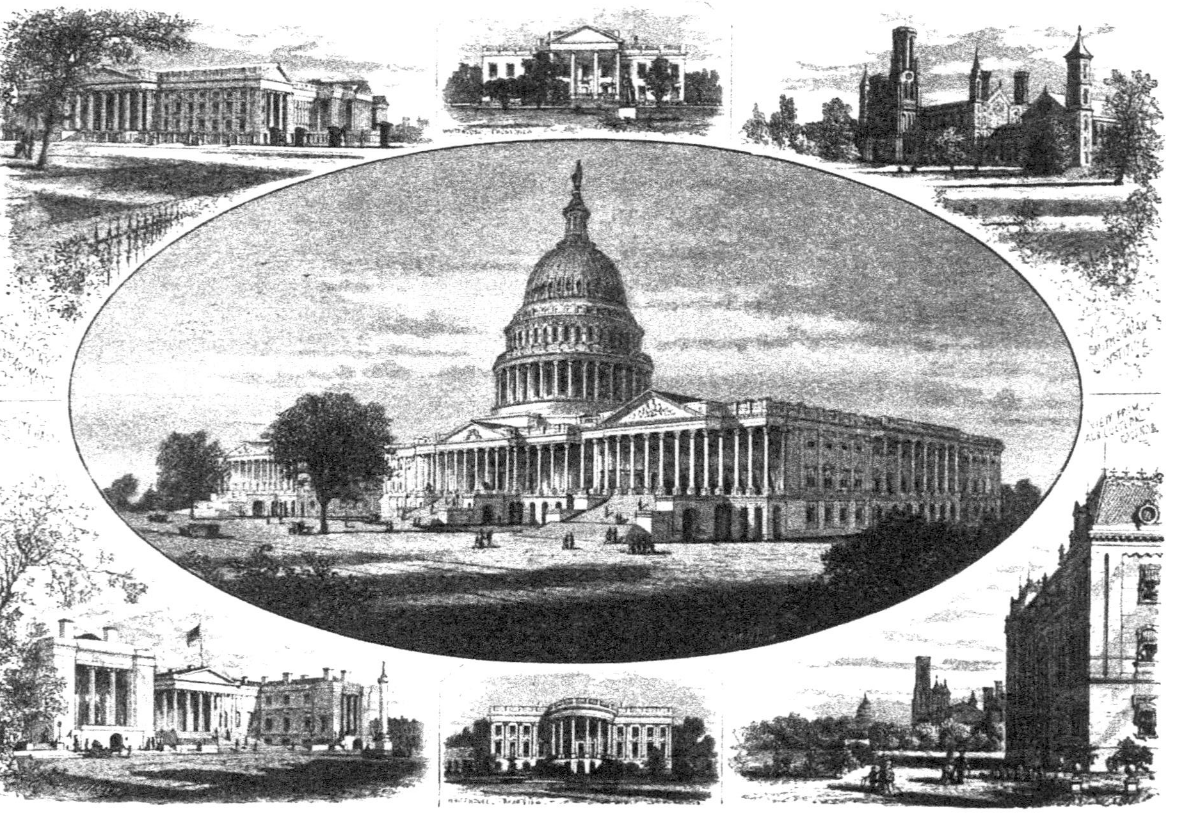

PUBLIC BUILDINGS IN WASHINGTON.

varied by artificial hillocks, and spread with closely-trimmed lawns, stretches off to a high-road separated from it by a high wall. This park is open to the public; and the chief magistrate and his family may enjoy its cheerful prospect from a handsome circular portico supported by high, round pillars, with solid arches below, and a broad stone staircase winding up on either side, fairly overgrown with ivy and other clinging parasites. The most prominent object seen from these "President's grounds" is the red Smithsonian Institution, which from here seems a very feudal castle set down amid scenes created by modern art. Beyond the presidential mansion and the cluster of department buildings Pennsylvania Avenue stretches over a flat and comparatively sparsely-settled district, until, by a sudden turn, it leads to the ancient, irregular, and now rather uninteresting town of Georgetown. Its former commercial bustle has departed from it; for Georgetown is older than its larger and more celebrated neighbor, and was once the third or fourth river port in the United States. It is still,

View from Red Hill, back of Georgetown.

however, a more picturesque place than Washington; built mostly on hills, which rise above the Potomac, affording really beautiful views of the river and its umbrageous shores. The town has many of those substantial old red-brick mansions where long ago dwelt the political and social aristocracy, and which are to be found in all Virginia and Maryland towns of a century's age, surrounded often with high brick walls, approached by winding and shaded avenues, sometimes with high-pillared porticoes, and having over the doors and windows some attempt at modest sculptured ornamentation. From Red Hill, which rises by pretty slopes at the rear of Georgetown, a fine view is had of the wide, winding river. The Potomac, just below, takes a broad sweep from west to east; and at the place where it is spanned by the famous Long Bridge, over which the troops

Glimpse of Georgetown, from Analostan Island.

passed from Washington to their defeat at Bull Run, it seems to form almost a lake. Washington itself is descried between the trees from the east of Red Hill; in the dim distance the shore of Maryland, lofty in places, and retreating southeastward; and, on the immediate right, the attractive Virginian shore, with a glimpse of the historic Arlington.

Now the Potomac is just below you, its stream not so turbidly yellow as it becomes farther down. The Capitol and the monument, both white and majestic, loom high above the city, the rest of which seems a confused mass of houses and spires; verdant meadows, pastures, and natural lawns, sweeping down by gentle inclinations beneath elms and oaks, are seen on the shore you are approaching; while quite near at hand the portico of Arlington rises on the summit of a higher slope, embedded in the richest Virginian foliage.

GREAT FALLS OF THE POTOMAC.

The Potomac, for several miles north as well as south of Washington, is bordered by attractive landscapes. One of the pleasantest walks in that vicinity is from Georgetown northward along the banks of the canal, with the artificial water-course on one side, and the broad, winding, and here rather rapid river appearing every moment on the other. A mile from Georgetown by this road, you never would imagine that you were in so close a proximity to one of the "centres of civilization." The scenery is wild, almost rugged. A profusion of brush and shrubbery mingles with the forest-trees along the banks, which rise in continual and irregular elevations; there are few habitations, and such as there are recall the former social status of the border States. After pro-

Arlington Heights, from Grounds in the National Observatory.

ceeding thus about three miles you reach Little Falls, which have no other pretensions to distinction than that they are surrounded by very attractive scenery, and form a modest cataract winding in and out among the rocks which here encounter the stream. Over Little Falls is a high bridge, by which one passes in a minute or two from Maryland into Virginia. Piled-up rocks line the shore, and anglers from the metropolis may often be found perched upon them, enjoying the very good fishing which the spot provides. Great Falls, as falls, are more pretentious than Little Falls; they are situated a short distance above. Here the water foams and rushes among jagged rocks, forming numerous cascades and pools as it hastens on. In this region the Potomac has become

LOOKING DOWN THE POTOMAC, FROM THE CHAIN BRIDGE.

a comparatively narrow stream, with limpid and rapid waters; and all along its course, as far as Harper's Ferry, its valley presents a varied, unkempt scenery, which makes the jaunt along its shores a thoroughly pleasant one.

But on the Potomac below Washington, where it is now broader and slower in motion, the aspects are perhaps more worthy of inspection, both because Nature here is more genial and more cultivated, and because at every step there is a reminder of some historical scene, old or modern. Passing down by the steamboat, less than an hour brings you, between verdant, sloping banks dotted by well-to-do-looking and for

Fort Washington.

the most part venerable country houses, to the landing-place, whence you reach Mount Vernon. It is unnecessary to describe this home of Washington, so familiar to every citizen by description if not by sight.

On either side of the river are Forts Washington, Foote, and other strongholds, familiar to the history of the war of the rebellion. The outlook northward from Fort Foote is especially fine, comprehending the view at its widest, bay-like expanse, and bringing into clear relief the city of Washington, with the bright dome still dominating all surrounding objects; while the shores in the immediate foreground are composed of gentle cliffs crowned with the rich growths of that Southern clime.

THE VALLEY OF THE CONNECTICUT.

WITH ILLUSTRATIONS BY J. DOUGLAS WOODWARD.

AMONG the hills of New Hampshire and Vermont the Connecticut River takes its rise. Flowing in a nearly southerly direction for four hundred miles, it forms the dividing line between the two States in which it had its birth. Crossing the States of Massachusetts and Connecticut, it empties into the Long Island Sound. Through

Saybrook.

this charming valley we now propose to pass, from the mouth of the river to its northern head, near Canada.

Leaving the train at the ancient town of Saybrook we make our way on foot to Saybrook Point, a distance of not over two miles. The walk will not be found a wearisome one, for the venerable elms beneath which we pass will remind us of the olden times, and there will be enough of the antique meeting our eye to carry us back to the times of Lord Say and Seal and Lord Brook, who lived in the unsettled period of the reign of Charles I. Our walk has brought us to a gentle rise of land, from which we get a distinct view of Long Island Sound. On our right is a cemetery, where we read the simple inscription, "Lady Fenwick, 1648;" and we are informed that she was the wife of General Fenwick, the commandant of the fort erected not far from this spot. Another item of historic interest also comes to our notice: Saybrook Point had the honor of being selected as the site for the collegiate school which afterward became Yale College.

Leaving Saybrook—a place around which cluster so many venerable associations—

we begin our ascent of the river. A sail of thirty miles brings us to Middletown, a partial view of which our artist has given us, the sketch having been taken above the city. As seen from the top of Judd Hall, one of the buildings of the Wesleyan University, Middletown presents a most attractive appearance, with its streets of generous width, adorned with shade-trees and many elegant mansions and public buildings. The Methodists have here one of their earliest and most flourishing seats of learning in the country, founded in 1831, and under the care of that denomination is taking high rank

Mouth of Park River.

among the colleges of the land. Some of its buildings—the Memorial Hall and Judd Hall—are especially fine.

Opposite Middletown are the famous freestone quarries, from which some of the most stately and costly buildings in New York and other cities have been erected. According to tradition, the rocks at the northern and principal opening originally hung shelving over the river.

The level tracts north of Middletown will not be overlooked by the tourist. These meadow-lands, which are found all along the Connecticut, are exceedingly fertile; and some of the finest farms in the New-England States have been formed out of this land. It was this very rich soil that attracted the attention of the early settlers of the State, and brought to Connecticut some of the best blood of the Plymouth and Massachusetts colonies. Above Middletown, a few miles, is Wethersfield, claimed by some to be the

THE CONNECTICUT, ABOVE MIDDLETOWN.

oldest settlement in the Commonwealth. It is a venerable, staid old place, long celebrated for the cultivation of the onion. It is also the seat of the State prison.

We are now approaching one of the most charming cities in our country—the city of Hartford. The scenery all about it is of a very picturesque character. Its banks are among the most beautiful levels on the river, and indicate at a single glance that they must be a mine of agricultural wealth to the cultivators of the soil. The original name of the place did not carry with it the euphony which usually characterizes the old Indian names, it being called Suckiaug. The story of the hardships of its early settlers is a familiar one. Dr. Trumbull tells us that "about the beginning of June, 1635, Mr. Hooker, Mr. Stone, and about one hundred men, women, and children, took their departure from Cambridge and travelled more than a hundred miles through a hideous and trackless wilderness to Hartford. They were nearly a fortnight on their journey." But we can not follow the fortunes of these adventurers.

Hartford, from East Side of the River.

HARTFORD. FROM COLT'S FACTORY.

The very irregularity of the laying out of Hartford adds to its charms. It is divided at the south part by Mill or Little River, two bridges across which are seen in

Stone Bridge, Hartford.

the accompanying sketches. We present also a sketch of Terrace Hill, in the City Park, one of the most beautiful spots in the city. Here has since been erected the

Terrace Hill, City Park, Hartford.

Capitol of the State of Connecticut, one of the most costly and elegant structures of its kind in New England. Hartford is celebrated as being the seat of some of the best charitable institutions in the United States, prominent among which are the Asylum for the Deaf and Dumb and the Retreat for the Insane.

Half a mile in a southwesterly direction from the centre of the city, on a most sightly spot, is the Retreat for the Insane. Its founders showed their good taste in selecting this place for an institution which, of all others, should be so situated as to

Main Street Bridge, Hartford.

secure for its inmates every thing that can charm and soothe a disordered mind. From the top of the building the eye ranges over a scene of rare beauty. The view of the Connecticut Valley in both directions, north and south, is very extensive, and embraces some of the choicest scenery on the river. Looking west, we see numerous villages, in which are found forest-trees and orchards, beneath whose grateful shade nestle cottages and farm-houses, the very sight of which awakens in the mind most gentle and soothing emotions, making us fancy, for the moment, that into such a paradise sin and sorrow have not found their way.

Any allusion to Hartford without reference to the famous "Charter Oak" would be like the play of "Hamlet" with the character of *Hamlet* left out. This famous tree, now no longer standing, occupied an eminence rising above the south meadows, not far from the ancient mansion of the Wyllys family. Like the great elm on Boston Common, its age is unknown, the first settlers of Hartford finding it standing in the maturity of its growth. The cavity in which the charter was hid was near the roots, and large enough, if necessary, to conceal a child. In December, 1686,

Windsor Locks, Connecticut River.

Sir Edmund Andros, who had been appointed the first governor-general over New England, reached Boston, from which place he wrote to the authorities of Connecticut to resign their charter. The demand was not complied with. The Assembly met in October, and the government continued according to charter until the last of the month. About this time Sir Edmund, with his suite and more than sixty regular troops, came to Hartford, where the Assembly were sitting, and demanded the charter, and declared the government under it to be dissolved. The Assembly were extremely reluctant and slow with respect to any resolve to bring it forth. The important affair was debated and kept in suspense until the evening, when the charter was brought and laid upon the table where the Assembly were

SCENES AT SPRINGFIELD.

sitting. Suddenly the lights were extinguished, and Captain Wadsworth, of Hartford, in the most silent and secret manner carried off the charter, and hid it in a large hollow tree fronting the house of Hon. Samuel Wyllys, then one of the magistrates of the colony. The candles were officiously relighted, but the patent was gone, and no discovery could be made of it or of the person who carried it away. The "Charter Oak"

Mount Holyoke.

was cherished as an object of veneration and affection by the inhabitants of Hartford for several generations. In 1856, weakened by age and decay, it fell before the blasts of a severe storm.

As we journey on up the valley of the Connecticut we do not lose our impression of the wonderful beauty of the extensive meadows and the indescribable charms of the neighboring and overshadowing hills. Had we time we would be glad to linger for a few hours in the ancient town of Windsor, settled as early as thirteen years after the landing of the Pilgrims at Plymouth. We must pause for a few moments at Springfield, and from the cupola that crowns one of the United States buildings, on Arsenal Hill, survey the scene. Rich alluvial meadows stretch far away in the distance along the river, rising gradually to quite an elevation, and terminating in a plain reaching several miles east. Lofty hills rear their heads in all directions, clothed in the summer with the richest verdure. Villages and farm-houses everywhere meet the eye, while the busy city is spread out like a map at our feet. And this is the Agawam of the olden times, when the wild Indian roamed over this splendid country, whose name—Springfield—was given to it as far back as 1640. It is the site of one of the most extensive armories in the country, and the Government has always a large force of men working in the arsenal. Springfield is full of enterprise and in many respects is the most thriving city on the Connecticut River.

The Connecticut Valley from Mount Holyoke.

Leaving, we pass rapidly over the level lands on the river, catching glimpses at every turn of scenes of singular natural beauty, and observing the many improvements made by man, pressing into service the immense water-power which he finds so useful as the propeller of the vast machinery here set in motion. Chicopee, and especially

The Oxbow—View from Mount Holyoke.

Holyoke, will not fail to attract the attention of the tourist, if with his love of Nature he combines an interest in works which give scope to human industry and minister to the comfort and add to the luxuries of life. The scenery along the river, if possible, grows more charming as we advance. The hills are nearer to the river, and begin to assume the name of mountains. We have reached Northampton, one of the most beau-

Mount Tom from Oxbow.

tiful villages of New England, on the west side of the Connecticut, on rising ground about a mile from the river, between which and the town lie some of the fairest meadow-lands in the world, covering an area of between three thousand and four thousand acres Like Hartford, the town is somewhat irregularly laid out, deriving from this circumstance the great charm of diversity. It abounds in shade-trees, the venerable appearance of which gives evidence of their great age. Few places of its size can boast of a larger number of elegant mansions and villas.

We will cross the river and view the scene from near the edge of a precipitous cliff on Mount Holyoke. The imagination can easily picture the exceeding beauty of the view. The illustration shows the river winding through the meadow-lands, which are of surpassing fertility. Changing our position, we are at the Mountain House, so distinctly seen in the preceding picture. Here we are nearly a thousand feet above the plain below. From this elevated point let us look about us. "On the west, and a little elevated above the general level, the eye turns with delight to the populous village of Northampton, exhibiting in its public edifices and private dwellings an unusual degree of neatness and elegance. A little more to the right, the quiet and substantial villages of Hadley and Hatfield; and still farther east, and more distant, Amherst, with its college buildings on a commanding eminence, form pleasant resting-places for the eye. Facing the southwest, on the opposite side of the river, the ridge called Mount Tom rises one or two hundred feet higher than Holyoke, and divides the valley of the Connecticut longitudinally. In the northwest, Graylock may be seen peering above the Hoosic; and, still farther north, several of the Green Mountains, in Vermont, shoot up beyond the region of the clouds in imposing grandeur. A little to the south of west, the beautiful outline of Mount Everett is often visible. Nearer at hand, and in the valley of the Connecticut, the insulated Sugar-Loaf and Mount Toby present their fantastic outlines, while far in the northeast ascends in dim and misty grandeur the cloud-capped Monadnoc."

The artist has given us another view of the valley from Mount Holyoke, showing a bend in the river which, from its peculiar shape, is known as the Oxbow. We have the same charming scene of meadow and winding river which we had in the previous picture. From Oxbow, also, we have a view of Mount Tom, the twin-brother of Mount Holyoke—not as much visited as the latter, but. worth climbing, and not disappointing the highly-raised anticipations of the tourist. The village of South Hadley lies on the east side of Mount Tom. This place has a national reputation, as being the seat of the famous Mount Holyoke Female Seminary. There are not a few spots in its neighborhood from which a spectator will get most picturesque views of the surrounding country.

South Hadley is in some respects the most beautiful village on the Connecticut.

Mount Holyoke from Tom's Station.

Looking toward the northwest from the bank of the river, Holyoke and Tom may be seen rising with boldness from the valley. They stand on either side of the river like

Titan's Pier, Mount Holyoke.

watch-towers, from whose lofty summits the observer may look out upon some of the most charming scenery in the world. Through the opening made between these twin-

Northampton Meadows.

mountains one can see two or three miles up the river, in which will be noticed one or two islands, looking peaceful enough to make another paradise on earth. Scattered over the meadows are the fine old trees whose summer shadows are so inviting, through whose foliage may be seen Northampton. Directly above the town the Connecticut, changing somewhat its usual course, turns northwest. Making a bend to the south again, it moves on for a little distance and then turns toward the east. South Hadley is famous as having been the residence of Whalley and Goffe, two of the regicides of Charles I., who escaped from England, and in 1664 came to America.

We should be glad to linger about these delightful regions of the Connecticut Valley. In no direction would it be possible for us to move without finding something most attractive to the eye and pleasing to a cultivated taste. A ride of about seven miles east of the river would bring us to Amherst, the seat of Amherst College, founded in 1821, and one of the most flourishing literary institutions in Massachusetts. In the distance, beyond Hatfield and Whately, rises a conical peak of red sandstone, reaching an elevation of five hundred feet from the plain, known as Sugar-Loaf Mountain. Although seemingly inaccessible, it may be ascended on foot without serious difficulty; and the tourist will be amply rewarded for the fatigue of the ascent when he reaches the summit. Rising some seven hundred feet above the plain on which the village of Deerfield stands is Deerfield Mountain. Standing on the western verge of this mountain, one may obtain charming views of the surrounding country. Other elevations, such as Mount Toby and Mount Warner, are worth climbing, and from their summits may be obtained magnificent views, each one of which will have some peculiar charm distinguishing it from all other scenes. We must hasten on, for other attractions are yet to be described before we reach our journey's end.

Table-Rock, Sugar-Loaf Mountain.

We have reached Greenfield, which combines the activity of a manufacturing with the quiet of a rural village of New England. The beautiful elm-shaded streets, and the neat and in many cases elegant and tasteful dwellings, are illustrative of the best style of a New-England village. The artist has given us a view of the valley of the Connecticut as seen from Rocky Mountain, in Greenfield. What images of summer repose are awakened in the mind as we gaze upon the scene on which the eye rests! We see before us a region the capabilities of which are far from having been fully developed, where future generations are to derive from the products of its fertile soil and its busy manufactures the means for their support.

We descend from this Greenfield eminence, therefore, and keep on in our northerly course, passing through Bernardston, and coming to South Vernon, from which we will take the few miles' ride required to bring us to that beautiful New-Hampshire village—Keene. The principal street, a mile long, is an almost perfect level, and is throughout

Sugar-Loaf Mountain, from Sunderland.

its entire length ornamented with what adds so much to the charm of our New-England villages—the fine old trees. Returning from our short circuit, it does not take us long to reach Brattleboro. We are approaching a more rugged portion of the country. We crossed the boundary-line of Massachusetts at South Vernon, and are in Vermont. Brattleboro has the well-deserved reputation of being among the most charming sites on the Connecticut. As a sanitarium it has been resorted to for many years by persons in search of health. The water here is said to be of remarkable purity, issuing cool and most refreshing from the hill-sides. The fine, invigorating air, and the roman-

Connecticut Valley, from Rocky Mountain, Greenfield.

tic scenery which in all directions meets the eye, make this village one to which invalids love to resort. We give a representation of Mount Chesterfield, which presents a singularly regular and unbroken appearance.

Our next stage is twenty-four miles, bringing us to the well-known Bellows Falls. In passing over this stage in our journey we have stopped for a few moments at Dummerston, one of the oldest towns in the State. Near the centre of the town is what is called Black Mountain—an immense body of granite, through which passes a range of argillaceous slate. Our artist has given us a sketch of an old mill in Putney, a few miles north of Dummerston. This village is pleasantly situated on the west bank of the Connecticut River, and embraces within its limits an extensive tract of river level known as the Great Meadows. Sackett's Brook is a considerable stream, which within a distance of one hundred rods falls one hundred and fifty feet. Our route has taken us through Westminster, whose soil has

MOUNT CHESTERFIELD.

made it a particularly fine agricultural region. A semicircle of hills encloses the place, touching the river two miles above and below the town.

Bellows Falls, of which we have three picturesque views, is well known as a railroad centre, and to some extent a place of summer resort. The falls, which afford the chief charm to the place, are a succession of rapids in the Connecticut. These rapids extend not far from a mile along the base of a high and precipitous hill, a partial view of which we have in one of the sketches, which bears the name of Fall Mountain. Standing on the bridge which crosses the river one looks down into the foaming flood below. The gorge at this point is so narrow that it seems as if one could almost leap

Mount Ascutney.

over it. Through this chasm the water dashes wildly, striking with prodigious force on the rocks below, and by the reaction is driven back upon itself for quite a space. In a distance of half a mile the water descends about fifty feet.

Keeping on in our northerly course, we come to Charlestown. At this point there are in the Connecticut River three beautiful islands, the largest—Sartwell's Island—having an area of ten acres, and well cultivated. The other two have not far from six acres each in them.

No lover of the picturesque will fail to see Claremont, a place watered by the Connecticut and Sugar Rivers, having a fine, undulating surface, and surrounded by hills with gentle acclivities, from the summits of which are obtained charming vistas of the surrounding country. From this spot we get fine views of Mount Ascutney, of which the illustration gives us an excellent idea. This mountain is in the towns of Wethersfield and Windsor, and is an immense mass of granite. It is well spoken of as "a brave outpost of the coming Green Mountains on the one hand and of the White Mountains on the other." It is sometimes called the Three Brothers, from its three peaks, which are so distinctly outlined as we look at the mountain from the point of view which the artist has selected.

Windsor is our next point of interest, situated on the elevated bank of the river, somewhat irregularly built, but in all respects one of the most attractive villages of Vermont. Its wide, shaded streets give it a peculiarly attractive appearance; and if one ascends the highlands in the neighboring town of Cornish, or climbs to the top of Ascutney, he will look out upon a scene which he will not soon forget. The location of Windsor is such that it has become the centre of trade both for the towns on the river and for the fertile interior country.

Whetstone Brook, Brattleboro.

We have reached White-River Junction, where the White River empties into the Connecticut, of which the artist has given us a view. It needs but a glance to indicate to us that we are in the midst of the mountains. We can almost feel the invigorating breezes as they blow pure and fresh from the "everlasting hills." The sketch is evi-

Bellows Falls from Distance.

dently intended to represent the evening hour. The new moon hangs over the valley which divides the two mountains in the left of the picture. The wind blows very gently down the mountain-gorge, bending a little to the right the smoke which ascends from the chimney of the cottage in the rear of the bridge. The whole scene is one of quiet beauty.

From White-River Junction we go to Hanover, New Hampshire, the great attraction of which is Dartmouth College, situated about half a mile from the Connecticut. The buildings are grouped around a square, whose area is twelve acres, in the centre of the broad terrace upon which the village has been built. The graduates of this institution have distinguished themselves in all the walks of professional life. Any college from which such men as Daniel Webster and Rufus Choate have gone forth may well pride itself on account of its sons.

The villages of Thetford, Orford, Bradford, and Haverhill may detain us for a few hours. We shall find in all this neighborhood excellent farms, and a busy, industrious population. In Orford, limestone is found at the foot of a mountain some four hundred feet above the Connecticut. Soapstone and granite abound, and some lead has been discovered. Brad-

Old Mill, Putney.

ford and Haverhill were so called because their earlier settlers came from towns of that name on the Merrimac, in Massachusetts. The town of Newbury is delightfully situated on the west side of the Connecticut River, and comprises the tract to which the name of "The Great Oxbow" has been given. This tract, on a bend of the Connecticut River, is of great extent, and is well known on account of its rare beauty and the fertility of its soil. Here we have one of the most charming of the many picturesque scenes which our artist has given us of the Connecticut. From the meadows of Newbury is seen the elevation called Moose Hillock. A few miles north of Newbury we

Bellows Falls.

reach Wells-River Junction. Not far from this point the waters of the Ammonoosuck empty into the Connecticut.

Our last sketch represents a scene in Barnet, Vermont, one of the best farming

The West Branch of Bellows Falls.

towns in the State, and abounding in slate and iron-ore. The water-power on the Passumpsic and Stevens Rivers is one of the finest in all this region. The fall in Stevens River, of which we have a view, is one hundred feet in the short distance of ten rods.

White-River Junction.

Moose Hillock, from Newbury Meadows.

Not very far from here the river Passumpsic discharges its waters into the Connecticut. From this point onward it bears the character of a mountain-stream. There are several pleasant villages on either side of the river, as we follow it up to its very source in the northern part of New Hampshire. The lover of Nature may be sure of finding abundant material to gratify his taste for the sublime and the beautiful all through this most picturesque region.

Stevens Brook, Barnet.

SCENES IN VIRGINIA.

WITH ILLUSTRATIONS BY WILLIAM L. SHEPPARD.

Interior of Natural Tunnel.

PICTURESQUE America may be said to find almost an epitome of itself in the State of Virginia. Her scenery—infinitely varied, beautiful exceedingly, and sometimes truly grand—repeats in her own boundaries features which would have to be sought in places widely separated. Here, indeed, are no Alps, no Matterhorn to tempt Whymper or Tyndall, and no glaciers to study; nor do those works of Nature find a parallel on this side of the Mississippi. But from Harper's Ferry to the farthest southwest corner of the State there is literally a world of scenic beauties, ravishing to the artist, and inviting to even the dullest traveller or sight-seer. Let us glance at a few of the more striking of these mountain-pictures.

The Natural Tunnel, in Scott County, is the first point to which we will conduct the reader. The variety and beauty of the forest-growths constitute the most striking peculiarity of this southwestern portion of Virginia—one might say the only striking peculiarity—and hence, no doubt, the surprise which the Tunnel excites when it is seen, albeit the spectator has been in momentary anticipation of the object of his quest. This

NATURAL TUNNEL

surprise recurs at every visit to the Natural Bridge, and the Tunnel is a similar formation, not so lofty in its arch, but longer and more tortuous in its course through the hill or shoulder of the mountain. In the one case there is a short and nearly straight tunnel; in the other the tunnel is long and very crooked; in both cases the country road runs over the tunnel, the traveller crossing it unawares. Stock Creek, a tributary of the Clinch, whose limpid waters have repeatedly wetted the hoofs of our horses in the zigzag course hither, has forced or found a passage through the ridge which stretches athwart the narrow, deep valley, and in so doing describes what railroad-men would call a "reverse curve," one hundred and fifty yards in length. Thus, although the arch is

New River.

seventy or eighty feet high, the light is intercepted, and, even when the sun is at its zenith, the passage of the Tunnel is attended with difficulties. At other times, when the rising or declining orb lends but a partial and imperfect illumination, the subterranean traveller, plunged in Cimmerian darkness, cannot repress a feeling of genuine horror as he toils through the central portion of the curve, and, as he emerges, hails the sunshine with rapture, exalted and prolonged by the precipices of naked rock ascending sheer three hundred feet above and around him; while higher yet rise the verdurous crests of the forest-crowned summits, and above all bends the intense blue of the welkin.

Leaving the Tunnel—which, after the Natural Bridge, is undoubtedly the most imposing *lusus naturæ* east of the Mississippi River—we retrace our way along the Norfolk and Western Railroad. Around us are mountains of every conceivable shape—all the rounded outlines, all the frightful angles, incident to such scenery; bays and nooks of greenery, reaching far off into coves; vales and chasms; bald knobs, dotted with the skeleton trees; jagged precipices, exposing the unhealed stumps of gigantic mountain-limbs torn off as by seismic violence; mountains lapped and dove-tailed within mountains, range above and beyond range, in seemingly endless succession, wooing us to stop, and flitting all too quickly past as the train flies on.

Debarking at Bristol, we start thence on horseback for a trip down New River, crossing it near the station. The river flows silently here, but with a subtile sort of

Sycamore on New River.

force, between banks lined with sycamores, which trail their branches in the water in many instances. Masses of brown-gray rock lift their heads above the foliage in many places, but the banks soon fall away, and the stream, gliding along the lowlands, divides with its silver breadth the alluvium which the plough has upturned to receive the corn.

We take a short cut athwart a bight or loop of the river, following a narrow path, the main road having been fenced quite across on account of some dispute as to the right of way. At last, regaining the main road, which goes over a ridge adorned with noble timber, we quicken our pace, observing, as we pass rapidly along, that even the local names are misspelled on the half-rotten sign-boards.

We go through five gates in two miles. Again we encounter the river, the road narrowing very much, and winding under steep bluffs; the river still flowing majestically, and the opposite banks getting higher, with no visible outlet for the stream. Now the road runs on the water's margin; and now it mounts far above, and the hoofs of our steeds are level with the tops of the white-and-brown-barred sycamores. Here the water

Great Falls, New River.

glides over ledges or eddies under willows; the mountains become higher and steeper—higher even than on the Hudson in the Highlands—and are thickly clothed with woods. Here and there great streams of loose stones—moraines, most likely—poured out as by a superhuman hand, extend away up the mountain-side. Mountains tower on every hand; there is seemingly no escape for the imprisoned waters, lake-like here, still as

NEW RIVER AT EGGLESTON'S SPRINGS

death, enchanted and asleep. The solitude and grandeur of the scene become oppressive; respiration is almost impeded. We push on. A murmur is heard; it becomes a roar; we turn a corner, and behold—the Great Falls!

The river, half a mile or more in width, foams and dashes over the ledges formed by the peculiar stratification, well shown on the mountain-side in the engraving, with

Anvil Cliff.

great but not unmusical violence in some places; while in others it slides between the huge rocks with a swift, treacherous look, which fascinates the observer. Boats equipped with oars at both ends shoot these dangerous rapids, guided with consummate skill by the boatmen, who are generally negroes. Getting back is a toilsome business, compelling the men frequently to plunge waist-deep in the powerful current in order to push

Purgatory Falls, Head-Waters of the Roanoke.

their boats up by main strength. The delighted visitor may linger long at the Falls; but, our sketching accomplished, we follow the course of the beautiful river, which soon resumes its placidity, although the actual velocity has not been greatly diminished. The scenery is literally magnificent, and of the character already noted, except that at intervals high crags tower above the stream, their gray, russet, and ochreous tints harmonizing admirably with the foliage.

At the point shown in the accompanying illustration, the river, lapsing once more into its lake-like aspect, composes itself into a picture which has an appearance that is almost *studio-like* if considered by the ordinary rules of composition, more striking in color than in form, but still most beautiful—the dreamless, perfect rest, after the strife and contention at the Falls.

Abruptly parting from the river, it being impossible to get along the banks, where cliff after cliff protrudes into the water, we make a circuit of several miles, and come suddenly in sight of the river again. The scene, viewed from the top of a lofty hill opposite Eggleston's or the New River White Sulphur Springs, is most remarkable. High hills enclose the place; back of these are mountains, and back of all the great Salt-Pond Mountain—a slumbering Titan. In the foreground, a hill-top, with gnarled and picturesque trees; beneath, the tranquil, gleaming river, shortly lost to sight in the sombre mountains; and immediately opposite the spectator, the rugged, riven, and weird Anvil Cliff lifts its awful but not repulsive front. Descending the winding pathway, under tall, fantastic rocks, we reach Eggleston's Ferry, and halt in mute admiration of the scene before us. The sketch leaves little to be added by way of description. The banks are lined with trees, mostly sycamores, but there are also some fine elms. Among the former we find a number of curious shapes, an example of which is given in the engraving.

Below the ferry, on the right, looking down the stream, rises the Anvil Cliff, the height of which, ascertained by triangulation, is stated to be two hundred and ninety-six feet—an over-estimate, probably. The cliffs are elevated in immense laminæ, and in a plane generally oblique to the stream; their color sombre gray, with brighter belts and dashes of dirty white; their summits black and riven, capped by twisted and storm-stained cedars. Mighty forest-trees are inserted between the crags; and in certain places the accumulated washings of the stream have formed, at the base of the cliffs, little levels in terraces of lively green, which afford foothold and nourishment to bright-leaved and gracefully-bending maples. At sunset the tops of the cliffs are illumined with brilliant gold or bathed in vivid red, as the character of the evening may be, while all below is enveloped in cool, purplish shadow—a noble and exquisite scene, worthy in form and coloring of the best master in the land.

Inconspicuous in itself and scarcely worthy of such august company, the "Anvil," which gives the name to this stately pile of rocks, is nevertheless much larger than it appears to the eye, being four by nine feet in actual dimensions. Near the foot of the cliff from which the Anvil juts, a stream gurgles between the fallen masses of rock. It is the outlet of a stream, which disappears strangely on the mountain-side in rear of the massive pile. Indeed, the behavior of the water hereabouts is very singular, for it sends up great bubbles continually, and on two occasions, in the last two years, threw itself, geyser-like, full twenty feet into the air.

A rough ride in a wagon brings us back to the railroad, and the train bears us eastward to Alleghany Station. Here the Roanoke River meanders so that it has to be crossed five times before we reach the Alleghany Springs, five miles from which one

Peaks of Otter.

of the streams which form the head-waters of the Roanoke, precipitating itself over a steep ledge, makes what is known as Purgatory Falls. Few approaches to a scene so beautiful are more picturesque. The detached masses of rock which impede and divide the stream are of enormous size, and out of all proportion to the volume of water, though that is by no means small. The large tree-trunks lodged against the huge rocks, which are not bowlders, but irregular solids, tell the fury of the torrent when at its height in rainy seasons. The water falls about seventy feet. Tall hemlocks and maples keep the gorge in a tender half-light, broken at mid-day by glaring rays, that give a magical charm to the place.

Still going eastward, we stop at Liberty, in Bedford County, in order that a sketch may be made of the famous Peaks of Otter. The view is taken a short distance from the village. The peaks have been made familiar by repeated descriptions. Ten miles distant from the village above named, the higher of the two is five thousand three hundred and seven feet above the level of the ocean; and the view from its top is truly magnificent. Eastward stretches an interminable plain, farther than the eye can reach;

Natural Towers.

while to the west a tumultuous sea of mountains extends on and on to the remote horizon. This grand panorama, once seen, can never be forgotten.

From Bedford County to the limestone region of the Valley is our next remove. Here caves and curious formations exist in numbers, surpassed only by the country around the Great Lakes. In Augusta County are the Natural Towers. A glimpse of them is caught in driving down the road that skirts the North River. No cliffs or mountains near at hand suggest the proximity of this wonder. Across the river is seen a plain skirted by a range of wooded hills of moderate height, and just at the foot of this range the Towers rise straight up from the cultivated field. The illusion is perfect; any one would mistake them for a ruined work of human hands. No other rocks are visible. From a distance, the ragged peaks of the Towers are transformed almost without an effort of the imagination into crumbling embrasures and machicolations. The first aspect is that of the large engraving, but, following the road, the observer is brought to the other face, and here the resemblance to a feudal ruin, the curtain-wall, with flanking towers, and low, central archway, is exact. A nearer inspection shows that the inner side of the pile is really attached to the hill-side. The colors are varied

Jump Mountain.

in horizontal bands, and, from the seams which appear at almost equal intervals in their height, the Towers are apparently the result of successive depositions.

Bidding farewell to the Towers, we proceed westward along the Chesapeake and Ohio Railroad to Goshen Pass. A stage hurries us through at night, for we are to sleep at the Rockbridge Baths, visit the Jump Mountain, and return to the Pass. We see the overhanging crags, the high, naked summits, the black masses of foliage, and

NATURAL TOWERS.

hear the melancholy winds soughing in unison with the invisible river rushing far below—that is all. It is simply grand, but we hasten on to the Baths, where we have things all to ourselves, the season not having begun.

Goshen Pass.

Early next morning we take a buggy and are off for Jump Mountain. Thunder-showers drag over the top of the "Jump" as we follow the road, prospecting for a good point of view, and the mountain appears to decide not to allow its portrait to be taken that day.

RAINBOW ARCH.

The western base of the Jump abuts on Goshen Pass, and the ascent on that side is so gradual that even ladies on horseback, during the Springs' season, ride to the edge of the cliff, five hundred feet perpendicular, which abruptly breaks the contour of the mountain. A prodigious stream of *débris*, the result of the forces which escaped the

Clifton Forge.

mountain's face, rolls from the base of the cliff nearly to the foot of the mountain, barring approach on this side.

On the morrow we are promptly at the Goshen Pass and through it—a narrow gorge, the like of which for length and depth is not in all Virginia, for it extends nearly nine miles between its frowning walls! At its southeastern entrance a spring of sulphur-water gushes out of a rock in the middle of the stream which traverses this

Cyclopean gorge. The river-waters, pure and sweet, flow around the Acherontic pool, as if shunning contact with a liquid of so infernal a savor that it is perceptible at a great distance.

And now we are fairly within the Pass. Words are of little use, and even the pencil fails, for that can give but one side at a time of this gigantic and horrible chasm. Overhanging crags, black and blasted at their summits, or bristling with stark and gnarled pines, tower in places into the very heavens, six, seven, eight hundred feet above the stream. Lower down, monstrous rocks threaten to topple and crush the foolhardy wayfarer who ventures beneath their dreadful masses. The roadway is in places walled up from the stream, which flashes deep down beneath him.

Quitting reluctantly the Pass, we are whirled along the new highway to the West by the Chesapeake and Ohio Railroad, over vast embankments, through yawning tunnels, and all along by delicious bits of scenery. The mountains close in as if to bar the way, then flit behind, displaying quiet meadows and charming vales. Clifton Forge is our destination. We arrive as the mists, slowly assembling in the hollows, begin to crawl to their rendezvous on the mountain-side. Looking into the gap, a single glimpse, the blue is of an intensity which the artist would hardly dare to put on his canvas.

Jackson's River, flowing between the sundered mountains, unites two miles below with the Cow-Pasture, to form the historic James. The stratification here is most rare and strange, describing the arc of a circle, and the contour of the opposing faces on the two sides of the river being so perfectly true that a projecting rock on the one side has, exactly opposite, the recess from which it was apparently torn.

The arch rises two hundred feet above the level of the stream, and is known as the Rainbow Arch. The whole scene is lovely. Graceful trees drooping over the clear water, an abandoned furnace, and the ruined piers of a long-swept-away bridge, add very much to the natural picturesqueness of the place. The view in the Forge Gap, combining the wreck of rocks and the ruins of man's handiwork, with foregrounds, middle distances, and horizon lines, finely balanced everywhere, is surpassingly beautiful. As you look up at the mountains, or along the stream which falls over the dam (built fifty years ago, when the forge was at work), the grandeur and loveliness of the picture bear an ineffaceable impression.

CAYUGA LAKE SCENERY.

WATER-FALLS AT CAYUGA LAKE.

WITH ILLUSTRATIONS BY J. DOUGLAS WOODWARD.

Taghanick Falls.

CAYUGA LAKE is noted for a great number of highly picturesque and beautiful waterfalls found at its southern extremity, in the vicinity of the town of Ithaca, famous as the site of Cornell University. The head of Cayuga lies nearly four hundred feet below the level of the surrounding country, while a remarkable feature of this elevation is a number of ravines and gorges, with an almost endless succession of waterfalls, formed by the primary streams which drain the middle portion of the northern slope of the watershed between Chesapeake Bay and the gulf of the St. Lawrence, their first point of rendezvous being Cayuga Lake.

VICINITY OF ITHACA.

The most northerly of those ravines is Fall Creek, in which, within a mile, there are eight falls, all of them exceedingly fine. The walls of the chasm are abrupt and high, fringed with a dusky growth of forest-trees. Four of the falls range from sixty to thirty feet in height, while a fifth—Ithaca Fall—attains one hundred and fifty feet. In the latter the foaming torrent leaps grandly between the fractured rock. Not far from here we also find the Triple Fall, which pours over the rock in threads, as in a veil of gauze, and is not woven into a mass, as in the Ithaca Fall.

In the vicinity are some curious formations which somewhat resemble the eroded sandstones of Monument Park, Colorado. Here is Tower Rock, a perfect columnar formation, about thirty-six feet high, with a sort of groove across the top. A still more extraordinary monument is Castle Rock, which has a certain regularity of form despite its unusual character. It consists of a massive wall, with a magnificent arched door-way. In the arch of the door-way there is a deep slit, whence spring two sturdy trees, their slender trunks appearing bleak and lonely in their exposed situation.

About a mile and a half south of Fall Creek is Cascadilla Creek, smaller than the former, but more delicate and harmonious in its scenery. Between the two ravines, its chimes mingling with their babble, the university is situated, on a fair expanse nearly four hundred feet above the level of the lake. The principal buildings are ranged on the summit of a hill, which slopes gently, and rises again in richly-scented fields of clover and wild-flowers.

Six miles from Ithaca, in a southwesterly direction, is Enfield Falls, a spot of great interest on account of the unusual depth which a stream has furrowed into the earth. The water reaches the main fall through a narrow cañon a hundred feet deep, and then tumbles down, almost perpendicularly, a hundred and eighty feet, into a chasm whose walls rise three hundred feet on each side. The torrent leaps six times over the protruding rock before it reaches the foot and proceeds on its way in comparative calm. The stream in the main fall of Buttermilk Ravine also issues from a deep channel, with jutting and somewhat steep walls. In this ravine there is another of those fanciful stone monuments which we have referred to.

But the most impressive of the water-falls about Cayuga Lake is the Taghanick. It is more than fifty feet higher than Niagara. The water breaks over a clean-cut table-rock, and falls perpendicularly two hundred and fifteen feet. Except in flood-time, the veil of water breaks, and reaches the bottom in mist and sheets of spray. The rugged cliffs through which the stream rolls before it makes its plunge are about two hundred feet in depth, and form a triangle at the brink of the fall.

A GLANCE AT THE NORTHWEST.

WITH ILLUSTRATIONS BY ALFRED R. WAUD.

WISCONSIN people submissively listen to a great deal of random talk about lone backwoods and prairie-wastes; but if you should feel weary of the more frequented routes of travel, you cannot do better than make a visit to the picturesque features of these backwoods and prairie-wastes. Go round the great lakes; break the trip at one of the lake ports—say Manitowoc, or Sheboygan—and find your way to the Wisconsin River by the railway

Near Kilbourn City, a sluggish town about half-way between the source and the mouth of the Wisconsin River, touched by the Lacrosse division of the Chicago, Milwaukee, and St. Paul Railroad, you will find Rood's Glen, a bit of scenery that will vividly recall to your memory Havana and Watkins Glens, the structure of which it resembles very closely, as will be seen in our illustration. It is deep-set between walls of soft-looking limestone and moist earth, fissured and wrinkled into many ledges and terraces, which are so near together in some parts as to almost form a cavern. The bottom is smooth

In Rood's Glen.

DEVIL'S DOOR-WAY, DEVIL'S LAKE, WISCONSIN.

Cleopatra's Needle, Devil's Lake, Wisconsin.

and sandy, covered with a shallow pool, which reflects the bright greenery of the trees and grass that are twisted and interlocked into a natural arch overhead.

And not many miles from this far-away city of Kilbourn are other scenes not less picturesque. In Barraboo County, in a basin for the most part walled in with abrupt hills, reposes the Devil's Lake, a sheet of water as pretty as its name is repellent. It is a gem of Nature; and, in the autumn, the contrast of its still, emerald-green waters with the rich colors of the foliage and the weird forms of its gray rocks is inexpressibly lovely. Its origin was, without doubt, volcanic, the surrounding cliffs bearing evidences of the action of great heat as well as of frost. Round about, too, are many extraordinary forms. The Devil's Door-way, of which we give an illustration, is characteristic, and from its portals we obtain an excellent view of a portion of the lake, and the serene vale of Kirkwood, with its orchards and its vineyards. Cleopatra's Needle is another

Lone Rock, Wisconsin River.

of the curious monuments of Nature's freaks to which we have alluded. It is an isolated column of rock, nearly sixty feet high, piercing a surrounding bosket at a point where the cliffs are sheer to the bosom of the lake.

Regaining the river, we travel southward, passing Lone Rock, a dot of an island in the mid-stream. It is nearly circular in form, with an area of not many square yards, and its sides have a streaky, corrugated appearance. A score or so of thin, repressed pine-trees do their best to shield its barrenness and be friendly; but it will not be comforted, and stands out bleakly, the current lapping and eddying sadly at its feet. At another point of the river the boundary rocks counterfeit the sterns of four or five steamboats moored together, with their several tiers of galleries one above another. Hereabout the stream straggles through a desolate, wild, melancholy reach of flat land, with low-lying forests of timber around; and the general inclination of the scenery to look like something artificial is again manifest in an opposite rock, the outlines of which hint at the paddle-box of a steamer. In the Dalles we pass through six miles of enchanting beauty. The forms are among the most picturesque that we have yet seen.

Steamboat Rock, Wisconsin River

Some of the rocks rise sharply from the water, and extend outward near their summits, so as to form a sort of shelter for the luxuriant grass that crops out in slender, wavy blades from the shoals. Others are perpendicular from their base to the table-land above, which is richly verdant with grass, and evergreen shrubs and trees. Here there is a narrow slope, bringing leafy boughs to the water's edge; and yonder a shadowy inlet, its entrance hidden by a curtain of delicately colored, seemingly luminous leaves. The shadows on the water are of exquisitely varied hues and forms. The sky, the

STAND ROCK, ON THE WISCONSIN RIVER.

clouds, the leaves, are mingled on the unruffled surface, save where the massive rock intervenes. At the Jaws we move from one spot which we think the most lovely to another that excels, and on through inexhaustible beauties, in a state of unalloyed rapture at the exquisite scenery.

Scattered over the plains of Wisconsin are found curious earthworks of fantastic

Dalles of the Wisconsin, "The Jaws."

and extraordinary forms, relics of a race that inhabited Wisconsin centuries ago. At Aztalan, in Jefferson County, there is an ancient fortification, five hundred and fifty yards long, two hundred and seventy-five yards wide, with walls four or five feet high. There are also numerous water-falls to be seen—the Chippewa, Big Bull, Grandfather Bull, and the St. Croix—all of them interesting and accessible; besides, Pentwell Peak, an oval

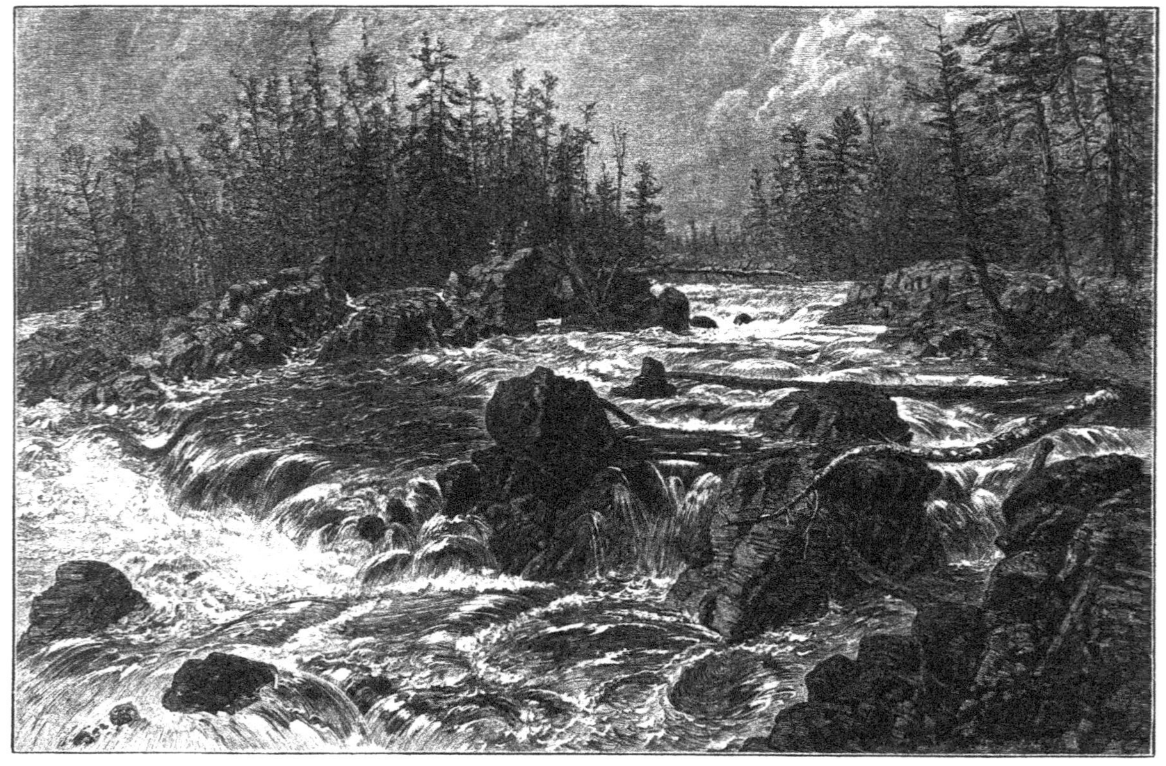

THE DALLES OF THE ST. LOUIS.

mass of rock three hundred feet wide, two hundred feet high, and nine hundred feet long; and Fortification Rock, a picturesque stroke of Nature, which towers one hundred feet high, and on one side is a sheer precipice, while on the other an easy descent is made to the plain by a series of natural terraces.

From Wisconsin we turn northward to Duluth and the St. Louis River, and visit the Dalles of the St. Louis, which are better known, but not more beautiful, than other places we have already seen. The sentiment of the scene is not inspiriting; Nature is

Red River, Dakota.

harsh, rugged, and sombre, tearing her way in a water-course four miles long, with a descent of four hundred feet. The banks are formed of cold, gray slate-rocks, clad with an ample growth of bleak pine, and twisted, split, and torn into the wildest of shapes. Through the dismal channel thus bordered the current surges with terrific force, leaping and eddying, and uttering a savage roar that the neighboring hills sullenly reverberate. Occasionally the spray leaps over the banks, and forms a silver thread of a rivulet, which trickles over the stones until its little stream tumbles into the unsparing current again and is lost. This continuous rapid of four miles is a grand, deeply impressive sight; but on a stormy day, when great white clouds are rolling downward, and the wind adds its voice to that of the turbulent waters, we shiver and sigh involuntarily as we contemplate it.

From Minnesota we cross to the Red River of the North, in North Dakota—a stream with an evil reputation for its sadness and loneliness. The names of its surroundings are far from encouraging—such as Thief River, Snake River, and Devil's Lake—but some of the scenery has a quiet, pastoral character, as will be seen in the accompanying sketches. The water is muddy and sluggish, and within Minnesota alone is navigable four hundred miles, for vessels of three feet draught, four months in the year. The banks are comparatively low, and are luxuriantly grassy and woody. There are "bits" of secluded landscape that transport us to New England, but we are soon recalled by a glimpse of an Indian trail through the grass, a canoe toiling against the stream, and a clump of decaying trees in withered, uncared-for desolation.

Indian Trail, Bank of Red River.

THE CATSKILLS.

WITH ILLUSTRATIONS BY HARRY FENN.

The Mountain House.

ABOUT one hundred and forty miles from the sea, on the western bank of the Hudson, the chain of mountains which, under various names, stretches from the banks of the St. Lawrence to Georgia and Tennessee, throws out a broken link toward the east. Clustering closely together, these isolated mountains, to which the early Dutch settlers gave the name of "Catskills," approach within eight miles of the river, and, like an advanced bastion of the great rocky wall, command the valley for a considerable distance, and form one of the most striking features in the landscape. Thus separated from their kindred, and pushed forward many miles in advance of them, they overlook a great extent of country, affording a wider and more varied view than many a point of far greater elevation. From few places in this world, even among the Alps of Switzerland, does the traveller see beneath him a greater range of hill and valley than from these heights. Nor are the Catskill Mountains famous only for this celebrated bird's-eye view. They contain some of the most picturesque bits of mountain-scenery on the continent. The beauties of the Clove and the Falls of the Kaaterskill have been immortalized by Irving and Cooper and Bryant, passing into the classics of American literature, and awakening in the genius of Cole its loftiest inspiration.

Twenty-five years ago the stage-routes were the only means of access to the mountains. At Catskill on the Hudson old-fashioned stages awaited the arrival of the tourist. Soon the huge vehicles started, and passed through the little village. Presently the bridge that spans the mouth of the Kaaterskill was passed, and the ride toward the

View of Mountains from Creek, Catskill-Mountain Road.

mountains fairly began. The road stretches ahead white and dusty in the sunshine. On either side the trees stand drooping, unstirred by a breath of air; and often, as the horses slowly pull their heavy burden up a rise in the road and stop a moment to rest, a locust, perched on a tree by the road-side, begins his grating cry. In the meadows the cows stand under the trees, switching away the buzzing flies; and the recently-cut grass breathes out its life in the soft perfume of new-mown hay. In the distance the clouds begin to gather on the tops of the mountains, and now and then a long rumble of thunder reverberates through them, and comes rolling down into the valley. It is at this point that our artist made his first sketch. Beside us, the little Kaaterskill, wearied with its rough journey down from the heights yonder, winds among the trees that line its banks, placidly smiling in the sun. Half a dozen cows are standing in the stream to cool themselves. In front, the valley rolls gradually (about a thousand feet in seven or eight miles) up to the base of the mountains, which rise in the distance like a wall.

Round Top and High Peak are buried in a dark cloud, but the scarred head of the North Mountain is in full view, and the Mountain House is clearly defined against a background of pines.

A ride of several hours across the fertile valley, climbing the ridges that lead like steps from the level of the river to the foot of the mountains, brings us at length to a toll-gate, from which we see the road straight before us, ascending steadily. We have now begun to climb in earnest. This excellent road takes advantage of a deep glen, or ravine, through which, in the winter, the melting snow finds its way into the valley. By clinging closely to the mountain—now creeping around a projecting rock, now crossing the beds of little streams, which in the midsummer heat trickle down the mossy rocks beneath the overshadowing trees—it brings us, at last, nearly to the highest point of the ravine. On every side huge trees overhang the road. On the right, the mountain towers straight up above our heads; on the left, the precipice plunges headlong down among the scattered rocks. As you climb up this steep road, you see, here and there, great bowlders lying on the slope of the mountain, covered with moss and fern, and in the perpetual shade of the forest-trees that interlace their leafy arms above you—catching a glimpse, every now and then, through some opening in the tree-tops, of the valley, a thousand feet below, and the river glistening in the distance. As we stop to rest the horses at a point where the road crosses the bed of a stream, from which we can look at the gorge and see a triangular piece of the

Rip Van Winkle's House, Catskill Road.

valley, set in the dark foliage on both hands like a picture in its frame, a sudden clap of thunder breaks on the peaks and echoes among the cliffs above our heads, rolling off slowly, fainter and fainter, till it dies away. Here, by the side of a little stream which trickles down the broad, flat surface of a large rock, is the shanty called "Rip Van Winkle's House," which is represented in the sketch. The artist is looking up the glen from a point on the left of the road. From this point the glen grows narrower and steeper, until it is finally lost among the crevices on the cliffs of the mountain.

The road now winds around the side of the North Mountain, creeping at times on the edge of the precipice, and steadily ascending. At a certain place it turns abruptly, and begins to climb in zigzags. At the first turn you suddenly see the Mountain House directly before you, apparently at the distance of half a mile. Perched upon a piece of rock which juts out far over the side of the mountain, in the bright sunshine glistening and white against the pine-clad shoulders of the South Mountain, the pile of buildings forms a singular feature of the view. On the left of the picture may be

South Lake.

noticed the opening of the Kaaterskill Clove, between the sloping side of the South Mountain and that of the more distant High Peak; and above the clouds, which are floating like bits of gauzy drapery about the sides of the mountains, see the valley of the Hudson fading off toward the south.

From this point there is a steady climb of three miles, the last part through a narrow gorge shaded by drooping hemlocks, when you have at last reached the plateau on which the hotel stands. Beneath it the cliff falls almost perpendicularly about eighteen hundred feet. Two or three trees, growing on the broken stones twenty or thirty feet below the level of the house, peep up above the rock in front; and between their waving tops the landscape for miles lies spread out before you. The Indian Ridge, and the smaller ridges beneath you, though in some places as much as seven hundred feet in height, are dwarfed into nothingness; and the hill-country, through which you have ridden from the river, looks like a flat and level plain. Through the centre of this, at a distance of eight miles, the Hudson winds along like a silver ribbon on a carpet of emerald, from the hills below Albany on the north to where, toward the south, its glittering stream disappears behind the Highlands at West Point. Directly beneath you, the fertile valley, dotted with farms, and broken here and there by patches of rich woodland, is smiling in the sunlight, constantly changing as the waves of shadow chase each other across the varied mass of green; and, beyond, an amphitheatre of mountains rises on the horizon, stretching, in jagged lines, from the southern boundaries of Vermont to Litchfield, in Connecticut—rolling off, peak after peak wave after wave of deepening blue, until they are lost in the purple of the Berkshire Hills.

The most famous beauty of the region is the Fall of the Kaaterskill. On the high table-land of the South and North Mountains lie two lakes, buried in a dense forest. Of one of these—the South Lake—Mr. Fenn has given us a sketch. It was taken from a high ledge on the North Mountain, looking southward. The shores are dark with pines, and the surface of the lake is dotted here and there with the broad leaves of the water-lily; but the most striking feature of the view is the summit of Round Top, reflected as in a mirror. A little brook, making its way from these lakes westward along the shoulder of the mountain, soon reaches the edge of a very steep declivity, over which it leaps into a deep pool in the centre of a great amphitheatre of rock.

First Leap of the Falls.

Gathering its strength again, the

UNDER THE CATSKILL FALLS.

peated fires, and the path winds among the rocks, half buried in long mountain-grass or blueberry-bushes, until it comes out to the eastern face of the mountain. You are high above the level of the Mountain House, which lies beneath you to the left, and the view over the surrounding country and the valley of the Hudson is even more extended than that from the piazza of the hotel. Farther along, still keeping southward, and occasionally climbing up steep steps, you find the cliffs exceedingly fine. Some of them are sharply cut, and overhang the tops of the tallest trees that grow from the *débris* at their base. On a promontory of high rock near the entrance to the Kaaterskill Clove lies "the Bowlder." It is a huge block of the pudding-stone, brought here doubtless by the ice in the glacial period, and left by some strange chance on the very verge of the precipice. A few feet farther and it would have toppled over the edge and crashed downward two thousand feet into the bottom of the Clove. Here the precipitous walls of rock hardly afford foothold for the weather-beaten pines that grow out of the crevices and wave their twisted arms from the dizzy heights. Sometimes, after passing through Pudding-Stone Hall, you keep straight along the path through the woods instead of turning eastward toward the face of the mountain. After a time you come to a point where the bits of rock have fallen from the ledge above and lie scattered along the hill-side, like the bowlders hurled about in the giant warfare of the Titans. The wood is dense and dark; the pines interlacing their arms above your head throw a perpetual twilight on the hill-side, and as you sit

Druid Rocks.

on the soft carpet of their fallen leaves and see these huge fantastic rocks scattered around you, you cannot but feel that the name of "Druid Rocks," which has been given to the place, is at once suggestive and appropriate. At times the path keeps close along the sloping hill-side, finding a doubtful way beneath the base of tall cliffs covered with moss; at others it climbs through some crevice, and, ascending to the top of the ledge, winds among the gray rocks in the full glare of a summer's sun.

Looking South from South Mountain.

A delightful walk brings you to Indian Head. This name is given to a bold promontory which juts out over the Clove until it overhangs the bed of the tumbling, tossing Kaaterskill. From this rock the mountain falls eighteen hundred or two thousand feet. Half a dozen tall pines growing out of the cliff, divided into two groups on either hand, form a sort of dark, rustic frame for the exquisite picture. The Clove at this place is very narrow, and along the bottom the Kaaterskill goes tumbling and foaming over the stones. Along the base of the cliff, on the left or southern side of the glen, winds the little road that leads from the village at its mouth up to the table-land beyond the famous falls. On both sides the mountains tower high above your heads, heavily wooded to the summits with chestnut and pine, through the rich green of which, here and there, you can see the rugged face of a huge precipice, scarred and broken by the frosts, and spotted with dark lichen and moss. The little rustic bridge which is seen in the view from Indian Head spans the stream at one of the most striking points in the Clove. The light structure, hardly strong enough, apparently, to bear the heavy stage that is about to cross it, hangs over the Kaaterskill where it comes tumbling over some huge rocks that have fallen in its path. The water boils and tosses into foam, and then dashes headlong down a succession of ledges beneath. On one side the cliff towers high into the air, sharp and smooth as masonry, looking like the walls of a great mediæval castle ; on the other, the spurs of the South Mountain, densely covered with trees, rise rapidly more than fifteen hundred feet. As you stand upon Sunset

Glimpse of Catskill Clove from Indian Head.

Bridge in Catskill Clove.

Rock and look westward up the Clove you have one of the most picturesque views in the range of mountain scenery. The rock is broad and flat, projecting far out over the precipice. A pine-tree stands like a sentinel upon its very verge. In front of and behind you, as you sit by the old tree on the dizzy edge, the mountain pushes two great, gray cliffs, bald and ragged, far out over the glen, and then falls in broken lines a scarred and frowning precipice.

The lines of the South Mountain and of the spurs of High Peak and Round Top blend so gently together, as they meet beneath, that it is difficult to trace the bed of the Kaaterskill or its tributary even by the shadows in the dense forest of green. Directly in front of you the table-land, which is formed by the shoulders of these mountains, rolls off toward the westward, where the sharp lines of Hunter Mountain are clearly defined against the sky among its sister peaks. Over the edge of this table-land leaps Haines's Fall.

One of the most beautiful of all the sketches made by Mr. Fenn is that of the Five Cascades, as they are improperly called. A stiff climb from

Sunset Rock.

the bottom of the Kaaterskill Clove—beginning at the point where the carriage-road leaves it and following the bed of the stream that comes down from Haines's, now clambering over bowlders and fallen trees, and again scrambling up the wet rocks or clinging to the vine-clad banks—brought us at last to the Five Cascades. The stream, after plunging over the cliff—as shown in the

The Five Cascades, Kaaterskill Clove.

view from Sunset Rock—like a far-off feathery vapor into a large shallow pool, jumps rapidly over a series of ledges from ten to forty feet in height, that lead like steps down into the Clove. Through the succession of the ages it has worn its way among the rocks, until for most of the distance its path is hidden from the sunshine. In many places the branches of the trees on the high banks above are intertwined across the ravine, down which the little stream dashes in hundreds of beautiful cataracts in a perpetual twilight. There are hundreds of these falls, but five of them are peculiarly striking, and three of these are represented in the engraving. As we sat upon a fallen tree and gazed upon the stream, dashing its cold, gray waters over the black rocks, a shaft of sunlight broke through the tree-tops above our heads and fell upon the middle fall. The change was instantaneous. Above it and below, the cataracts were still in shadow, but the central one, in the bright sunshine, threw over the glistening rock a myriad of diamonds.

The last illustration is a distant view of Stony Clove—a pass in the mountains famous for the wildness of its scenery. It is always dark and cool, and even in mid-August you may find ice among the crevices of the rocks that have fallen in great numbers from the cliffs above. The sketch was made as we drove toward the northern entrance. A thunder-storm was gathering about the southern gate of the pass, and a rainbow seemed to rest upon the mountains hovering above the Clove.

Such are a few of the attractions of this charming region. Of course there are drives over fine roads among the hill-tops, and countless walks through the forests and over the ledges, with the usual results of torn clothes, sunburnt faces, and hearty appetites. The natural beauties of the Catskills still remain, and the walks and the drives are there, but the pleasant stage-ride is a thing of the past. At first the railway

Stony Clove.

extended from the steamboat landing at Catskill to the base of the mountain. Within a year or so the old picturesque road by which the hotels on the summit were reached has given way to a rapid railway, that ascends, elevator-like, abruptly up the face of the mountain. Now the modern tourist leaves the station in the distant metropolis, and, at lightning speed almost, is transported to the junction where connection is made with the elevating railway, and the distance of one mile and a quarter up the mountain is made in exactly ten minutes. To the dweller in a city of the plain, weary of work and worn with the tumult of its life, there are few places in the whole range of American scenery so attractive and refreshing as the Catskill Mountains.

THE VALLEY OF THE GENESEE.

WITH ILLUSTRATIONS BY J. DOUGLAS WOODWARD.

THE Genesee River in its early course is not marked by any exceptional beauty or peculiar charm of surroundings. It is not till the falls at Portage are reached that the river asserts its claim to recognition as one of the most beautiful and picturesque of all our Eastern streams.

Railroad-Bridge, Portage.

The tourist, on leaving the train at Portage Village, will be first attracted by the great bridge that spans the ravine and river at this point. The viaduct crosses the river at a spot hardly a stone's-throw above the brink of the First or Upper Fall; and its lightly-framed piers, with their straight lines reaching from the granite base to the road-way above, contrast strangely with the natural chasm over which it extends.

We turn away from a hasty survey of the bridge to the contemplation of the rough-hewn, rugged walls of the fissure it spans. Divided for an instant by the stone buttresses of the bridge, the waters of the river unite again, just in time to present a bold and unbroken front upon the brink of the first fall. As the body of water which passes over these falls is comparatively small—except in seasons of flood—and as the first precipice is but sixty-eight feet in height, the effect would be of little moment were it not for the striking character of the surroundings.

Middle Falls, Portage.

Entering the gorge a short distance above the brink of this Upper Fall, the river has cut for itself a wild, rugged channel, the walls of which rise in a perpendicular height of from two to six hundred feet, each successive fall resulting in a deepening of the chasm, and a consequent increase in the height of the rocky barriers.

It is this chasm that constitutes the distinctive feature in the upper course of the Genesee. Beginning abruptly at a point not far above the Upper Fall, it increases in depth and wildness until the village of Mount Morris is reached, where the stream makes its exit from the rocky confines as abruptly as it entered them, and, as though to atone for the wildness of its early course, settles at once into a gentle and life-giving current, gliding through rich meadows and fertile lowlands, its way marked by a luxuriant growth of grass and beautiful woodland.

Having recovered from their first bold leap, the waters unite and flow onward in gentle current, with an occasional ripple or miniature rapid, for the distance of half a mile, when the brink of the second and highest fall is reached. Over this the waters pour in an unbroken sheet, a depth of one hundred and ten feet. At the base of

Lower Falls, Portage.

this fall the waters have carved out, on the western side, a dark cave, which may be approached by a wooden stairway, standing at the foot of which we see the sky as from the depths of a crater.

Ascending again to the plateau that reaches out on a line with the brink of this fall, we come in sight of Glen Iris, a rural home, the owner of which is the possessor of an appreciative taste for the beauties with which he is surrounded.

Upon the lawn that divides Glen-Iris Cottage from the brink of the precipice there is a rude log-cabin that is a monument of peculiar interest. It is an old Indian council-house, and stands alone, the only ruin of what was once a village of the Iroquois. It was here that the chiefs of the Seven Nations were wont to hold their councils of war. This ancient building of Caneadea stood originally upon a bluff of land overlooking the Genesee, about twenty-two miles above its present site. It was the last relic of aboriginal sovereignty in the valley. During the Indian wars all the white captives brought in from the South and East were here received and compelled to run the gauntlet before this council-house, its doors being their only goal of safety. There is no record of the

Indian Council-House.

date of its construction, but upon one of the logs is the sign of a cross, the same as that which the early Jesuit fathers were known to have adopted as the symbol of their faith. When the Indians took their departure to more western reservations, the old council-house came into the possession of a white squatter, who guarded it against decay, and made it his home for fifty years.

Turning again to the river, we follow down a wild mountain-road for the distance of two miles, at which point a narrow, winding foot-path leads down a steep and rugged defile. Descending this, and guided by the rush of waters below, we suddenly come upon the Lower Falls. Here the waters of the river are gradually led into narrower

High Banks, Portage.

channels, until the stream becomes a deep-cut canal, which, rushing down in swift current between its narrow limits, widens out just upon the brink of the fall, that more nearly resembles a steep rapid than either of the others. Standing upon one of the projecting rocks which are a feature of this fall, we can only catch occasional glimpses of the cavern's bed, so dense and obscuring are the mist-clouds. A second and more hazardous pathway leads from these rocky observatories to the base of this the last of the Portage falls; and the course of the river now lies deep down in its rock-enclosed limits, until the broad valley is reached, to which the general name of High Banks is given—a name rendered more definite by a prefix denoting their immediate locality. Thus we have the

High Banks at Portage, the Mount-Morris High Banks, and at the lower end of the valley the High Banks below the lower fall at Rochester.

A journey along the river's shore from the lower falls to the valley will reveal wonders of natural architecture hardly exceeded by the cañons of the far West. Here, hidden beneath the shadows of the overhanging walls of rock, it is hard to imagine that, just beyond that line of Norway pines that forms a fringe against the sky above, lie fertile fields and quiet homes. A just idea of the depth of this continuous ravine can best be secured by an ascent to one of the projecting points above, where, resting on a ledge of rock, the river is seen at one point six hundred feet below—a distance which changes with the varying surface of the land above.

Although the point where the river enters the ravine at Portage is but twelve miles in a direct line from that of its exit at Mount Morris, the distance, following its wind-

High Banks, Mount Morris.

ing course among the hills, is much greater. Emerging through what is literally a rocky gate-way, the whole mood of the river seems to have changed with that of its surroundings. In order to make this change as conspicuous as possible, we ascend to one of the two summits of the terminal hills. Standing there, and shaded by the grand oaks which crown it, we have but to turn the eye southward to take in at a glance the whole val-

Elms on the Genesee Flats.

ley below, which is a grand park, reaching far away to the south. The sloping highlands are dotted here and there with rural villages, whose white church-spires glisten in the rich, warm sunlight. Below and around are the meadows and alluvial places known as the Genesee Flats.

The illustration shows broad, level fields, marked out by well-kept fences, enclosing areas often one hundred acres in extent. To the right are the celebrated nurseries, with their lines of miniature fruit and shade trees; the distant slopes are dotted with the golden wheat-harvests, while reaching far away to the south are the rich meadow-lands of the Genesee. In the midst of all flows the river, its waters giving life and beauty to the numerous groves of oaks and elms which shadow its course. It is, in fact, a broad lawn, unbroken save by an occasional hillock, with here and there groves of rare old oaks, beneath whose shade droves of cattle graze at leisure.

Flats of the Genesee

This valley, like all others watered by rivers taking their rise in neighboring mountain districts, is subject to frequent and occasionally disastrous inundations. Fortunately, however, the moods of the river are oftenest in accord with those of the varying seasons; for this reason freshets seldom come upon the ungathered harvests. The possibility of this event, however, leads the landholders to reserve their meadows upon the flats for grazing purposes, and hence the damage from a flood is never great.

Passing northward by means of the railway, we soon leave behind us the valley villages of Dansville, Mount Morris, Geneseo, and now Avon, the first a flourishing town seated upon one of the tributaries of the Genesee, and thus being entitled to a place among this beautiful sisterhood. At Avon are the justly-famous sulphur springs. Continuing our journey twenty miles farther, following the line of the river along its eastern shores, we enter Monroe County and approach the city of Rochester.

This city stands in the same relation to the valley as does a storage and distributing reservoir to the streams from which the supply is received. In its early days the

ROCHESTER, FROM MOUNT HOPE CEMETERY.

East Side, Upper Falls of the Genesee.

life of the city was dependent upon the harvests of the valley; when these were abundant then all went well. We can therefore readily understand the need and consequent prosperity of the city, and how it was long known as the "Flour City of the West."

West Side, Upper Falls of the Genesee.

Although it now ranks as the fourth city in the State, there are yet living some persons whose childhood dates back of that of the city in which they dwell. From a brief historical sketch on the subject, we learn that, in expressing astonishment at the career of

Lower Falls.

Rochester, De Witt Clinton remarked, shortly before his death, that when he passed the Genesee on a tour with other commissioners for exploring the route of the Erie Canal, in 1810, there was not a house where Rochester now stands. It was not till the year 1812 that the "Hundred-acre Tract," or "Allan Mill Tract," as it was then called, was planned out as the nucleus of a settlement under the name of Rochester, after the senior proprietor, Nathaniel Rochester.

Having previously referred to the second series of falls and high banks, we will now return to the guidance of the river as it enters the city limits at its southern boundaries. Its course lies directly across or through the centre of the city, the main avenues, running east and west, being connected by several iron bridges, with the exception of that known as the Main-Street Bridge, concealed beneath the street, which is of stone, and the several railway-bridges.

It is at the city of Rochester that the Erie Canal encounters the Genesee River, which it crosses upon the massive stone aqueduct that has long been regarded as one of the most important works of American engineers. In its present course the river has rather the appearance of a broad canal, save that the current is rapid, and at times

Light-house, Charlotte.

boisterous. The shores are lined by huge stone mills and factories, the foundation-walls of which act the part of dikes in confining the water to its legitimate channel. Not far from the Erie Railway station the river is crossed by a broad dam, from either side of which the water is led in two mill-races, which pass under the streets and conduct the water to the mills along the route. At a point somewhat below the centre of the city, and yet directly within its limits, are the First or Upper Falls. These are ninety-six feet in height, and it is thus evident that, with such a cataract in the centre of the city, the facilities for obtaining water-power could hardly be excelled. The mill-races conduct the main supply along the two opposite shores, and as the mills are mainly situated below the level of the falls the full force of the water can be utilized. The illustrations of the Upper Fall have been so designed that the two combined present a full view of the whole front as viewed from the chasm below, the darkened channels through which the water from the races is returned to the river being shown to the right and left.

The brink of this fall marks the upper limit of a second series of high banks similar in general character to those that lie between Portage and Mount Morris. The height of these walls at certain points exceeds two hundred feet. At the distance of about a mile from the Upper Fall, a second descent of about twenty-five feet is followed, at the distance of a few rods only, by the Third or Lower Falls, which are nearly one hundred feet in height. It thus appears that, within the limits of the city, the waters of the Genesee make a descent, including the falls and the rapids above them, of two hundred and thirty feet; and the water-power, as estimated for the Upper Fall alone, equals forty thousand horse-power.

As the river has now reached the level of Lake Ontario, it assumes the character of a deep-set harbor, and the vessels engaged in lake-traffic can ascend it five miles to the foot of the Lower Falls. The port of entry, however, is now at the mouth of the river, where stands the village of Charlotte, which in recent years has become a summer resort. It is within easy reach of Rochester by means of a fine broad road, along the side of which passes an electric railway. A large hotel faces the lake and is provided with ample facilities for bathing and boating. Summer cottages have been erected in the vicinity, where many of those whose business interests prevent a long absence from the city spend the summer. It is likewise the headquarters of the yachting interests of the neighborhood. Here are also wharves, a light-house, and a railroad-station, which road leads direct to Rochester.

SCENES IN NORTHERN NEW JERSEY.

WITH ILLUSTRATIONS BY JULES TAVERNIER.

Scene on the Passaic.

THE picturesque features of New Jersey lie almost entirely in the northern section of the State, and are within easy reach of the great metropolis. Her territory includes every variety of scenery, from the picturesque hills and lakes of her northern to the broad sand-wastes of her southern counties.

A ride of seven or eight miles brings the traveller from the valley of the Hudson to the valley of the Passaic, the latter being bounded, at some distance inland, by the abrupt, precipitous range of hills known generally as Orange Mountain. Thirty years ago this mountain was a wild, uninhabited region. The Dutch farmers who originally settled in this vicinity were content to nestle in the grassy valleys, preferring for their homes the quiet plains, rather than seeking for picturesque nooks on the frowning hill-side. They built solid one-story houses of gray-stone, covering them with overhanging roofs, and caring in their domestic arrangements rather for comfort than for elegance. Many of these simple yet substantial structures are standing at this day, giving shelter

EAGLE ROCK, ORANGE.

WASHINGTON ROCK.

to the descendants of those who built them. Others have passed into the hands of city folk, and have been decked out with verandas, furnished with larger windows, and even provided with Mansard roofs, so that it is difficult to recognize in these reconstructed edifices the solid old farm-houses of a hundred years ago. In recent years railway communication has increased to such an extent that almost every farm in Northern New Jersey enjoys the advantage of being "near the station"—a privilege which only those who live in the country can fully appreciate.

One of the first and most successful attempts at landscape-gardening on a large scale, in this country, was made by Llewellyn S. Haskell, who was especially fond of rural life. He purchased a tract of land on Orange Mountain and laid it out as a park, in which he and his friends built a variety of elegant private residences. No attempt was made to deprive this region of its wild primeval beauty. Roads were laid out, winding in gentle curves amid the rugged rocks and through the rich and picturesque forests. Near Eagle Rock he built his own home, which commands a view more extensive than any other in the vicinity of New York. Beneath lies the cultivated valley, covered with villages, and partially bounded by the Bergen Hills. To the south can be seen the gleam of the waters of the bay of New York and of the Atlantic Ocean, and, under favorable atmospheric circumstances, the spires of the great city. The whole eastern slope of the mountain, for several miles in length, is dotted with residences, most of which share this delightful view, that increases in diversity and beauty, though not in extent, as you go northward into the prosperous town of Montclair.

At the foot of the mountain there is a well-kept road, which is a favorite drive for the residents of the vicinity, affording, as it does in the warm summer afternoons, that "shadow of a great rock in a weary land" of which the scriptural poet spoke so many thousand years ago, and at the same time offering a goodly view of the level plain. From this road—though it is at a much lower elevation than the point of view suggested in our engraving—Eagle Rock is seen towering in majestic grandeur, as bold and rugged as when only the red-men inhabited this charming region. The eagles, which gave it its name, are now but seldom seen; yet the hoary, scarred projection seems to the eye as distant and as desolate as when it was indeed the home of the king of birds.

Still more striking in appearance, and more picturesque in formation, is Washington Rock, on the same range of hills. This rock is divided by a deep chasm into two parts, one of which has evidently been cleft from its fellow by some great convulsion of Nature, and has fallen several rods down the slope of the hill, where it stands firm and upright. From this rock it is said that George Washington viewed the land below, eager to trace the course of the British army. At that time the plain was cultivated, it is true; but the pretty little village of Dunellen, which now forms so pleasing a feature of the scene, was unthought of, and the mountain itself was as wild and uninhabited as

RAMAPO RIVER.

the distant Sierras. Washington Rock is a favorite resort for picnic-parties, and for the tourist who seeks to gratify his taste for the picturesque.

Farther to the north of the State is the Ramapo River, a stream which finds its way between high hills, and is frequently used for manufacturing purposes. Over one of the dams which obstruct its course the water flows in a graceful cascade, which, but for its prim regularity, would equal in its beauty of motion the natural falls which are

Terrace House and Thorn Mountain.

ever such a source of delight to the lover of the beautiful. To such, indeed, the Ramapo offers many attractions. The stream, in its numerous curves, constantly presents fresh points of view. The hills—sometimes abrupt, sometimes rolling—here and there recede from the river's edge, leaving grassy fields or rocky plateaus, on either of which it is a pleasure to stroll. The sails on the river add to the variety of the scene; the fisher-

BREAKWATER, RAMAPO.

man's row-boat imparts to it notable life and vivacity; and the wreathed smoke of the locomotive does not seem wholly inharmonious.

graceful beauty, will remind the traveller of the famed English lake.

Among the hills and streams of the section of country to which these pages are

Little Falls.

with the mountains of New Hampshire. But, as a general thing, the scenery of Northern New Jersey is on a less extensive scale. The hills, rugged and wild as they may be, after all cannot fairly be called mountains. The lakes are small, and the narrow rivers find devious paths among their rocky barriers. Principal among these streams is the Passaic, not only in its historic interest, but in its great length, breadth, and commercial importance, a notable exception among the rivers of New Jersey. From its extreme source, in the upper part of Morris County, it flows between the hills of that county and Essex, taking toll of Dead River as it passes the base of Long Hill, and thence stealing its way, with scarcely a ripple, through narrow vale and broad valley, for

THE PASSAIC, BELOW LITTLE FALLS.

THE PASSAIC BELOW THE FALLS.

twenty miles among the defiles of the Horseshoe Mountain, till it receives the tribute of the vivacious Rockaway. Stimulated apparently by the instillation of this lively little rock-stream, or perhaps awakened to the scene of an impending crisis in its fate, it emerges from the last defile with a sudden start, and almost rushes for a few miles toward its first leap over the rapids of Little Falls, nearly opposite the pleasant village of that name. The fall is more than three hundred feet broad, and is formed with an obtuse angle opening down-stream, over which the river, just pausing to smooth its ruffled surface on the brink, leaps in two broad sheets of foam-capped, spray-clouded water, and then glides away serenely to perform a similar feat a short distance lower down, at the Second Fall—the two being possibly in the nature of rehearsals for the final struggle at the Great Passaic Falls, some six miles below. The scenery along the river, during its leisurely loiterings through the mountains, and its scarcely more hurried voyage along the valleys of its upper course, is of that peculiar character which belongs to such regions. Tall masses of rock rise abruptly, at intervals, on its banks, like great buttresses, or still more like the massive and forest-grown ruins of mighty rock-structures, such as are found here and there along the water-courses of the wondrous Southwest. The river-bed is rocky; yet the flow is hardly fretted into ripples by these up-cropping barriers, but seems to hold the even tenor of its way with a quiet disregard of obstacles that is eminently suggestive of a serene philosophy. At Little Falls the Morris Canal crosses the river by a handsome stone aqueduct; and from the summit of this the artistic loungers may obtain a charming view of the stream, winding down between overhanging hills of greenery and jutting escarpments of cedar-crowned trap-rock and sandstone, toward Great Falls, and the more level reaches of the Paterson plains and the salt-marshes of Newark. Before reaching this point, however, the river undergoes a second tribulation in the shape of another fall and rapid, which rouse its sluggishness into momentary and picturesque fury, and over and down which it roars in foamy wrath, scarcely subdued in time to collect itself for the struggle five miles beyond. But it does subside, and, assuming once more a tranquil air of unconsciousness, rolls smoothly to the verge, and then plunges boldly, in one unbroken column, over the precipice of the Great Falls, dropping sheer ninety feet into a deep and narrow chasm of less than sixty feet in width, through which it dashes and foams in short-lived madness, to rest and glass itself upon a broad, still basin, hollowed by its own labors from the solid rock. After leaving this basin the river is vexed no more, but flows pleasantly past many thriving towns and hamlets, giving of its tide to turn the wheels of industry, spanned by bridges of many forms and used for many purposes, from the elaborate iron arch of the railway to the rude rusticity of the wooden foot-bridge. Its path now lies amid rich uplands and orchards, teeming fields, and the dwellings of a prosperous agricultural community. But there are still many picturesque glimpses of a wilder nature along its course, and many a spot known to the disciples of the "gentle Izaak" as giving and

PASSAIC FALLS.

fulfilling the promise of excellent sport and the added charm of attractive scenery. From Paterson to Newark the shores spread like an amphitheatre covered with verdure, dotted thickly with dwellings and the monuments of successful enterprise and industry, giving it the appearance of a watery highway through a picturesque succession of close-lying

Near Greenwood Lake.

villages and centres of busy life. In Newark the river loses itself in the lake-like bay of Newark, which is the beginning of the end, for the waters of the bay soon find their path to the great mother ocean; first, however, passing through the Kill van Kull to the bay of New York, or by the Arthur's Kill into Raritan Bay.

THE PLAINS AND THE SIERRAS.

WITH ILLUSTRATIONS BY THOMAS MORAN.

Witches' Rocks, Weber Cañon.

THE present banishes the past so quickly in this busy continent that to the younger generation of to-day it already seems a very dreamy and distant heroic age when men went out upon the great prairies of the West as upon a dreaded kind of unknown sea.

The route of the Union Pacific Railway is the most familiar path across the Plains, not only because it passes nearest to the well-known emigrant-road of former days, but because it is the road which, though it misses the nobler beauties of the Rocky Mountains, shows the traveller the prairie itself in its truest and most characteristic aspect. It passes through almost every change of prairie scenery—the fertile land of the east and the alkali region farther on; past the historic outposts of the old pioneers; among low *buttes* and infrequent "islands;" and over a country abounding in points of view from which one may take in all the features that mark this portion of the continent. To the south, the great level expanse is hardly interrupted before the shore of the Gulf of Mexico is reached, and the Mexican boundary; to the north, the hills and high table-land of the Upper Missouri are the only breaks this side of the Canadian border.

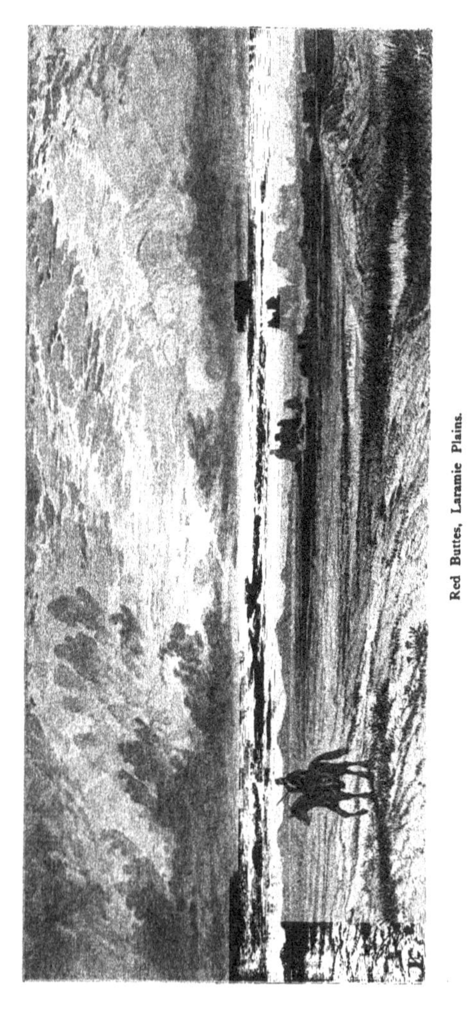

Red Buttes, Laramie Plains.

Omaha, our starting-point, stands looking out upon the muddy water of the Missouri, and watching with interested eyes that transient traveller whom it generally entices in vain to linger long within its precincts—a city that has been all its life a starting-place; and yet, in spite of the fact that every man seems to arrive only with the thought of departing, a prosperous, thrifty place, not without a look of permanence, though as yet only a little over a quarter of a century old.

The valley of the Platte lies before one almost immediately after he has left the Missouri behind him. The word "valley," in this apparently unbroken plain, seems a misnomer; but it is everywhere used—as in regions where its significance is truer—for

DIAL ROCK, RED BUTTES, LARAMIE PLAINS.

Buttes, Green River.

the slight depression that accompanies the course of every stream; and an old traveller of the Plains will tell you that you are "entering the valley of the Platte," or "coming out of the Papillon Valley," with as much calmness as though you were entering or leaving the rockiest and wildest cañon of the Sierras. There is only a short reach of railway to the northwest, a sharp turn to the westward, and the clear stream of the river is beside the track—a free, full channel

CLIFFS OF GREEN RIVER.

if the water is high, a collection of brooks threading their way through sandy banks if it is low. For more than a whole day the railway runs beside the stream, and neither to the north nor south is there noteworthy change in the general features of the scenery. A vast, fertile plain, at first interrupted here and there by bluffs, and for some distance not seldom dotted by a settler's house or by herds of cattle; then a more monotonous region, still green and bright in aspect; farther on—beyond Fort Kearney, and Plum Creek, and McPherson, all memorable stations with many associations from earlier times—a somewhat sudden dying away of the verdure, and a barren country,

Church Butte, Utah.

broken by a few ravines. This again gives place, however, to a better region as the Wyoming boundary is approached.

The border crossed, a new region is entered. The Plains do not end, but they are already closely bordered, within sight, by the far-outlying spurs of the Rocky Mountains. Beyond the civilized oasis of Cheyenne the scenery takes on a darker look, and if one chances to come to the little station of Medicine Bow when the sunset begins to cast long shadows from the black mountains on the southern side of the North Fork of the Platte, there is something almost sombre in the aspect of the shaded plain. The Laramie plains have just been passed; indeed, they still lie to the northward. Hills break the monotony of their horizon, and here and there the regular forms of castellated *buttes*

Castle Rock, Echo Cañon.

stand out sharply against the sky. The far-off Red Buttes are most noteworthy and most picturesque of these; grouped together like giant fortresses, with fantastic towers and walls, they lift ragged edges above the prairie, looking lonely, weird, and strong. Among the singular shapes their masses of stone assume, the strangely-formed and pillar-like Dial Rocks tower up—four columns of worn and scarred sandstone, like the supports of some ruined cromlech built by giants. About them, and, indeed, through the whole region, the country is a barren, unproductive waste. The curse of the sage-brush, and even of alkali, is upon it, and it is dreary and gloomy everywhere save on the hills.

With the approach to Green River the verdure comes again, but only here and there, and generally close by the river-bank. Here the picturesque forms of the buttes

WEBER RIVER—ENTRANCE TO ECHO CAÑON.

reappear—a welcome relief to the monotony that has marked the outlook during the miles of level desert that are past. The distance, too, is changed, and no longer is like the great surface of a sea. To the north, forming the horizon, stretches the Wind-River Range—named with a breezy poetry that we miss in the later nomenclature of the race that has followed after the pioneers. To the south are the Uintah Mountains.

At some little distance from the railway the great Black Buttes rise up for hundreds of feet, terminating in round and rough-ribbed towers; and other detached columns of stone stand near them—the Pilot, seen far off in the view that Mr. Moran has drawn of the river and its cliffs. And through all this region fantastic forms abound everywhere, the architecture of Nature exhibited in sport. All about one lie "long, wide troughs, as of departed rivers; long, level embankments, as of railroad-tracks or endless fortifications; huge, quaint hills, suddenly rising from the plain, bearing fantastic shapes; great square mounds of rock and earth, half-formed, half-broken pyramids. It would seem as if a generation of giants had built and buried here, and left their work to awe and humble a puny succession."

The Church Butte is the grandest of the groups that rise in this singular and striking series of tower-like piles of stone. It lies somewhat farther on, beyond Bryan, and forms a compact and imposing mass of rock, with an outlying spur that has, even more than the main body, the air of human though gigantic architecture. It "imposes on the imagination," says Mr. Bowles, in one of his passages of clear description, "like a grand old cathedral going into decay—quaint in its crumbling ornaments, majestic in its height and breadth."

It is a characteristic of this whole portion of the Rocky-Mountain chain, and one that disappoints many a traveller, that there are here no imposing and ragged peaks, no sharp summits, no snow-covered passes, and little that is wild and rugged. All that those who remember Switzerland have been accustomed to connect in their minds with great groups of mountain-masses must be sought elsewhere. The Plains themselves rise; one does not leave them in order to climb. Over a vast, grass-covered, almost unbroken, gradual slope, extending over hundreds of miles of country, the wayfarer has come imperceptibly to the great water-shed. It is scenery of prairie, not of hills and peaks, that has surrounded his journey.

For the last fifty miles, indeed, before the arrival at Sherman, which is eight thousand two hundred and thirty-five feet above the level of the sea, the rise has been barely appreciable. A new circumstance makes the descent from the great height much more perceptible and enjoyable through a new sensation. It is then that the traveller over duller Eastern roads, who has flattered himself that the "lightning express" of his own region was the highest possible form of railway speed, first learns the real meaning of a "down grade."

We have reached the Church Butte, beyond Bryan, and have crossed Green River, on the old overland stage-route and the emigant-road, travellers ›rd the stream—no unwelcome task, with that great Bitter-Creek esh in the memories and hardly out of their view.

the great railway goes on beyond Green River through the valley ws down from the Uintah Mountains, and, leaving at the south ›ssing the old Mormon road, enters Utah. A little farther, and we st scenes of the journey this side the far-away Sierras.

the Union Pacific Railway that lies between Wasatch and Ogden, corner of Utah, will some day be that part of the journey across ntinent that will be especially regarded by the tourist as necessary ll others. It does not in grandeur approach the mountain scenery ast, but it is unique; it is something the counterpart of which you the world; and long after the whole Pacific journey has become nmonplace, this part of it will keep its freshness among the most ie journey.

:e west from Wasatch the road passes through a tunnel nearly eight gth. The preparation for what is to come could not be better; and, bleak and dreary region that has been passed over adds so much to picturesqueness of these Utah scenes that it may very possibly have little to the enthusiasm they have called forth. From the darkness suddenly, and, tunnel and cutting being passed, there lies before the the green valley before the entrance to Echo Cañon. Through it River, bordered with trees, and making a scene that is suddenly de- weirdness and look of dreary devastation that has marked the country miles of this long journey. The valley is not so broad, so pastoral which comes after the wild scenery of the first cañon is passed, but and valley of home lying here in the wilderness.

id of Echo Cañon stands Castle Rock, one of the noblest of the great that are passed in all the route—a vast and ragged pile of massive y cut, by all those mighty forces that toil through the centuries, into ce of a mountain fortress. A cavernous opening simulates a giant door een its rounded and overhanging towers; the jagged points above are battlements left bristling and torn after combats of Titans; the huge rn sides seem to have been builded by skilful hands; and the great ions, from which the sandy soil has been swept away, would appear ry central earth. It surmounts a lofty, steep-sided eminence, and frowns wesome strength and quiet on the lonely valley below it.

the road enters the Echo Cañon itself. It is a narrow gorge between t tower hundreds of feet above its uneven floor, along which the river

DEVIL'S GATE, WEBER CAÑON.

runs with a stream as bright and clear as at its very source. Not simply a straight cut between its precipices of red-and-dark-stained stone, but a winding valley, with every turn presenting some new variation of its wonderful scenery. On the mountains that form its sides there is little verdure—only a dwarfed growth of pine scattered here and there, leaving the steeper portions of the rock bare and ragged in outline. Now and then there are little openings, where the great walls spread apart and little glades are formed; but these are no less picturesque than the wilder passages.

Devil's Slide, Weber Cañon.

There are memorable places here. Half-way down the gorge is Hanging Rock, where Brigham Young spoke to his followers after their long pilgrimage, and pointed out to them that they approached their Canaan—preached the Mormons' first sermon in the "Promised Land." Full of all that is wild and strange as is this rocky valley, seen even from the prosaic window of a whirling railway-car, what must it have been with the multitude of fanatics, stranger than all its strangeness, standing on its varied floor and looking up at the speaking prophet, whom they half believed, half feared?

The cañon is not long; the train dashes through it at sharp pace; and suddenly, without passing any point of view that gives the traveller a warning glance ahead, it turns and dashes out into the beautiful and broad valley beyond, halting at Echo—most picturesque and bright of little villages.

TERRES MAUVAISES, UTAH.

is a really
ad plain left
ountains is
river winds
anse in pleas-
awl or rush;
se in a famil-
de. Only the
surrounding
there the ap-
izon of some
distant peaks,
is whereabouts,
rom the quieter
about him.
est gorge is still
road enters it
ter crossing the
Weber Cañon,
se Utah ravines.
are grander by
Echo; the forms
lges and the carv-
ices are more fan-
deep, dark aspect
rrow valley gives
nobler scene. It
ved on a cloudy,
realize its whole
randeur. The little
t the left of the
ider of the locomo-
om the high preci-
les; the rush of the
motion adds a cer-
wildness to the shad-

Salt Lake.

follows the nomen-
Veber Cañon would

PALISADE CAÑON.

line and plummet. On either side of this smooth, white line is what appears to the eye to be a well-laid white stone wall, varying in height from ten to twenty feet. This white spectacle on the red mountain-side has all the appearance of being made by man or devil as a slide from the top of the mountain to the bed of Weber River."

And now we are nearing the very centre of Mormondom; for only a little beyond the Devil's Gate, which, though first named, is farther toward the western extremity of the cañon than the "Slide," we come to Uintah, glance at the Salt-Lake Valley, and are hurried on to Ogden, whence the trains go out to the City of the Saints itself. Ogden lies in the great plain of the valley, but from the low railway-station you see in the distance long ranges of mountains, more picturesque than almost any distant view you have had thus far; and all about the town are green fields—yes, positively fenced-off fields—and beyond them the prairie, but here no longer without trees.

From here a branch road may be taken down to Salt-Lake City to see the curious civilization there. "It lies in a great valley," says an accurate description of this city of the Mormons, "extending close up to the base of the Wasatch Mountains on the north, with an expansive view to the south of more than one hundred miles of plains, beyond which, in the distance, rise, clear-cut and grand in the extreme, the gray, jagged, and rugged mountains, whose peaks are covered with perpetual snow. Adjoining the city is a fine agricultural and mining region, which has a large and growing trade. The climate of the valley is healthful, and the soil, where it can be irrigated, is extremely fertile. . . . The city covers an area of about nine miles, or three miles each way, and is handsomely laid out. The streets are very wide, with irrigating ditches passing through all of them, keeping the shade-trees and orchards looking beautiful. Every block is surrounded with shade-trees, and nearly every house has its neat little orchard of apple, peach, apricot, plum, and cherry trees. Fruit is very abundant, and the almond, the catalpa, and the cottonwood-tree grow side by side with the maple, the willow, and the locust."

So it will be seen that even a city on the Plains has elements that entitle it to a place in this record of the picturesque, and that it is not as other cities are. But after doing the few sights, and when you have enjoyed the magnificent view from the back of the city of the valley and the snow-capped peaks which lie on the other side, you have seen all of Salt-Lake City, and have time, if you have risen early, to bathe at the sulphur spring.

The great inland sea should be seen. It is a remarkable sight from any point of view, and as you come suddenly upon it, after the long days of travel, in which you have seen only rivers and scanty brooks, it seems almost marvellous. A great expanse of sparkling water in the sunshine, or a dark waste that looks like the ocean itself when you see it under a cloudy sky, it is an outlook not to be forgotten in many a day.

West from Ogden lies the second great reach of the long overland journey. Salt-

e River.

ɔf human-
ıigh order
to mark the half-way
crossing of the Plains.
t Ogden, and the con-
·d the western coast is
ral Pacific Railway. It passes westward
station which derives its life and pros-
its communication with the Utah silver-
Promontory—properly, it seems, called
" which appears a strange bit of tautol-
oteworthy place, and one which all his-
·e ought to celebrate, each after his manner. Close by the station—
ches after skirting the shore of the great Salt Lake for a little time
curving away, the great iron line, pushed westward from the east, met
ich for many months had grown slowly toward it from the west—the
ron chain were riveted. There were jubilant ceremonies when the
g the road came at last, on the 10th of May, 1869. A rosewood
st rails; and solemnly, in the presence of a silent assembly, a golden
·ith silver hammer—the last of the thousands on thousands of fasten-

ings that held together the mightiest work made for the sake of human communication and intercourse in all the world.

Beyond the memorable Promontory comes a dreary waste—the dreariest that has yet been passed, and perhaps the most utterly desolate of all the journey. Nothing lives here but the hopelessly wretched sage-brush and a tribe of little basking lizards; yes, one thing more—the kind of gaunt, lank animals called "jackass-rabbits," that eat no one knows what on this arid plain. The horizon is bordered by bare, burned mountains; the ground is a waste of sand and salt; the air is a whirl of alkali-dust. Kelton; and Matlin, and Toano, dreariest of Nevada stations! Could any man wish his direst enemy a more bitter fate than to be kept here in the midst of this scene for a decade?

There is some mineral wealth farther on, hidden near the route of the railway; but apart from this there would seem to be nothing useful to man obtainable from all this region. We dash across the sterile space in a few hours; but imagine for a moment the dreary time for the old emigrant-trains, which came on to these gusty, dusty levels in old days, and found neither grass, nor water, nor foliage, until they came to the wells at Humboldt, blessed of many travellers, lying close together within a few hundred yards of the present road, and surrounded with tall, deep-green herbage. There are nearly a score of these grateful springs scattered about in a small area; and they are of very great depth, with cool, fresh, limpid water.

They herald the approach of another and a different district, for now we soon come to the Humboldt River itself, and for a time have all the benefit of the growth of trees along its sides, and the fertility that its waters revive along its course. The soil here is really arable; but go a little distance away from the river, and the few water-pools are alkaline, and the land resumes the features of the desert-soil. The scenery here, in the upper part of the Humboldt Valley, is for a time varied, and in many places even wild and grand. The road winds through picturesque cañons, and under the shadow of the northernmost mountains of the Humboldt Range, until the important station of Elko is reached.

For now we have a little more of sage-brush and alkali, ant-hills and sand. Let him who passes over the Humboldt Plains on a hot August day, and feels the flying white dust burning and parching eyes and mouth and throat, making gritty unpleasantness in the water wherewith he tries to wash it away, and finding lodgment in every fold of his clothing, be sufficiently thankful that he is not plodding on with jaded horse by the side of a crowded emigrant-wagon, with days of similar journeying behind him, and some of it still to come.

He will be glad to come near to the refreshing grandeur of scenery of the Palisades—though the finest of this is not seen without leaving the established route and penetrating a little into the mountains at one side. It is here that you come upon such glimpses and vistas as the one Mr. Moran has drawn—breaks in the rocky wall,

ne
lly
ic-
iot
in
of
petual steam, of sul-
he ground is tinged
s, as at the geysers
around us, too, are mining dis-
n old and exhausted, some still

nd the emigrant-road have long
se of the Humboldt River, but
in sight after Battle Mountain
old Indian combat—is passed;
ost to view altogether, and the road runs by the fresh, bright-looking
Humboldt itself; past Golconda, and Winnemucca, and Lovelocks, and

DONNER LAKE, NEVADA.

ave histories; and finally Wadsworth is reached. And here the
Sierras fairly begin. The monotony of the view begins to change;
ıbout us, as we enter the well-named Pleasant Valley through
flows, and at last, passing through well-wooded land again, reach
city in the wilderness, standing among the very main ridges of the
rn—the first of the stations within the actual limits of California—
t place of several thousand inhabitants; a place that has had its
l, its riots, and adventures not a whit behind those of the larger
toward the interior of the State.
shores of its river lie the noblest scenes; the tall cliffs are ragged
ee-crowned; the rock-broken water ripples and thunders through
tches of fertile plain; and the buzzing saw-mills of an incipient
a homelike, New-England sound on its banks. From the town
ges of luxury and civilization, too—carry the traveller to the beau-
own Donner Lake, only two or three miles away. The great sheet
l water lies high up in the mountains, between steep sides, and in
ldest and most picturesque of the scenery of the Sierra summits.
e is very great, but its waters are so transparent that one can look
into them; they are unsullied by any disturbance of soil or sand,
formed almost entirely of the solid rock.
d have more perfect beauty than this mountain-lake, and its even
or, Lake Tahoe, some fifteen miles farther to the south. The scene
me. Though it lies under the unbroken sunlight through a great
weather, there is perpetual variation in the great mountain-shad-
nd calm on the surface. There is a climate here that makes almost
. It is neither cold to chilliness nor warm to discomfort, but always
, inspiring with a kind of pleasant and energetic intoxication.
ere in the Sierras, the rock-forms are picturesque and grand at all
Castellated, pinnacled, with sides like perpendicular walls, and sum-
platforms, they give a strangely beautiful aspect to every shore and
The road, twelve miles in length, by which Lake Tahoe is reached
s some of the most remarkable and memorable views of these forma-
singularities of outline, that can be obtained in any accessible region
range; and it would be impossible to find a more glorious drive
the edge of the river-bed, over a well-graded path, through the very
noblest groups of the Sierra chain.
dred miles," says Clarence King, who knows these mountains better,
other American, "the Sierras are a definite ridge, broad and high, and
of a sea-wave. Buttresses of sombre-hued rock, jutting at intervals

LAKE TAHOE.

Giant's Gap.

from a steep wall, form the abrupt eastern slopes; irregular forests, in scattered growth, huddle together near the snow. The lower declivities are barren spurs, sinking into the sterile flats of the Great Basin.

"Long ridges of comparatively gentle outline characterize the western side; but this sloping table is scored from summit to base by a system of parallel, transverse cañons, distant from one another often less than twenty-five miles. They are ordinarily two or three thousand feet deep—falling at times in sheer, smooth-fronted cliffs; again, in sweeping curves, like the hull of a ship; again, in rugged, V-shaped gorges, or with irregular, hilly flanks—opening at last, through gate-ways of low, rounded foot-hills, out upon the horizontal plain of the San Joaquin and Sacramento. . . .

"Dull and monotonous in color, there are, however, certain elements of picturesque-

›ne. Its oak-clad hills wander out into the great plain like coast
ng yellow, or, in spring-time, green, bays of prairie. The hill-forms
:tch in long, longitudinal ridges, broken across by the river-cañons.
red earth, softly-modelled undulations, and dull, grayish groves, with
owns, dotted ranches, and vineyards, rise the swelling middle heights
road, billowy plateau, cut by sharp, sudden cañons, and sweeping up,
·b growth of coniferous forest, to the feet of the summit-peaks. . . .
›per limit, the forest-zone grows thin and irregular—black shafts of
firs clustering on sheltered slopes, or climbing, in disordered proces-
and rocky faces. Higher, the last gnarled forms are passed, and be-
ɜ rank of silent, white peaks—a region of rock and ice lifted above

The San Joaquin River.

fe. In the north, domes and cones of volcanic formation are the summit
t three hundred miles in the south it is a succession of sharp granite
crags. Prevalent among the granite forms are singularly perfect conoidal
symmetrical figures, were it not for their immense size, would impress one
artificial finish.

lpine gorges are usually wide and open, leading into amphitheatres, whose
ıer rock or drifts of never-melting snow. The sculpture of the summit is
ly glacial. Beside the ordinary phenomena of polished rocks and moraines,
eneral forms are clearly the work of frost and ice; and although this ice-
nly feebly represented to-day, yet the frequent avalanches of winter, and
d mountain flanks, are constant suggestions of the past."

OAKS OF OAKLAND

e of the whole Sierra—of all of it save that which lies above the
e life—is its magnificent forest covering. It may well be doubted
:sts of pine is ever seen in greater perfection than is found here.
oble shafts are the very kings of trees. Covering the great slopes
of sombre green, they lend a wonderful dignity to the peaks, as
1 from a distance; and, to one already in the forest, they seem the
the mountain-sides. They are magnificent in size, as they are ad-
on. No mast or spar ever shaped by men's hands exceeds the
: of their straight, unbroken trunks. They are things to study for
s individual trees, apart from their effect upon the general landscape,
them would be wild and picturesque enough.

isses on from Truckee, climbing a gradual slope to Summit, fifteen
iighest station on the Central Pacific, though still lower than Sher-
:y Mountains. Summit, standing at the highest point of this pass
is at an altitude of seven thousand and forty-two feet above the level
ɔ reach it, the track has ascended twenty-five hundred feet in fifty
hundred and four miles between this and Sacramento, on the plain
it must again be made to a point only fifty-six feet above sea-level.
the journey—the western descent from Summit—is one that is most
glorious period of sunrise. There can be no more perfect scene. The
the edges of great precipices, and in the deep cañons below the shad-
;. Those peaks above that are snow-covered catch the first rays of the
th wonderful color. Light wreaths of mist rise up to the end of the
id then drift away into the air and are lost. All about one the aspect
s is of the wildest, most intense kind; for by that word "intense"
to be expressed of the positive force there is in it that differs utterly
if such a scene as lies passive for our admiration. This is grand; it is
is no escaping the wonder-working influence of the great grouping of
avines, of dense forests, and ragged pinnacles of rock.

ie mountains seem to fade away, and before we realize it we are among
hose oak-clad or bare brown hills that "wander out into the great plain
ontories, enclosing yellow, or, in the spring-time, green bays of prairie"—
on the plain of the San Joaquin. We might fancy ourselves back again
s were it not for the still farther range of heights before us. These are
npicturesque, outlying hills, and we dash through them by Livermore's
assed Sacramento, and speed on our way toward the coast.

resque is over; and already the hum of the still distant city seems almost
ars as we dash in under the great green oaks of Oakland.

THE EASTERN SHORE, FROM BOSTON TO PORTLAND.

WITH ILLUSTRATIONS BY J. DOUGLAS WOODWARD.

Pulpit Rock, Nahant.

THE coast of New England between Boston and Portland is for the most part irregular and rocky, and in many spots picturesque. Nature seems to have supplied it with every variety of sea-coast aspect and beauty, from the jagged mass of frowning and rough-worn rock overhanging the waters to the long, smooth reach of broad, curving beaches, and the duller landscape of green morass extending unbroken to the water's edge.

The picturesqueness of the Eastern shore betrays itself as soon as you have steamed away from the Boston docks. Eccentric and irregular peninsulas of land, abruptly widening and narrowing, now a mere thread between water and water, now a wide, hilly

at once. The steamboat is forced to make many a curve and ıfter leaving East Boston, passes through a straitened channel be- ›w Point Shirley, a mere needle of a peninsula, and the irregularly-

'und Point Shirley, the broad stretch of Chelsea Beach comes into

Swallows' Cave, Nahant.

; from the lower part of the peninsula to Lynn Bar. Beyond Pine's the strip of land at the northern end of Chelsea Beach, the sea makes ıpt invasions into the line of coast, and has scooped out there a minia- .th uneven coast borderings, called Lynn Bar. This is the inlet to the r city," which stands just by, intent on supplying mankind with shoes.

Lynn Bar is bounded on its eastern side by the long and slightly curved western side of the peninsula of Nahant. From this point of view you can form no conception of the noble picturesque beauties and architectural decorations which this bold and strangely-shaped promontory affords.

Nahant is about eight miles northeast from Boston, and is easily reached from the city by boat. The peninsula, as it stretches out from the main-land, is at first a narrow neck, crossed by a few steps, for some distance almost straight. On one side is the pretty harbor of Lynn; on the other a noble, wide beach, sweeping in a direct line for some distance, then curving, in a short semicircle, round the rocky cliffs beyond which lies the lovely and famous Swampscott. This narrow neck begins anon to thicken irregularly, with here and there a sudden eruption of ragged rock, and finally broadens into a rocky, uneven eminence. This promontory is shaped like a horseshoe. On the

The Old Fort, Marblehead.

two sides the shore is rocky, with its Black Rocks, West Cliff, Castle Rock, Saunders's Ledge, Natural Bridge, and so on; while in the convex side of the horseshoe are several exquisite diminutive beaches lying below the jagged eminences. A writer, describing the rocky beauty of Nahant, says: "The rocks are torn into such varieties of form, and the beaches are so hard and smooth, that all the beauty of wave-motion and the whole gamut of ocean eloquence are here offered to eye and ear. All the loveliness and majesty of the ocean are displayed around the jagged and savage-browed cliffs of Nahant."

The artist has reproduced two of the most striking of the many natural wonders which the eternal lashing of the waves has wrought out of the obstinate rock-masses about Nahant. Pulpit Rock lies just by the lower eastern shore of the horseshoe, between the Natural Bridge and Sappho's Rock. It is a huge, jagged mass, rising some

water,
sides,
w, but
ito an
legrees
tle dis-
appears
which
s Bible
re been
—hence
, if one
venture
s-grown
, whence
sea, sit-
its wash
Swallows'
, at the
eastern
shoe. It
y cavern,
me of ir-
ed togeth-
ing layers.
feet high
long, and
om having
d by colo-
which built
ts sombre
w in and
multitudes;
n of their
ity-seekers
em thence.
other won-
more strik-
John's Peril,

Salem, from the Lookout on Witches' Hill.

a great, yawning fissure in one of the cliffs; the huge, oval-shaped mass called Egg Rock; Castle Rock, a beautiful structure, which might almost be taken for a mediæval fortress, with battlements, embrasures, buttresses, and turrets; a boiling and seething Caldron Cliff; a deep-bass Roaring Cavern; and a most grotesque yet noble arch, with a cone-like top, and leading to a natural room in the rock, to which the romantic name of Irene's Grotto has been given.

Beyond the broad Long Beach, which sweeps from the promontory of Nahant in almost a straight line to Red Rock, is the fashionable sea-side resort of Swampscott, with its Dread Ledge, and pretty beach, and clusters of charming marine villas; while just northeastward of Swampscott juts out far into the sea the rude and uneven and

Norman's Woe, Gloucester.

historic peninsula of Marblehead. The sea penetrates the peninsula with a narrow and deep little harbor, and it is around this that the town has clustered. Once on a time Marblehead was famous for its fishermen; and it is the scene of Whittier's poem, "Skipper Ireson's Ride." The Old Fort is a plain, hoary-looking edifice, standing on the rugged slope of the promontory looking toward the sea.

Just around the extremity of Marblehead are the harbor and the still more ancient Puritan settlement of Salem. Of all New-England towns, it bears most plainly the stamp of a venerable antiquity. It is a grave and staid place, and there are still streets largely composed of the stately mansions of the colonial and marine aristocracy; for Salem was once not only a metropolis, but a port teeming with lordly East-Indiamen, and warehouses packed with the choicest fabrics and spices of the Orient. Salem was

and it
ented by
h a fine
que and
–that the
suspected
and spells
anged by

coast, after
Iarbor, you
reach the
ing penin-
which is
1 forms the
f Massachu-
d between
1 the south,
rcular basin
the spacious
The coast
1 Gloucester
ots at once
and histor-
The rocky
pears in full
Opposite to
nd, is Bever-
he old town
a few years
alem, in the
A little to
is Wenham,
arming lake;
Ipswich, with
," and its an-
minary, where
udents, says a
"are wont to

Bass Rocks, Gloucester.

GLOUCESTER AND ROCKPORT.

ves wives of the daughters of the Puritans." The quaint village of Man-
the rugged shore, and, soon after passing it, Gloucester is reached, long
of the northern fisheries.
is a characteristic New England sea-coast town. Its harbor is one of the

Point of Cape Ann, from Cedar Avenue, Pigeon Cove.

esque and attractive on the coast; and the town rises gradually from the
senting at once the aspect of venerable age and of present activity. All
e fine points of view seaward, beaches, and rocky cliffs, with a greater share
of verdure than along the more southerly coast.

PORTSMOUTH AND ISLES OF SHOALS.

PORTLAND HARBOR, AND ISLANDS.

Cushing's Island.

watering-place; for here is not only a noble view of the waters, but the opportunity to enjoy its wide avenues and promenades, with groves of oak and pine, which lead to striking landscape-views—among them the Breakwater, which forms the outer wall of the snug little cove, and Singer's Bluff, which overhangs the sea.

Passing from the varied beauties of Pigeon Cove, the northern side of Cape Ann is crossed by an ancient road, which at times enters beneath an arching of willows, and again emerges in sight of the waves and sails. The coast thereafter for a while lacks any peculiar characteristics of picturesqueness, until the broad, bay-like mouth of the "great" river Merrimac is approached. From its entrance is espied the old, historic town of Newburyport, surmounting an abrupt declivity some three miles up the broad and rapid river. Few places more abound with old traditions and family histories, and few inspire more pride in their annals and past glories in the breasts of the natives.

The shore between Newburyport and Portsmouth is

ght and even. The abrupt eccentricities of boulder and storm- nearly disappeared. Long and sunny beaches have taken the ıs and yawning fissures, sinuous inlets and shapeless projections. Rye, occupying the larger portion of the brief coast which New long stretches of sand, interspersed with rocks, but presenting ·e cheerful than the rugged aspects of marine Nature.
wburyport, is on a river-bank, some three miles from the open ːs bay between it and the Maine shore, with an island directly in a singularly venerable and tranquil-looking old place, with many hich look as if they had been slumbering for generations.
:raction in the vicinity of Portsmouth is the Isles of Shoals, a rugged islands lying about nine miles off the coast. The isles largest—Appledore—containing only about three hundred and ıin-land they appear shadowy, almost fairy-like, in their dim out- cipal island of the group, rises in the shape of a hog's back, and)earance. Its ledges extend some seventy-five feet above the sea, .rrow, picturesque little valley, wherein are here and there timid where are the few buildings on the island. The solitude and to be enjoyed to the fullest on these gaunt rocks, in whose in- ook may be discovered where, fanned by cool breezes of pure ıpe may be contemplated amid a surrounding stillness broken r, and trickling in and out of the waves. Just by Appledore ow, flat, and insidious, on whose black reefs many a stalwart destruction. A quarter of a mile off is the most picturesque Island, with its odd little village of Gosport, the quaint tow- of which crowns the crest of its highest point; and just by is west, toward the main-land, is Londoner's, jagged and shapeless, while two miles away is the most forbidding and dangerous k Island—many of whose ledges are hidden insidiously beneath nd at low tide are often seen covered with the big, white sea- ıabited isles.
the coast runs tolerably even for some distance northward, ·r, bends gradually to the northeast, until the isle-crowded en- :eached. It is dotted all along with marine hamlets and fishing- a bit of broken beach, and now and then a slight promontory 'ork Beach is the principal sand expanse between Portsmouth ; gently to the water from the eminences behind. The coast :auty north of York, and affords ample opportunities for fisher- rs by the ocean.

Nothing could be more strikingly picturesque, however, than the marine scenery about Portland. One of the largest and most attractive spots in its harbor is Cushing's Island, the edges of which are bordered by high bluffs crowned with shrubs and turf, with here and there a low, rocky shore or a graceful inlet. The view from here is perhaps more various and extensive than from any other point, for it includes the harbor, ship-channel, and city, on the one hand, and the towering ledges of Cape Elizabeth on the other. Forts Preble, Scammel, Gorges, and Portland Light loom in the near distance; the busy wharves of Portland are seen crowded with their craft of many climes; the neighboring islands present each a novel and contrasted aspect of shape and color; the heavy sea breakers may be seen settling themselves into the smooth, blue ripple of the bay; and sometimes a glimpse is had of the snowy summit of Mount Washington and its sister eminences, dimly outlined on the far northwestern horizon.

Portland, from Peak's Island.

.LLEY OF THE HOUSATONIC.

'H ILLUSTRATIONS BY J. DOUGLAS WOODWARD.

Housatonic.

few New-
vers of any
gth which do not present, in
ieir flow, not only a great va-
a striking contrast of aspects.
River is no exception to this
s in the beautiful Berkshire re-
chusetts, where its first ripples
ts of grand hills; and, after flowing for a century of happy miles amid
not suffer it to quite forget its mountain-cradled laughter, it glides gravely

Housatonic Valley, near Kent Plains.

enough through the plains of old Stratford, on the Connecticut shore, and is lost thereafter in the expanse of Long Island Sound.

The beauties of the Housatonic Valley were little known, and still less pictured, before the opening of the railway that connects the sea-coast of Connecticut with the mountains of Massachusetts. It begins at the thrifty city of Bridgeport, and enters the valley of the Housatonic only above Brookfield; thence it traverses the valley closely through nearly all its remaining extent; and there are few stations beyond at which the tourist might not tarry, and, with brief excursions to the right or left, fill his eye with the charms of mountain outlines, valley reaches,

r water-falls. There is, therefore, quite a long interval of the
c which the tourist cannot, if he would, follow by the railway.
ꞓ it, for its first half-score of miles, from Stratford, on the rails
and this will afford him pleasing glimpses of the river where it
Naugatuck, and where the busy manufacturing interests of such
rmingham subsidize and utilize the water-power of the streams,
ɹresqueness of appliance or effect.

The Housatonic at Derby.

ıt span the rivers here, one at least is pretty enough to have
tist; and, with the accessories of fine old elms, and the placid,
tream, it can hardly fail to renew its fascination on the page.
ew Milford there is too little, perhaps, of the romantic in this
tempt any one but the determined pedestrian to follow the

Old Furnace at Kent Plains.

he is traversing, girt closely on the west by almost abrupt hill-sides, and, on the other hand, spreading out into sweet pastoral reaches and green undulations.

The old furnace, which the artist has so faithfully reproduced with his pencil, will suggest to the mind one of the industries of the Housatonic Valley—the working of the iron which is found in many localities.

A day or two may be well spent between Kent and Canaan—a northward reach of twenty-five miles, which brings the valley of the Housatonic close upon the dividing line between Connecticut and Massachusetts. This interval is rich in picturesque delights. The lofty ridge has now assumed a true mountain aspect, and lifts up, here and there, noble crowns to the sky.

Falls Village is the centre of some of the chief attractions of the section under notice. Close at hand are the falls of the Housatonic—the most prominent, perhaps,

Old Bridge, Blackberry River, near Canaan.

of the cataracts in Connecticut. They are worthy of attention, but it is difficult to avoid some feeling of vexation on finding that near views of them are blemished by the unsightly encroachments of that barbarism which, under the misnomers of "civilization" and "progress," clutter our water-falls and rapids with ugly shanties and shops. These falls are commonly known as the Canaan Falls, and fill up the whole breadth of the

umultuous dash and roar over a steep, terraced ledge of dark rock. bly exceeds fifty feet; and seen at a distance, and especially under c of the moonlight, they inspire no small degree of admiration in

Old Mill, Sage's Ravine.

rises about two miles from these falls, in a northwesterly direction; nay be reached in a carriage, by the road which the woodmen fol- When gained, it opens to the view of the tourist such a scene few other mountain crests in the valley, though some are of more ıe great bosom of the interval between the east and west ranges of hills is heaving with its green billows beneath him. A thousand wavy crests are in his view; and, threading its way near and afar, the silvery line of the river stretches amid picturesque homesteads, which now and then cluster into villages. A deep, dark, and ugly fissure into wild, outlying rocks, at the foot of this mountain, bears the ap-

Silver Cascade, Sage's Ravine.

propriate but not attractive name of the Wolf's Den.

Within an hour's walk of the Great Falls lies the pretty village of Salisbury, which is the social centre of the beautiful and populous county of Litchfield. Lying close under the deep shadows of the great Taconics, Mount Riga may be said to be its especial guardian, whose noble crest, known as Bald Peak, alternately smiles upon it in sunshine and frowns upon it in storm.

It would carry the reader quite out of the Housatonic Valley to press him beyond Bald Peak on to the Dome, and westward still, a dozen miles, until we came to the

Bashbish, and its grand but gloomy water-fall, closely overlooking g village of Copake, in New York.
ısing the ridges of the Taconic, and quite within the legitimate com- it is proper for us to explore a mountain gorge less known than of the terrible but with far more of the beautiful in its aspect. ıt an easy walk—or a delightful drive, if preferred—of four miles

f this noble ravine there are a fine old mill, and a picturesque bridge which comes dashing and foaming down the wild cleft. Leave the

Mount Washington, from Sheffield.

the clatter of the mill-gear behind, and go up the ravine, with ou the possible paths. There is hard climbing before the Twin ıre reached. Your feet will sink in clumps of moss and decayed if you are not wary. You must cling to birch-boles, and often to u swing round opposing barriers of rock. You may get a foot- ıu cross the foaming torrent to find an easier path on the other there, all along the wild way, are pretty cascades, tortuous twists lichened or dark-beetling rocks, mossy nooks or gloomy tarns, and, overhead, maples and birches, mingling their rare autumnal splendors of red and gold with the sombre greens of hemlocks, and cedars, and pines.

Prospect Rock, East Mountain, Great Barrington.

A week in Salisbury would be none too much time for the leisurely enjoyment of the many charming views to be found in its neighborhood. There, very near to the iron-smelting hamlet of Chapinville, spread the sweet waters of the Twin Lakes—the Washinee and Washineën—encompassed by winding drives, with ever-shifting visions of the kingly Taconic crests, and these, on the nether slopes, displaying, in the bright autumn days, such splendors of variegated color as would intoxicate with delight the heart of a devotee of illuminated missals. These pretty lakes lie in enticing proximity to a limestone cave, into which the tourist may be induced to venture by the promise of rare visions "of stalactites and stalagmites."

Canaan, near the outgoing of the river and valley from the Connecticut border, is an important station on the intersection of two railways. A pretty village in itself, it has its special picturesqueness along the pleasant little valley of the Blackberry River, on whose banks it lies.

Leaving it, the tourist crosses almost immediately the southern boundary-line of the renowned Berkshire County, a region not surpassed, in picturesque loveliness, throughout its whole longitude of fifty miles and its average latitude of twenty miles, by any equal area in New England, and perhaps not in all this Western world. Off the railway, in village nooks, in glens and by-ways, upon near crests and remote hill-tops, the lover of the beautiful will find innumerable views to gaze upon, to sketch, or haply to impress only on his memory.

Sheffield is a good lingering point for those who do not wisely shun, amid Nature's

the engine and the sharp click of the electric hammer. From
ɔunt Washington—one of the Taconic giants—is easily made,
ill be a cheap purchase of "far prospects," exchanged for the
.t its foot. Mount Washington was once a part of the great
ts summit commands a view of the rich and lordly domain
r half-forgotten name.

Green River, at Great Barrington.

hardly do better than to take up his abode for a little while in
w miles east of the railway, and just under the lofty crest of
ce, at his own sweet will, he may go and climb or ramble. He
in, by way of "its vast, uncultivated slope, to a height of two
—to his astonishment, if not before informed—he would find a
twelve score of inhabitants are literally mountaineers, and whose
aily outlook, with such a panorama as a sensitive valley or sea-
ɔ into ecstasies to behold. It is not finer, perhaps, though far
ainable from Prospect Mountain; but then it takes in half the

Monument Mountain.

whole stretch of the Housatonic River, and below the eye lie lakes and woodlands, lawns and villas, gleaming spires, and little rifts and puffs of smoke from furnaces and creeping engines; and all this so far away, so still, that it is more like a picture on canvas than a real scene. East and west, the eye has broad extent of vision into Connecticut and New York. The Catskills make a blue and wavy western horizon; and the Hudson, in the interval, twins the nearer Housatonic in its sparkling flow.

The practical man, who shuns the toilsome clamber to Mount Everett's crest, may go afoot, or in his light wagon, from his inn, to see the famous marble-quarries of Egremont, whence were hewn the white columns and walls of the Girard College, more than half a century ago.

Great Barrington is a most attractive point in the valley of the Housatonic. The river, losing all the while in volume, is gaining in picturesqueness. Its narrowing banks

›velier fringes, and its tones ring more musically in the swift, broken,
es of its waters. Barrington has many summer charms—in its splen-
its streets, in its attractive drives over fine roads, and in its pleasant

the road that leads to the two Egremonts—North and South—will
› a charming bit of land-and-water view at the rural bridge over Green
stream that flows along as if in sweet and delighted consciousness of
and there discloses.

a' great mistake of the explorer of Berkshire to go from Barrington
rail. The highway is shorter by nearly two miles, and not a furlong
or tedious, for it is thick set with those sweet surprises that charac-
in Berkshire.

wonder is the renowned Monument Mountain, which Stockbridge num-
›le pride, among her special attractions. This mountain was called by
:w Indians—the old Stockbridge tribe—"Maus-was-see-ki," which means
est." Its present appellation was given to it, perhaps, on account of a
its southern crest, which has connected with it an Indian myth of a
ho, disappointed in love, jumped from the precipice and was killed—a
e which the braves commemorated by flinging a stone upon the fatal
hey passed by it.

ument Mountain to the village of Stockbridge is less than half an hour's
carriage-road has been regained. There are few—if, indeed, there are
our land that can rival it in rare and fascinating aspects of rural beauty,
urroundings of unwonted charms, in worthy and precious historical asso-
beauties of Stockbridge lie in many directions. To the north, the pretty
c—more familiarly known as the "Stockbridge Bowl"—spreads its trans-
shapely, in its outline, as a gigantic basin, on whose margin Hawthorne
a succession of seasons. A mile or more from the village is found that
ture, the Ice Glen, which pierces the northern spur of Bear Mountain;
g and awsome corridors and crypts, formed by massive and gloomy rocks
prostrate trees, the explorer may find masses of ice in the heart and heat
r. The passage of this glen, though not perilous, requires nerve and pa-
e cheer of glowing torches withal. The heights that overhang the village
for situation," and studded with pleasant villas, whose fortunate possessors
will over the fair interlocking valleys of the Housatonic and the Konkapot.
ng Stockbridge, the tourist may scarcely venture to promise himself a beauty
he has already enjoyed; and this may be suggested without disparagement
scenery of Northern Berkshire. It may hardly be doubted that the rare
is attractions of this whole region—so aptly called "the Palestine of New

Housatonic River at Stockbridge.

England"—are crystallized, in excess of loveliness, around Stockbridge as a nucleus. If this verdict had gathered something of weight to the judgment from the acknowledged union in Stockbridge of all the forces—natural, historical, social, intellectual, and religious, alike—which have given to Berkshire its enviable renown, the influence would be, nevertheless, legitimate and just.

Lee and Lenox are the two villages that lie in the Housatonic Valley between Stockbridge and Pittsfield, which latter village is rapidly growing into the rank of a city, and is the metropolis of all the Berkshire region.

Lee has a pretty lake, within a pleasant half-hour's walk on the road to Lenox; but, for heavier charms, its summer guests make excursions to quaint old Monterey and to Tyringham, on the east, and to Lenox and Stockbridge, between which places it is about equidistant.

Of Lenox—reached by a drive of constantly-increasing picturesqueness—these chronicles can make but inadequate mention. Professor Silliman designated it, in his enthusiastic admiration of its pure, exhilarating air and its lovely views, "a gem among the mountains." It deserves the praise. Till recently it was the shire-town of the region, and term-time gave it a measure of importance and influence which it has since lost. But it cannot lose its beauty, and the summer doubles its population with hundreds of happy pilgrims from the cities, some of whom occupy their own villas, while more crowd its hotels and the numerous boarding-houses.

Ice Glen, Stockbridge.

the crests and slopes of its constituent and outlying hills are ıd villas, which one might remember for their architectural indi- ıot always eclipsed by the surpassing breadth and beauty of the ding vicinity.

:h the "Ledge" contributes to the embellishment of this paper commentary on the breadth and manifold charms of the Lenox landscape. The summer guests of Lenox find great delight in. gazing out from its noble "coignes of vantage." For still wider range of vision they go to Perry's Peak, a bald and lonely summit on the west, easily reached in an hour's ride, and standing like a grim sentinel on the New York border. There is a scientific interest, also, about

Lenox Station.

Perry's Peak, in that it is strewed with the fine boulders which are traced in seven parallel lines across the Richmond Valley, intervening between the peak and Lenox Mountain. These stones attracted the careful notice and diligent review of that eminent English geologist, Sir Charles Lyell.

Among the attractive points included in the magnificent overlook from the peak

View from the "Ledge," Lenox.

are the Shaker villages of both Lebanon, in New York, and Hancock, in Massachusetts, the former being, perhaps, the metropolis of the sect of Shakers. The Boston and Albany Railway passes close by the village of the Hancock Shakers, and has a station there. The town of Hancock is itself one of the outlying characteristics of the Housa-

Banks of the Housatonic, at Pittsfield.

together mountainous, being only a long and narrow tract on the
f the Taconic range, with a single hamlet crouching in a beau-
near the northern end of it. The roads which cross this attenu-
r romantic and very rough, except, perhaps, those from Lebanon
direct, which are fine in summer, and much travelled.

ell may be—is that which takes in and overpasses the exquisite
.ke, two miles to the west. This view, besides its immediate love-
sheen of its waters, and the sweet variety of the pastoral and
ıviron them, has for its central but remote background the splendid

> "Graylock, cloud-girdled on his purple throne."

the outlet of a contiguous pond, is a wild mountain-torrent that
of the mountain in a rugged cleft known as Tories' Gorge. This
ıf the eastern branch of the Housatonic. To the eastward, also,
lton, with its busy paper-mills; and beyond it, on the acclivity of
:he village of Hinsdale, from which point, as also from Dalton, the
Falls may be reached by a brief carriage-drive. These falls lie at
the review which this article will make of the Housatonic Valley.
nding waters" narrow into shining becks and brawling brooks.
l, beyond Onota already named, a mountain-road leads across Han-
ɔn Springs, and to the village of the Lebanon Shakers, affording,
ɔspects, but, from its highest point, a scene never to be forgotten.
expanse of the sweet vale of Lebanon, and, beyond this, stretches
vague and violet-hued.
ıota, on the slopes of the Taconics, are found delightful bits of
u Cascade, a much-frequented haunt of those who fain would find
ıs" hides; there, Rolling Rock, a huge and nicely-poised bowlder;
the table of a giant crest, as pretty a mountain-lake as the eye
ılled Berry Pond, but not for the profusion of raspberries to be
er. The lakelet has crystal waters, a sparkling, sandy beach, is
:vergreen and deciduous trees, and to these charms adds that of a
ı all sounds upon its margin.
tsfield lie Pontoosuc, a populous mill suburb, and a lake bearing
iles beyond, old Lanesboro' is reached by a delightful drive. Here
fail to make a slight circuit, and gain, either afoot or in a carriage,
ıtion Hill, lying just west of the village and the iron-furnace.
ı unique and delightful village, with a green park for its main
ng, hurrying Hoosac singing along its borders. It is a fit place
ing one for summer life and recreation, though hardly for fash-
which, indeed, its vigilant wardens evermore oppose their classic
Villiamstown, who are familiar with Swiss scenery, are wont to say
's and wonderful atmospheric effects they see there more nearly
ɛs than those of any other mountain recesses in this land.
the railway from Pittsfield to this region passes twenty miles

through a country contrasting strangely with the deep rural isolation of that just glimpsed along the by-road through New Ashford. It is a tract of new activities and industries, of glass-furnaces and sand-quarries, of lumber-mills and cotton-looms, of woollen-mills and populous hamlets—in succession, Berkshire, Cheshire, South Adams, until he comes at last to North Adams.

Indeed, North Adams is the upper "metropolis" of Berkshire, and is more thickly

Hoosac River, North Adams.

studded about with wild and romantic spots than its southern sister. Graylock, the loftiest mountain in Massachusetts, is within easy distance by means of a fine road, though not visible from its streets.

When Graylock, and the Hopper, and Money Brook have been explored—or between these explorations, as separate adventures—there are dainty and most compensat-

Natural Bridge, North Adams.

›rth Adams which should not be left unseen. Some of these lie
·ious object, the Natural Bridge, a rare freak of the waters of a
the rocks. It is a vast roof of marble, through and under which
et contrived, with incessant, fretting toil, to excavate a tunnel—a
rds wide, and ten times as long. This wonderful viaduct is loftily

Hoosac Mountain and Tunnel Works.

arched over the torrent, and displays its marble sides and ceiling sometimes of a pure white, but oftener with strange discolorations. Through this weird corridor the brook flows with thunderous echoes, booming up to the ear and filling the mind of the beholder with strange, wild fancies. In the ravine of this brook there are many picturesque points to arrest the tourist's attention, but next in interest to the bridge itself is a strange, columnar group of rocks, which at its overhanging crest assumes, to a facile imagination, the aspect of gigantic features, and bears, therefore, the appellation of Profile Rock. These and other scenes are within a mile or two of the village, where there will be found inducements for more than ordinary lingering, and still more reluctant leave-taking, on the part of the visitor. Those who have enjoyed the magnificence and varied charms of the eight-mile coach or carriage drive from North Adams to the east end of the great Hoosac Tunnel during its long working will doubtless almost lament that it is now an accomplished fact, because the splendid road across the great Hoosacs will now be no more needed, and will very likely fall into disuse, thus spoiling a most unique and almost unparalleled mountain-ride.

The Hoosac Tunnel is a bold and fortunate feat of engineering skill. Second in

amous
under
es the
ate of
e with
ly five
d thus
ble toil
tlay, a
between
Hudson
miles
preëxist-

busy and
ebb of
ld Gray-
peers of
Hoosac
ok down
they did
noil and
aster with
rn end of
is wrought
mpleteness,
ing to the
oral, if not
beauty and
e Berkshire
ns are natu-
d this brief
the Hoosac
ords a con-
ortunity to
scription of

Profile Rock, North Adams.

of the Housatonic" to a close. In beauty of scenery it compares favor-
y section of the country, which feature is greatly enhanced by its historic

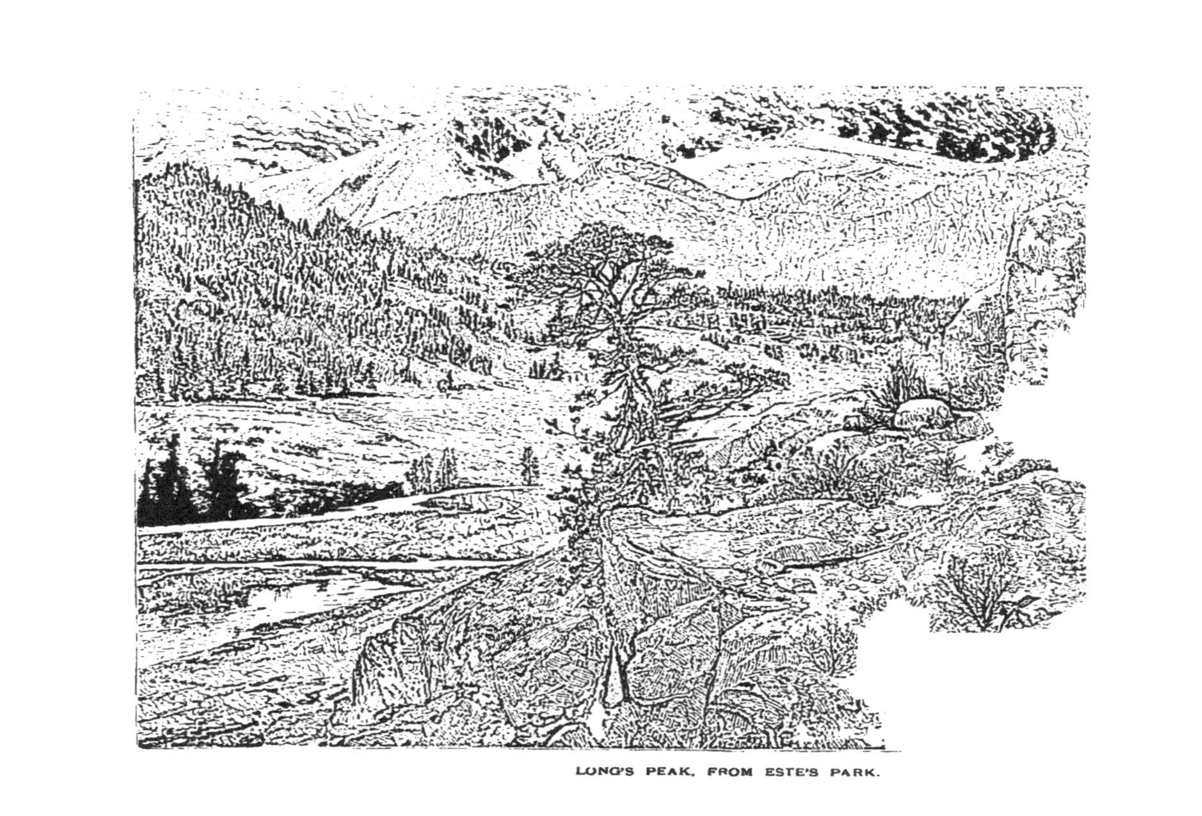

LONG'S PEAK, FROM ESTE'S PARK.

BOULDER CAÑON.

Gulf of California. On one side are the famous Gray's and Evans's Peaks, scarcely noticeable among a host of equals; Long's Peak is almost hidden by the narrow ridge; Pike's is very distinct and striking.

Until 1873 only small areas of our vast Territories had been surveyed and accurately mapped; the greater space had been unnoticed and uncared for; but in that

Gray's Peak.

year a geological and geographical survey of Colorado was made, under the able direction of Dr. F. V. Hayden, and the results have exceeded all expectations. The position of every leading peak in thirty thousand square miles was fixed, including the whole region between parallels 38° and 40° 20′ north, and between the meridians 104° 30′ and 107° west. The ground was divided into three districts, the northern district including the

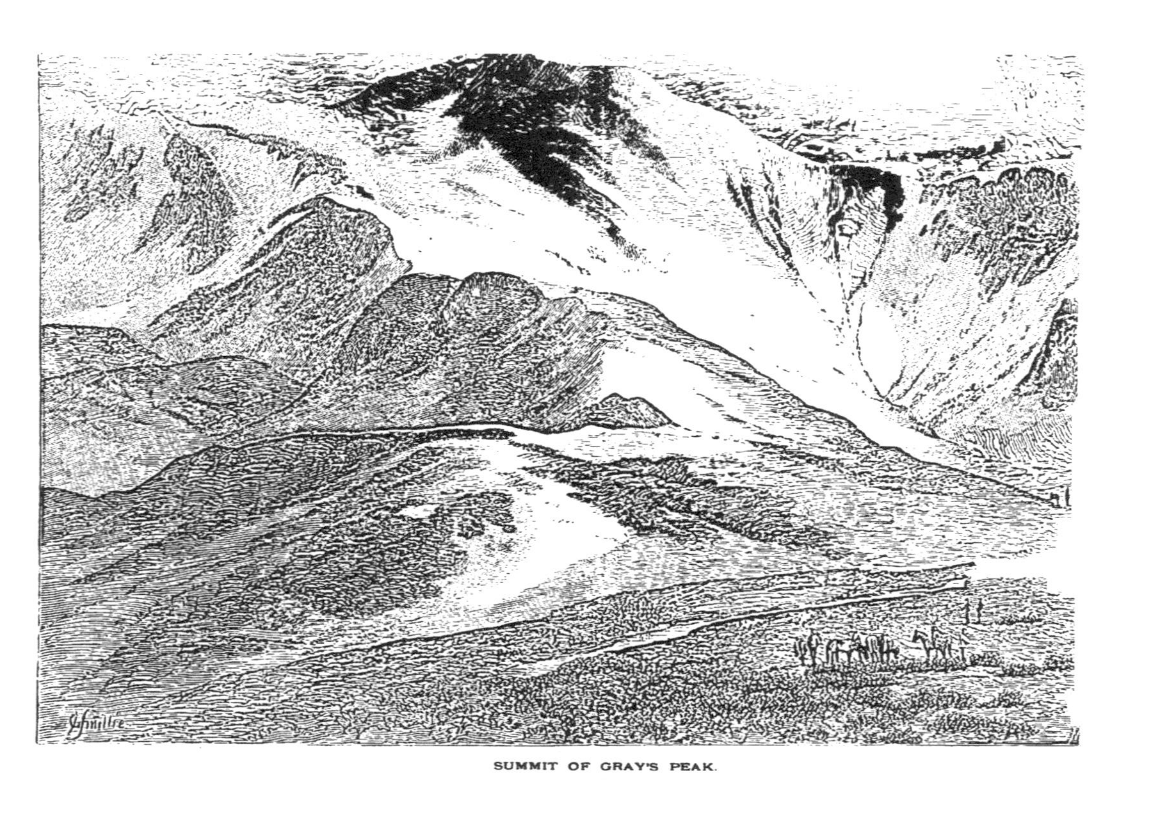

SUMMIT OF GRAY'S PEAK.

Middle Park, the middle district including the South Park, and the southern district the San-Luis Park. In these three districts the range reveals itself as one of the grandest in the world, reaching its greatest elevations, and comprising one of the most interesting areas on the continent.

Early in May of 1873, with a detachment of the Hayden expedition, we encamped in the Estes Park, or Valley. The night is deepening as we pitch our tents. We are at the base of Long's Peak—about half-way between Denver City and the boundary-line of Wyoming—and can only dimly see its clear-cut outline and graceful crests as the last hues of sunset fade and depart. In this lonely little valley, with awful chasms and hills around, in a wilderness of glacier creation, scantily robed with dusky pine and hemlock, the hearty voice of our expedition breaks many slumbering echoes in the chilly spring night. The shabby hut of the squatter, and straggling mining-camp, deep set in a ravine, are an inexpressible relief; and so our white tents, erected on the fertile acres of the Estes Park, throw a gleam of warmth among the snowy slopes, and impart to the scene that something without which the noblest country appears dreary, and awakens whatever latent grief there is in our nature.

The park is in a lovely spot, sheltered, fertile, and wooded. It is an excellent pasture for large herds of cattle, and is used for that purpose. A few families are also settled here; and as the valley is the only practicable route for ascending the peak, it is destined, no doubt, to become a stopping-place for future tourists. It is seven thousand seven hundred and eighty-eight feet above the level of the sea, and six thousand three hundred feet below Long's Peak, which is said to be about fourteen thousand and eighty-eight feet high. The peak is composed of primitive rock, twisted and torn into some of the grandest cañons in this famed country of cañons. While we remain here we are constantly afoot. The naturalists are overjoyed at their good fortune, and the photographers are alert to catch all they can while the light lasts.

Soon we are on the march again, tramping southward through stilly valleys, climbing monstrous boulders, fording snow-fed streams, mounting perilous heights, descending awful chasms. Then, from cañons almost as great, we enter Boulder Cañon, cut deep in the metamorphic rocks of foot-hills for seventeen miles, with walls of solid rock that rise precipitously to a height of three thousand feet in many places. A bubbling stream rushes down the centre, broken in its course by clumsy-looking rocks, and the fallen limbs of trees that have been wrenched from the sparse soil and moss in the crevices. Boulder, at the mouth of the cañon, has a small population, and is the centre of rich gold, silver, and coal mining districts.

James's Peak comes next in our route, and at its foot we see one of the pretty frozen lakes that are scattered all over the range. It is a picturesque and weird yet tenderly sentimental scene. The surface is as smooth as a mirror, and reflects the funereal foliage and snowy robes of the slopes as clearly. The white dress of the moun-

tain hereabout is unchanged the year round, and only yields tribute to the summer heat in thousands of little brooks that gather together in the greater streams. The lakes themselves are small basins, not more than two or three acres in extent, and are ice-locked and snow-bound until the summer is far advanced.

We encounter civilization, modified by the conditions of frontier life, in the happily-situated little city of Georgetown, which is in a direct line running westward from Denver, the starting-point of tourist mountaineers. It is locked in a valley surrounded by far-reaching granite hills, with the silver ribbon of Clear Creek flashing its way through, and forests of evergreens soaring to the ridges. Roundabout are wonderful "bits" of Nature, and from the valley itself we make the ascent of Gray's Peak. We toil up a winding road, meeting plenty of company of a rough sort on the way. There are silver-mines in the neighborhood, and we also meet heavily-laden wagons full of ore, driven by labor-stained men. The air grows clearer and thinner; we leave behind the forests of aspen, and are now among the pines, silver-firs, and spruces. The forest still grows thinner, the trees smaller. Below us are the successive valleys through which we have come, and above us the snowy Sierras, tinted with the colors of the sky. Twelve thousand feet above the level of the sea we reach the highest point in Colorado where mining is carried on, and then we pass the limit of tree-life, where only dwarfed forms of Alpine or arctic vegetation exist. Higher yet! Breathless and fatigued, we urge our poor beasts on in the narrow, almost hidden trail, and are rewarded in due time by a safe arrival at our goal.

Foremost in the view are the twin peaks, Gray's and Torrey's; but, in a vast area that seems limitless, there are successive rows of pinnacles, some of them entirely wrapped in everlasting snow, others patched with it; some abrupt and pointed, others reaching their climax by soft curves and gradations that are almost imperceptible. We are on the crest of a continent. There is a sort of enclosure some feet beneath the very summit of Gray's Peak, or, to speak more exactly, a valley surrounded by walls of snow, dotted by occasional boulders, and sparsely covered with dwarfed vegetation.

Pike's Peak was an El Dorado to immigrants, who, in adventurously seeking it, often fell victims on the gore-stained ground of the Sioux Indians. Foremost in the range, it was the most visible from the plains, and was as a star or beacon to the travellers approaching the mountains from the east. Thither we are now bound, destined to call, on the way, at the Chicago Lakes, Monument Park, and the Garden of the Gods. Chicago Lakes lie at the foot of Mount Rosalie, still farther south, and are the source of Chicago Creek. Mount Rosalie, ridged with snow and very rugged in appearance, is over twelve thousand feet high. Monument Park is probably more familiar than other points in our route. It is filled with fantastic groups of eroded sandstone, perhaps the most unique in the Western country, where there are so many evidences of Nature's curious whims. If one should imagine a great number of gigantic sugar-loaves,

ERODED SANDSTONES, MONUMENT PARK.

Upper Twin Lake.

irregular in shape, but all showing the tapering form, varying in height from six feet to nearly fifty, with each loaf capped by a dark, flat stone, not unlike in shape to a college-student's hat, he would have a very clear idea of the columns in Monument Park. They are for the most part ranged along the low hills on each side of the park, which is probably a mile wide, but here and there one stands out in the open plain. On one or two little knolls, apart from the hills, numbers of these columns are grouped, producing the exact effect of cemeteries with their many white-marble columns.

Once more we are on our way, and still in the mountains. We linger a while in the Garden of the Gods, which is five miles northwest of Colorado Springs, among the magnificent forms that in some places resemble those we have already seen in Monument Park. At the "gateway" we are between two precipitous walls of sandstone, two hundred feet apart and three hundred and fifty feet high. Stretching afar is a gently-sloping foot-hill, and beyond that, in the distance, we have a glimpse of the faint snow-line of Pike's Peak, which is still about ten miles off. It forms, with its spurs, the southeastern boundary of the South Park. It offers no great difficulties in the ascent, and a good trail for horses

PIKE'S PEAK, FROM GARDEN OF THE GODS.

has been made to the summit. More recently a cogwheel railway has been built to the top of the mountain, similar to that up Mount Washington.

Now we bear away to Fairplay, where we join the principal division of the expedition, and thence we visit together Mount Lincoln, Western Pass, the Twin Lakes, and

Teocalli Mountain.

other points in the valley of the Arkansas; cross the National or Mother range into the Elk Mountains; proceed up the Arkansas and beyond its head-waters to the Mount of the Holy Cross.

The course of the Arkansas River is southward hereabout, touching the base of the central chain of the mountains. So it continues for one hundred miles, then branching

SNOW-MASS MOUNTAIN.

eastward toward the Mississippi. In the lower part of the southward course the valley expands, and is bordered on the east by an irregular mass of low, broken hill-ranges, and on the west by the central range. Twenty miles above this point the banks are closely confined, and form a very picturesque gorge; still farther above they again expand, and here are nestled the beautiful Twin Lakes. The larger is about two and a half miles long and a mile and a half wide; the smaller about half that size. At the upper end they are girt by steep and rugged heights; below they are bounded by undulating hills of gravel and boulders. A broad stream connects the two, and then hurries down the plain to swell the Arkansas. Our illustration does not exaggerate the chaste beauty of the upper lake, the smaller of the two.

We advance from the Twin Lakes into the very heart of the Rocky Mountains, and sojourn in a quiet little valley while the working-force of the expedition explores the neighboring country. Two summits are ascended from our station, one of them a round peak of granite, full fourteen thousand feet above the level of the sea, and only to be reached by assiduous and tiresome scrambling over fractured rocks. This we name La Plata. The range is of unswerving direction, running north and south for nearly a hundred miles, and is broken into countless peaks over twelve thousand feet high. It is penetrated by deep ravines, which formerly sent great glaciers into the valley; it is composed of granite and eruptive rocks. The northernmost point is the Mount of the Holy Cross, and that we shall visit soon. Advancing again through magnificent upland meadows and amphitheatres, we come at last to Red-Mountain Pass, so named from a curious line of light near the summit, marked for half a mile with a brilliant crimson stain, verging into yellow from the oxidation of iron in the volcanic material. The effect of this, as may be imagined, is wonderfully beautiful. Thence we traverse several ravines in the shadow of the imposing granite mountains, enter fresh valleys and contemplate new wonders. We pitch our tents near the base of an immense pyramid capped with layers of red sandstone, which we name Teocalli, from the Aztec word meaning "pyramid of sacrifice." The view from our camp is, we should say, surpassing, could we remember or decide which of all the beauties we have is the grandest. Two hills incline toward the valley where we are stationed, ultimately falling into each other's arms. Between their shoulders there is a broad gap, and, in the rear, the majestic form of the Teocalli reaches to heaven.

In the distance we have seen two mountains which are temporarily called Snow-Mass and Black Pyramid. The first of these we are now ascending. It is a terribly hard road to travel. The slopes consist of masses of immense granitic fragments, the rock-bed from which they came appearing only occasionally. When we reach the crest, we find it also broken and cleft in masses and pillars. It is this snow-field which forms the characteristic feature of the mountain as seen in the distance. There is about a square mile of unbroken white, and, still lower down, a lake of blue water. A little to

MOUNTAIN OF THE HOLY CROSS.

the northward of Snow-Mass the range rises into another yet greater mountain. The two are known to miners as "The Twins," although they are not at all alike, as the provisional names we bestowed upon them indicate. After mature deliberation the expedition rechristen them the White House and the Capitol, under which names we suppose they will be familiar to future generations. Not a great distance from here, leading down the mountain from Elk Lake, is a picturesque cascade, that finds its way through deep gorges and cañons to the Rio Grande.

Elk-Lake Cascade.

The Mountain of the Holy Cross is next reached. This is the most celebrated mountain in the region, but its height, which has been over-estimated, is not more than fourteen thousand feet. The ascent is exceedingly toilsome even for inured mountaineers. There is a very beautiful peculiarity in the mountain, as its name shows. The principal peak is composed of gneiss, and the cross-fractures of the rock on the eastern slope have made two great fissures, which cut into one another at right angles, and hold their snow in the form of a cross the summer long.

Elsewhere we have given a description of "The Plains and the Sierras," so that with the account of Colorado's most picturesque peak our article on "The Rocky Mountains" properly comes to an end. That full justice has been accorded to all the marvellous scenes of this region is perhaps too much to claim, but certainly no characteristic feature has been knowingly omitted, and we feel assured that this brief glance into Nature's wonderland will be appreciated.

ɅOHAWK, ALBANY, AND TROY.

TH ILLUSTRATIONS BY MESSRS. FENN AND WOODWARD.

a part of New-York State around which the spell of the pastoral ages y been thrown, and which gives to it a sentiment of extreme antiquity ry refuses to account. It is the valley of the Mohawk, a river whose signation is unknown, but which has preserved the name of the abo- elt upon its banks.

ıwk rises in Oneida County, about twenty miles north of Rome; flows east, falling into the Hudson, after a stretch of one hundred and thirty-

The Mohawk at Utica.

five miles, ten miles above Albany. It is but a petty stream near its origin, nor is it fed by important tributaries until it has passed the city of Utica. It meanders placidly past that place, travelling very slowly, and with many turns and bends. On each side are long, tranquil meadows, studded with trees that mount up from the water's edge with a most gradual ascent. Beyond the meadows rise gentle hills, whose sides are thick with trees that glance and gleam in the sunlight as the frolicsome winds display the upper and the lower sides of the leaves. The cattle graze close to the river, near the bulrushes; and the sheep feed higher up, where the grass is shorter and less rank. And the river goes murmuring on through this scene of quiet happiness until it comes to a place where the Adirondack Mountains have thrown out a line of skirmishing rocks,

quillity of
ught to an
This is at
ust be con-
rmishers of
pursuance
waged be-
ınd the riv-
ade a most
determined
place is liter-
rocks. They
-cropping up
ses, over the
ens; bursting
of the green
become really
ıd starting up
ıe river in the
manner. The
ːes a descent
ːet, accomplish-
in three small
'e been turned
by the people
for they furnish
ı a great many
se, for the most
the island which
the river below
and this island
ıe rockiest part
settlement.
other side rises
not so precipi-
gher, and terraced
h grand, curving
show clearly the
er of the Mohawk

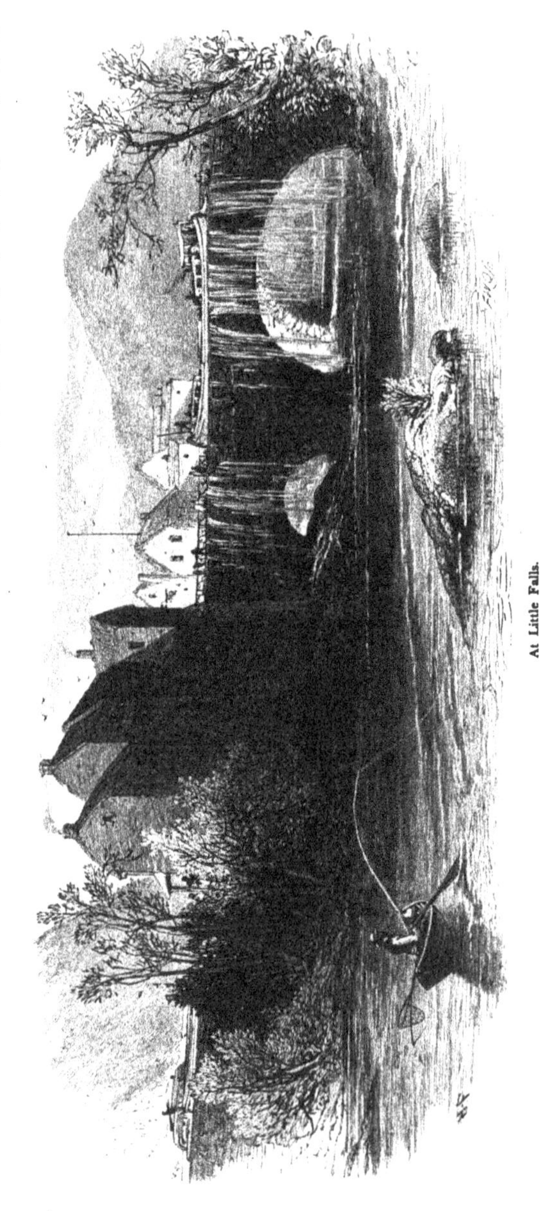

At Little Falls.

LITTLE FALLS.

Profile Rock.

s. It had its turbulent youth also; and the day was when it swept these
fierce current that laughed at such puny obstacles. Now it glides peace-
l, and sings with a pleased murmur to the fat cattle and the impudent birds
its waters and toss their heads half disdainfully.

ere are witnesses still extant of what the waters did in the remote past; for

The Mohawk Valley.

here is Profile Rock, where the hard stone has been so mauled, and had its stratification so handled, that the very fair likeness to a human profile has been washed out. That tow-path, where the canal-horses tug and strain so, is the favorite drive of the townspeople; and, indeed, the good folks have nowhere else to drive, being circumvented and hemmed in by their rocky girdle. The view along the canal tow-path is exceedingly interesting. The side of the Rollaway runs along the canal for several miles, and is clothed with a fine growth of trees—stately, dark pines, white beeches with gleaming, silvery trunks, and bending aspens here and there. On the other side is the Mohawk, once more united, for the rocky island terminates at the end of the town. The rocks, however, continue, and though of no height, are strangely varied in shape, and mingled with bosky shrubs and thick bushes, waving grasses, and delicate hare-bells. But gradually the Rollaway dwindles to a bank, and the rocks to pebbles; and after the Suspension Bridge is passed, the Mohawk is itself again, and the pastoral era is renewed.

ıt to Sche-
ermed the
wk Valley
say which
picturesque
ʏ—the val-
awk from
king west-
Suspension
.ittle Falls
Both have
ral beauty;
ne low hills,
'ering trees.
larity about
ormer which
tself to the
ry, and there
e looseness
which many
artistic. At
e capital of
find that it,
the quiet, un-
eristics of the
city has this,
with the sur-
. it appears
ın it really is.
it has many
he title of the
e State. This
ırely upon the
rst settlement
ıich some de-
taken place in
's in 1623; but
confusion about
cause there was

Schenectady, from the West.

Cohoes Falls.

undeniably a time when the Indians called both *Skaunòghtada*, which means "town across the plains." However that may be, in those remote times it is certain that Schenectady proper was more flourishing than Albany. It was at the head of the rich Mohawk Valley, and did an immense business in dairy produce and Indian peltries. The Indians seem to have lived in harmony with the Dutch settlers for many years, and it was not until 1690 that they suddenly became enemies. On this occasion the whole population, save sixty souls, was annihilated; and the town was destroyed by fire. It was burned

TROY AND VICINITY.

Albany, from East Albany.

again in 1748, which gives it quite a history; and the most astonishing thing about it is, that it looks as if it had been existing for untold generations. The Mohawk, at this point, is broad and deep, and the old wooden bridge that formerly spanned it was a long one; for the stream has been recruited by several large tributaries since it swept by the city of Utica, the chief contribution coming from the West Kanahta Creek, which, after dashing down the wildly-beautiful Trenton Falls, glides peaceably enough into the placid bosom of the Mohawk, and remembers its past furious excitement only in dreams.

Beyond Schenectady the river sweeps on with a majesty obtained from its increased volume, but the country is not so pastoral as it was. The soil is shaly, and the hills are low. At Cohoes there is a great fall; about a mile above the falls the river, broad and deep as it is, has been hemmed in by a dam, and a portion of its waters drawn off. The town of Cohoes is entirely manufacturing. Here are the great cotton-mills; and here, also, are many woollen-mills, besides paper - factories and other industries.

SCENES IN AND AROUND ALBANY.

Albany, from Kenwood.

Below the falls the river is divided by a green island, the favorite resort of pleasure-seekers from the neighboring city of Troy. This is a great manufacturing centre, especially of metals, and therefore abounding in tall chimneys vomiting forth black smoke. For this reason the inhabitants, who love to call themselves Trojans, prefer to dwell upon the other side of the river, which is only a mile or so from Cohoes. It is here that the junction of the Mohawk and the Hudson takes place, between East and West Troy. There is here, also, a large island, on which the Troy Bridge finds a support for its central part. The view here of the bustling place is inspiriting, and makes one as eager to be up and doing as the pastoral scenes of the Mohawk Valley made us wish to live and die shepherds. Troy is a city of some sixty-five thousand inhabitants, situated at the mouth of Poestenkill Creek, six miles above Albany, and a hundred and fifty-one miles above New York—an active, enterprising, and bustling city.

Albany, which now numbers over a hundred thousand souls within its limits, is an important railway centre, and the main point of departure for Western travellers. It is the terminus of nearly all the great steamboat lines of the Hudson; but its chief importance is that of being the capital of the grand Empire State. Albany is the oldest settlement in the original thirteen colonies, except Jamestown, Virginia. Henry Hudson, in the yacht Half-Moon, moored in September, 1609, at a point which is now in Broadway, Albany. In 1617 a fort was built at the mouth of the Normanskill; and in 1628 another was erected near the present steamboat-landing in the south part of the city, and named Fort Orange. The place was called, by the Dutch, New Orange, and retained that name until the whole province passed into possession of the English, in 1664, when New Orange was changed to Albany, in honor of the Duke of York and Albany, afterward James II. Peter Schuyler was its first mayor. The Schuyler family possessed the good-will of the Indians to such a degree that, while other settlements were desolated by Indian forays, Albany was never attacked by them. Besides its ancient importance as a centre of the Indian trade, Albany afterward became the point where the great military expeditions against Canada were fitted out. Here assembled the first convention for the union of the colonies. It was held in 1754, Benjamin Franklin being presiding officer.

There are two views of Albany which are specially good; one is from the other side of the river, where the city rises up from the western bank in irregular terraces, the culminating point being crowned with the capitol, embowered amid the foliage of old trees. Now a more palatial building takes the place of the structure shown in the illustration, and gives to the heights of Albany a magnificent apex. Above, the hills of the town rise, covered with fine old houses, and towering churches, and massive legislative halls. The other view shuts out the river almost—at least, all the activity along the western bank—and gives to the eye a wider stretch of vision. Looking from Kenwood, one sees the city foreshortened and gathered into a huge mass; while the two bridges across the Hudson, and the labyrinthine railway-lines of East Albany, become very prominent. One can see the masses of foliage of the trees in Washington Park, and the brown sedges of the flats above the town. Far in the distance lie quiet hills, on whose sides the reapers are at work on the browned wheat; while at the base are serried lines of trees that may have stood there in the old days, when the Mohawks ruled the land. From the summits of those hills, looking northward, one can see, with the utmost distinctness, the junction of the broad Hudson with the quiet Mohawk.

THE SUSQUEHANNA.

WITH ILLUSTRATIONS BY GRANVILLE PERKINS.

THE Susquehanna is considered with justice one of the most picturesque streams of America. The river is not only beautiful in itself, but its attractions are greatly enhanced by the soft, silvery haze through which they are presented. This gives to its scenery an indescribable charm, which defies alike the pencil and the pen, but which never fails to make itself felt by the heart.

All of the Susquehanna scenery is not beautiful. The ending is dull and prosaic; and the long stretch south of Columbia, in Lancaster County, Pennsylvania, to Havre

, is unattractive. Above Columbia begins the beautiful land. .lows the path of the river, which is due northward. Here we —waves of the main ranges of the Blue Mountains, so called to the very summits, an unusual amount of the cerulean haze is istance, and the hills appear intensely blue. The road skirts the ıs, running along the eastern bank of the river, and affords, from ırs, ample opportunities for inspection and admiration. To the

Above Columbia.

ıins rise up in grand, rounded masses, with an inexhaustible wealth of ı their sides. Nowhere can one see such superb forms of vegetation as mountain, for here they are fully developed, whereas in the forests they having excessively tall, thin trunks, and a head of small branches, but middle. They are choked for want of air; and so they aspire toward no marked development save that which is upward. On the boulder- is a superbly colored carpet of many kinds of undergrowth, convolvuli,

Harrisburg, from Brant's Hill.

and creepers, wild grape-vines and huckleberries, flowers of a hundred different kinds, and humble strawberries that cling to the ground as if to hide themselves and their delicate points of crimson fruit. Scattered over the surface of the gleaming waters of the river are islands, too small to be habitable, covered with the densest vegetation, that fairly glows with vivid hues of green. Onward rushes the train with its living freight, and soon reaches Harrisburg, the capital of Pennsylvania, and a thriving manufacturing town. It occupies the ground between the river and the hills, which here retreat considerably. The foot-hills, or low spurs, are close to the city, and are already built upon.

Brant's Hill is almost in a direct line with the crest of ground, in the centre of the town, on which the capitol is built; and the city, therefore, can be seen most excellently from this point—lying, indeed, spread out before one like a panorama. But the best view is from the cupola of the capitol. From here, more elevated than Brant's Hill, not all the city, with its climbing spires and its massive manufactories, ·y to the northward comes into view, and one has a distant though Hunter's Gap and the range of mountains through which the Sus- ;ht its way. Opposite Harrisburg the river is unusually wide, and shallow, which increases the brown appearance of its waters; for in ɜam is not a foot deep, and the sandstone bed is plainly visible, the ll the lines of its cleavage. In the centre of the river, straight in a :ring, whitewashed cottages of the village, are three islands covered of such a size that picnics are possible on them.

Glimpse of the Susquehanna, from Kittatinny Mountains.

ırns and twists, writhing like a fever-burned mortal, or some animal try- ɔm a trap. The mountains compass it about on every side; they hem ınd, east, west, north, and south. But though the general aspect is ter- ɜ quiet sylvan nooks, where the mountains show their gentler sides, and, nting their fronts, turn to us huge, undulating flanks, covered with glo- ɟ noble oaks, spreading hickories and dark hemlocks. These are the c trout-streams come singing through the ravines, murmuring their thanks ɔr their shelter and companionship. Hunter's Gap is the last gate-way er through the hills; but there is, in fact, a succession of gaps, through

SCENES ON THE SUSQUEHANNA.

North Point.

Northumberland, however, these foot-hills become larger, higher, and less pastoral in character, until, at the actual point of junction of the two rivers, those on the east bank are actually precipitous; and, moreover, they are ruder in appearance than elsewhere, being almost entirely denuded of timber. The West Branch at this point runs due north and south, and receives the North Branch, running nearly due east. The latter is very nearly as large a stream as the former, but the majesty of its union is somewhat marred by a large, heavily-timbered island, which occupies the centre of the current.

Everywhere around Northumberland are strong hints that the tourist is getting into the lumber-region; and the next place of importance—Williamsport—is the very head-

'INE FOREST ON WEST BRANCH OF THE SUSQUEHANNA.

quarters of the lumber-trade in the eastern part of the United States. The West Branch of the Susquehanna at this point has taken a bold, sweeping curve due west, and has left behind it a spur of the Alleghanies. Here comes in the Lycoming River, down which thousands of logs float. But down the Susquehanna come hundreds of thousands of oak and hemlock, and, above all, of pine. Close by the opposite bank of the river the hills rise up very grandly, but on the other side of the town they are far away, for the valley of the Susquehanna at this point is quite broad. It begins to narrow a little as we approach Lock Haven, which is also a lumber-place. It is a very charming little town, very bustling, very thriving, and more picturesque than Williamsport. The canal at Lock Haven is fed with water from the Bald-Eagle-Valley Creek, which falls here into the big river, after traversing the whole valley from Tyrone, not far from the head-waters of the Juniata, the principal tributary of the Susquehanna. At North Point the mountain-forms fairly arrest the eye of the most phlegmatic. In one direction, one mountain proudly raises itself like a sugar-loaf; in another, the side is presented, and it is not unlike a crouching lion; in a third the front is shown, and the mountain then turns in so peculiar a fashion as to uncover its great flanks, giving it the appearance of an animal lying down, but turning its head in the direction of the spectator. Close by is another pyramidal-shaped mass, whose body meets the flank of the former, forming a ravine of the most picturesque character, where the tops of the pines, when agitated by the breeze, resemble the tossing waves of an angry lake.

Renovo, which is the only stopping-place of importance between Lock Haven and Emporium, is a favorite summer resort, being located at a most picturesque point on the river, in the immediate vicinity of many beautiful mountain-streams, in which the trout shelter during the hot weather. The valley of the Susquehanna at Renovo is nearly circular in shape, and not very broad. The mountains rise up almost perpendicularly from the south bank, which is most picturesque, the other bank being low and shelving. The hotel, surrounded by beautifully-kept lawns adorned with parterres of brilliant flowers, becomes a marked point in the landscape, although in the early summer its blossoms are put to shame by the wild-flowers of the surrounding mountains; for at this time the slopes of the giant hills are everywhere covered with the pale-purple rhododendrons, which, when aggregated into large masses, fairly dazzle the eye with the excess of splendid color. Just opposite the hotel a mountain rises to a height of twenty-three hundred feet in one vast slope of living green, ascending without a break in a grand incline right up from the water's edge, whose brown flood is not here broad enough to reflect the entire outlines of the stupendous mass; for here the river narrows considerably, and is very deep under the mountain-side, becoming shallower as the bed approaches the northern bank. The town of Renovo is stretched along the Susquehanna side, its breadth being inconsiderable, although the valley here must be nearly half a mile wide. The hills on the other side are not so high as the one that bids

FERRY AT RENOVO.

SCENES ON THE NORTH BRANCH OF THE SUSQUEHANNA.

North Branch of the Susquehanna, at Hunlocks.

sitors in the hotel, daring, as it were, their utmost efforts to climb up
other side of the valley the mountains are easily accessible, and, in fact,
rt of tourists. There is a mountain-road here which penetrates through
e southward, and the teams cross the river in a curious rickety ferry.
of flat-boat, which is propelled across by a man hauling on a rope sus-
high south bank to a huge pole on the other shore. The view from

Canal at Hunlocks.

the centre of the stream is exceedingly beautiful. One gets a better idea of the circular shape of the valley, and the manner in which the hills have retired to let the little town have a foothold. And there are islands in the channel covered with beautiful mosses, and stretches of shallow water where rocks peep up, on which gray cranes perch with solemn air, busily engaged in fishing. The shadows of the mountain's bank, too, are thrown into relief by the sunshine reflecting on the water, and the mountains to the westward form a brilliant background, with their tree-laden slopes brightened with golden tints.

The Susquehanna, after receiving the cold waters of Kettle Creek, begins to incline southward, and, from its junction with the Sinnemahoning, makes an abrupt turn due southward toward the town of Clearfield. From this point it ceases to be a river, branching off into numerous creeks that rise from the mountains of this region, where it is all either hill or valley, and where a plain is a rarity. The land here is cultivated with care and success, but the prevailing industry is mining, all the mountains here containing iron-ore. There is some considerable difficulty in floating down logs to

PILLSBURY KNOB

the main stream of the Susquehanna below Clearfield, and most of the timber cut is used for the purpose of smelting or for forges where the charcoal hammered iron is made. The scenery is not so wild as might be imagined, the forms of the mountains seldom varying from somewhat monotonous grandeur, relieved by the beauty of the forest-trees upon their sides.

To describe the North Branch of the Susquehanna, it will be necessary to retrace our steps to Northumberland, the point of junction. The North Branch runs here almost

Below Dam at Nanticoke.

due east, rushing right through a majestic range of mountains, which pass under the generic title of "Alleghanies." The mountains here are far bolder, more rocky, and with far less timber, exhibiting huge crags of a picturesque character, very unlike the small fragments that cover the hills of the Western Fork. The many chimneys vomiting black smoke at Danville, the first place of importance the tourist reaches, remind him forcibly that he is not out of the iron-region; and the coal-cars which pass him on the road tell him that he is approaching the very centre of the famous Pennsylvania coal-

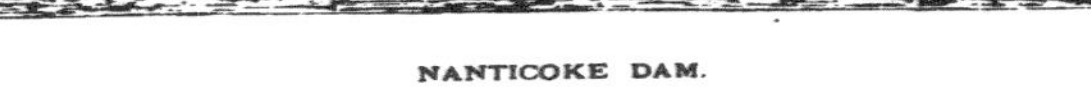

NANTICOKE DAM.

mines. Beyond Danville the river makes a bend away from the overhanging mountains of the northern side, and approaches somewhat closely to the southern, which are densely wooded, and have consequently many more runs brawling and bubbling down their sides. The hills on the northern bank are distant, but there are foot-hills that come down to the river. At the foot of these hills runs the railway. In immediate proximity comes the canal—a quiet, peaceable, serviceable servant of commerce, vexed with few locks. Between the canal and the river there is only an artificial dike of little breadth. Beyond the dike, some feet lower in level, the river rushes vigorously onward to join its waters with those of the West Branch. Its stream is more rapid, and its waves are of a clearer hue, than that which glides past Renovo, Williamsport, and Lock Haven. Rising from the southern bank are wood-covered mountains, boasting fewer oaks and hickories than we have seen in our progress hitherto, but having a sombre grandeur of tone from the more numerous evergreens. The extreme background is veiled by a soft haze, through which the river looks silvery and the mountains an ethereal blue. The finest points to the artist are the places where the rushing, tumbling, foaming creeks from the mountains come raging down to join the river, and to frighten the canal from its staid propriety, necessitating great enlargements of the dike and beautiful bridges. These swellings of the dike gladden an artistic eye; for they are often covered with fine, large trees, aud produce all the effects of islands hanging, as it were, over the brink of the river. There are several places where these bits of scenery exist—at Mifflin, Shickshinny, but above all at Hunlocks. Hunlocks Creek is not very long, but it has a commendable breadth, and so precipitous a course that it is more like a cataract than a creek, and its turbulent, shallow stream carries down boulders of a most respectable size. There is a coal-mine at Hunlocks, close upon the brink of the creek, and the miners down the shaft can hear the growling of the water-course in the spring, like distant thunder; for then its waters are swollen from the mountain snows, and it carries away, encumbered with its ice-masses, tons upon tons of rocks, which go hurtling down the stream, dashing against each other, and crashing with as much noise and fury as if an avalanche had been precipitated by the melting of a glacier. Our artist has given a group of illustrations characteristic of this region, showing the furnace on Hunlocks Creek, Nanticoke ferry, Danville, the hemlock-gatherers, the stone-quarry, and similar scenes.

After passing Pillsbury Knob, a remarkably bold promontory on the northern bank, the tourist arrives at Nanticoke, where the river expands considerably, becoming very shallow. Here there is a dam erected for the lumberers, though the business is yearly decreasing in this part. There are on the southern side broad stretches of fertile land below the bank, and these are cultivated with profit. The hills here rise in three several ranges upon the northern side and two upon the southern, and the effect from the lowlands on a level with the river is very grand. The majority of the hills to the northward are not well wooded, and their prevailing hue is a dull, purplish brown. To the south the mountains are better wooded, but the slope is very considerable and the height not very great. Between these the river winds in a serpentine form, creating a thousand different views of transcendent loveliness. Here we are actually entering the famous Wyoming Valley, so renowned for its beauties. The hills are not high, never

Wyoming Valley.

exceeding two thousand feet, but the banks of the river and the river itself present such combinations of form and color as kindle the admiration of the most apathetic. The hills on the northern bank, which is the higher, are more picturesque, and it is impossible to get the best view until the river is crossed.

There is an island in the river just opposite Kingston, of which the bridge takes

WYOMING VALLEY, FROM PROSPECT ROCK, WILKESBARRE.

WYOMING VALLEY, FROM CAMP HILL.

MONUMENT.

VIEW FROM KINGSTON.

advantage. From the centre of this there is a lovely view. At this point the river makes a superb curve, like the flashing of a silver-sided fish, and disappears, showing, however, through the trees, broad patches of gleaming white. But this is only a slight glimpse. The real place for a striking view is from Prospect Rock, about two miles behind the town, nearly at the top of the first range of hills on the southern side of the river. This post of observation is on the summit of a jutting crag, and from its picturesquely-massed boulders one can survey the whole of the Wyoming Valley, which, from Nanticoke westward to Pittston eastward, lies stretched before the eye of the visitor like a lovely picture. It is not broad; for, from Prospect Rock to the topmost crest of the first range of opposing hills, the distance, as the crow flies, is not more than four miles, and the farthest peak visible not six. But this is a gain rather than a loss; for the views that are so wide as to be bounded by the horizon are always saddening. Step by step the landscape leads you beyond the winding river, and beyond the swelling plain, to vast distances, which melt by imperceptible gradations into the gracious sky.

Wyoming Valley, with its narrow boundaries of northern hills, tossing their crests irregularly like a billowy sea, steeped in clear, distinct hues of a purplish brown, and having every line and curvature plainly in sight, compel the eyes to rest on the green and smiling country, dotted with countless houses, ever scattered sparsely or gathered thickly into smiling towns. Through the points of brilliant light with which the sun lights up the white houses the Susquehanna glides like a gracious lady-mother, making soft sweeps here and noble curves there, but ever bordered by fringes of deep, emerald green. The whole valley is green, save where the towns toss up to heaven their towers and spires from numberless churches, and where behind, as if in hiding, black mounds and grimy structures mark the collieries. Too far to be vexed with details, too near not to see distinctly, the gazer on Prospect Rock views the landscape under just such circumstances as will delight him. Therefore, all who have stood upon these masses of sandstone and have watched the cloud-shadows sweeping over the broad plain, and have seen the sun go down in beauty and the stillness of twilight overstretching the happy valley, have gone away with hearts satisfied and rendered at ease. But this was not always a happy valley, and the time has been when this fair stretch of smiling green was smoking with the fires of burning homes, and the green turf was gory with the blood of men defending their families from the invader and his savages; when the Susquehanna shuddered at the corpses polluting her stream, and the mountains echoed back in horror the shrieks of wretches dying in torture at the Indian's stake. Out of misery came bliss; out of defeat, bloodshed, burning homes, and captured wives and daughters, came tranquil happiness and a material prosperity almost unequalled. The whole valley is one vast deposit of anthracite coal, and is now only in the dawning of its prosperity. What it will be in the full sunlight of fortune it passeth here to tell.

LAKE MEMPHREMAGOG.

WITH ILLUSTRATIONS BY J. DOUGLAS WOODWARD.

Owl's Head Landing.

GRANDER scenes there may be, but they tire and oppress, and the tourist comes back to the Connecticut Valley year after year loving it the more, and deriving from it the solace that empowers him for renewed toil at the treadmill of city life. You may start out from your home in the metropolis for Memphremagog direct; but when the path-way leads through the Connecticut Valley, you will linger, inhaling the breath of the daisy-scented fields, resting the wearied mind with the tranquil sentiment of the

LAKE MEMPHREMAGOG, SOUTH FROM OWL'S HEAD.

Arcadian life that dreams in the brook-side villages on your way. Loitering in these pastures a while, you arrive at the foot of Lake Memphremagog in a fit state of mind to appreciate its beauties, not so fagged-out as you would have been had your journey been unbroken. Disembarking at the little Vermont town of Newport, you submit yourself to the regimen of a fashionable hotel, sleep well, and dream of peace. The morning breaks on a bracing day in the season of Nature's most gorgeous transformation; the autumn foliage is crowned with the richest hues; your fellow-tourists have less of the jaded expression that is almost habitual on their features, and so all circumstances are propitious for a voyage over the lake.

Some people say that it rivals Lake George, but this admits of difference of opinion; yet it is almost impossible that there should be any thing more picturesque, in the exact sense of that word, than this beautiful expanse with the awkward name. It is overshadowed by mountains and bordered by dense forests and grassy reaches. At one point it is in Quebec, and at another in Northern Vermont. It is thirty miles long and two miles wide; the basin that holds it is deep and narrow; numerous islands spring from its depths, where speckled trout, of enormous size, dart and glimmer. There is one object already in sight, as the white steamer leaves the wharf for the other end of the lake, that must not be missed—the Owl's Head, a mountain surpassing others around the lake in form and size. It is twelve miles distant, and in the mean time our eyes are attracted by other enchantments that the shore sets forth.

Here is a narrow cape jutting out, the shimmering ripples tossing in play around; and yonder the land inclines into two bays, one of them sheltering the boats of some lazy boys, who are stretched on the thwarts, with their vagabond faces raised to the unclouded sun. The shore varies in character: for a mile it is high and craggy, and then the banks are low and rolling, girt by a belt of yellow sand. The deep water readily imprints the colors on its smooth surface, and duplicates the forms of earth and sky. Past Indian Point there is a village, and farther on are the Twin Sisters, two islands, thickly wooded with a growth of evergreens. Beyond is another village, and soon we are abreast of Province Island, a garden of one hundred acres. Nearer the eastern shore is Tea-Table Island, with many cedar-groves.

Now we enter British waters, with British soil to the right and to the left of us. The scenery of the lake does not diminish in beauty. The banks are always picturesque, almost invariably fertile and under cultivation. Here is Whetstone Island, so named because of the stone found there for axe-grinding. A little farther in the course is Magoon's Point, a grassy slope coming to the water's edge.

Owl's Head is not far distant. The boat winds in and out between the cedar-robed islands, and the golden haze vanishes into the clear and breezy day. No landing is made during the journey down the lake, but Owl's Head is passed then, with only a glimpse at its magnificent height. Round Island, cap-like in shape; Minnow Island,

LAKE MEMPHREMAGOG, NORTH FROM OWL'S HEAD.

the most famous fishing-place; and Skinner's Island, once the haunt of an intrepid smuggler, are left behind. North of Skinner's Cave is Long Island, covering an area of about a square mile, with a rugged shore. At one place the shore is almost perpendicular, and on the southern side there is an extraordinary granite boulder, balanced on a natural pedestal, named Balance Rock.

Owl's Head is the most prominent mountain, and is cone-shaped; but in the passage to the head of the lake are noticed other heights that do not fall far below it. Mount Elephantus, faintly resembling an elephant's back, afterward changing, as we proceed farther north, into a horseshoe form, is conspicuous. Yonder is Mount Oxford, not unlike Owl's Head; and here is a landing, toward which the steamer's prow inclines. The drowsy little town of Magog, at the foot of the lake, is reached.

In the morning the ascent of Owl's Head is made. The pathway is in good condition, overarched by pines and cedars, bordered by pleasant fields. Occasionally, through the green curtain that shelters us from the mounting sun, a glimpse of the untroubled, azure sky is caught. On the way a shelving rock is passed, under which shelter is afforded during a passing shower; and proceeding farther, a mass of stone, plumed with ferns and covered on the sides with a velvety moss, is reached. From the summit a superb view is the reward for the tiresome climb. Looking south, the lake from end to end—its islands and villages, the near rivers flashing in the sunlight—is seen. Looking north, the picture expands into other beauties; and to the east and west there are more lakes, plains, islands, and mountains. The summit itself is riven into four peaks, silent ravines intervening between them.

Mount Elephantus, from the Lake Steamer.

THE UPPER DELAWARE.

WITH ILLUSTRATIONS BY J. DOUGLAS WOODWARD.

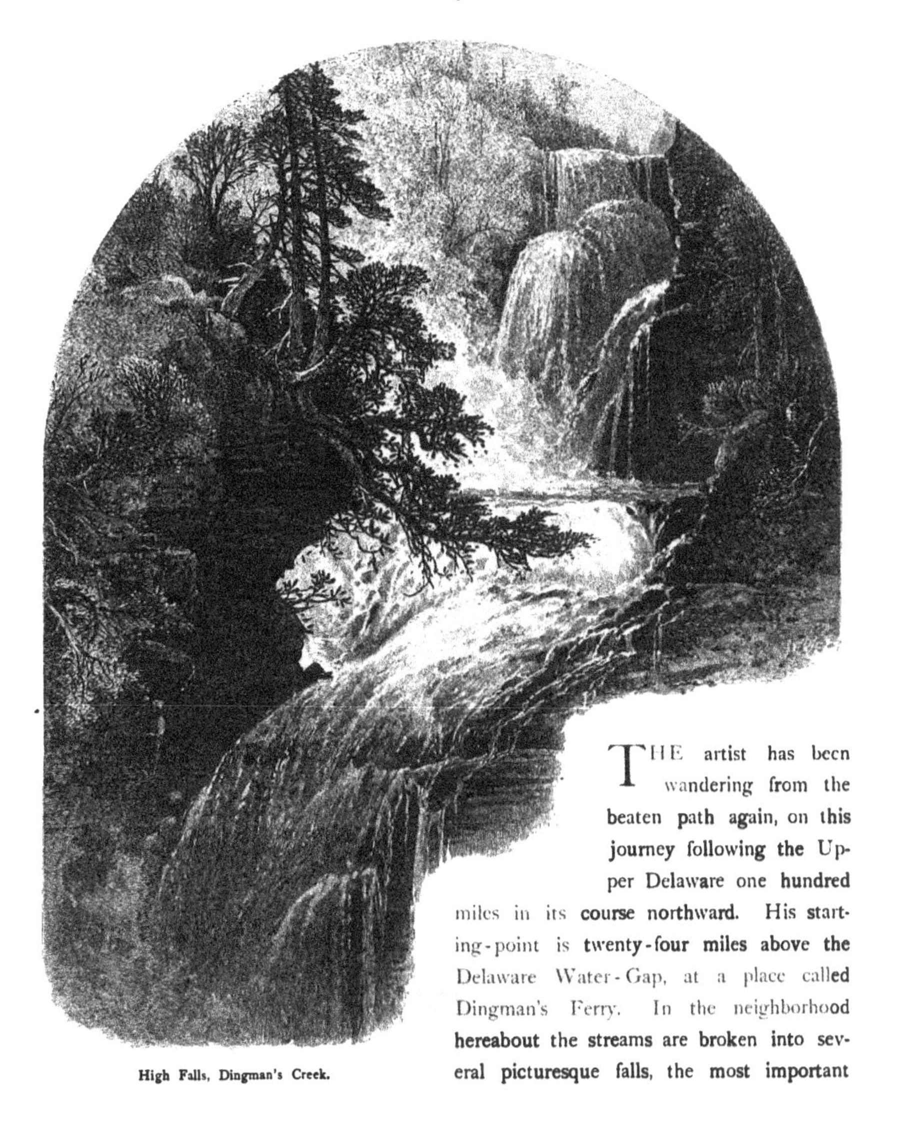

High Falls, Dingman's Creek.

THE artist has been wandering from the beaten path again, on this journey following the Upper Delaware one hundred miles in its course northward. His starting-point is twenty-four miles above the Delaware Water-Gap, at a place called Dingman's Ferry. In the neighborhood hereabout the streams are broken into several picturesque falls, the most important

SCENES IN AND ABOUT MILFORD.

of which are the High Falls, shown in our first sketch. It was in the morning when we first rambled through the bosky approaches to this cascade; and, after leaping down slippery, moss-covered rocks, we reached the foot, only to find a thin stream of water trickling down, with very little music, and less spray. The weather had been dry—but that fact scarcely consoled us—and we could only admire the tints of the rocks, and the foliage that seemed to grow out of the basin into which the waters made their first leap before rushing through a narrow bit of hill and descending to a lower level. The artist was content, thankful for the smallest share of Nature's bounty; but the literary soul was disappointed and growling.

We were retracing our steps to the hostelry leisurely, when the premonitions of a storm urged us into a quicker pace. Gusts of wind soughed among the trees, and heavy drops of rain pattered fast on the trembling leaves and parched earth. The sunshine was hidden beneath the gray clouds that came rolling from the east. We considered ourselves in for a wet day, and we dozed near the veranda, puffing at our brier pipes in a mood of bachelor meditation.

But in the afternoon there was clearer and warmer weather, and we again tramped to the foot of the High Falls. If the spirit of the artist was content before, it was aglow now. The scene had changed, and instead of a mere thread of water, there was a bubbling, foaming, boisterous torrent, echoing its voice in the walls of the hills through the veins of which it found a sparkling way. The moss in the crevices held glittering drops on its velvety surface; and the branches of overarching trees looked as though they, too, were crystallized. The changing position of the clouds threw shadows across the water, varying its tints, and first giving it the appearance of a pure white, then of a faint green, afterward of a soft blue. The artist drew our attention this way and that—one moment toward yonder darkling hollow in the rocks, as the spray dashed itself into the brown seams; next toward the water, as the light played ever-new tricks with it; and then to a little pool formed in the cup of a boulder. That keen eye of his discovered effects in the smallest nooks, underneath the fronds of the tiniest fern, among the grains of sand that lodged in the crevices, and in the swaying shadows of the forms around. He occupied us constantly for more than two full hours, and was even then inclined to linger, although our journey was long and the time short.

From the ferry we proceeded toward Milford. The stage-road runs along the base of a mountain, so precipitous as to resemble the Palisades of the Hudson. Atoms of rock, rolling down, have made the bed as hard as concrete; and they have been spread so evenly that travelling is smooth and comfortable. The outlook is magnificent. The sheer wall of the mountain is on one side of us, protecting us from the scorching rays of the sun; and undulating meadows reach afar in the opposite direction, dotted with many a snug farm-house, painted red or white, that shows its thatched roof over the tops of the orchard. The river glistens through this green expanse, and is spanned, here and

PORT JERVIS AND VICINITY.

THE UPPER DELAWARE.

there, by a picturesque bridge. Still farther away are the purple lines of more hills, mysterious in the haze of a warm autumn morning.

Some distance below the village of Milford we reach the falls of the Raymondskill, in which the artist finds more beauties and wonders. The torrent tumbles from among a mass of foliage down a rock, and is broken several times by projections, which cause it to surge and foam in a grand tumult. Three miles farther in our course we enter the village, which is prettily situated in a valley, and divided through the centre by a romantic glen. Glens always are romantic, for lovers invariably choose to make love in their shade and quiet. The Sawkill, scarcely more than a brook, trembles over the pebbles, and glints vividly as a stray shaft of sunlight breaks through the boughs overhead. Ferns, mosses, and wild-flowers are sprinkled on the path, and strive to hide the decay of a felled hemlock that rests between two sturdier brothers. It is a lovely spot, picturesque in the extreme, a most suitable retreat for the shepherds and shepherdesses of the Pennsylvania Arcadia.

Not more than two miles farther north are the principal falls of the Sawkill, which in general characteristics much resemble the High Falls and the Raymondskill. As in the latter, the water dashes against some projecting rocks in its downward course and is broken into clouds of spray, which the sunshine colors with rainbow hues. The volume of water is, in reality, divided into two separate falls by an elbow of the rock; but, before the two reach the level below, they mingle in one mass.

Following the windings of the river, our next stopping-place was Port Jervis, which borders on New York, New Jersey, and Pennsylvania. Near here the Neversink River enters the Delaware from a valley of great beauty. We followed the artist to a place called Mount William, from which there is a superb view—a wide, extended plain, through which the winding river can be traced for many miles. The afternoon was far advanced, and the sun was declining westward. The whiteness of the light was subdued, changing into a pale yellow, that soon again would deepen into crimson. You see how he has expressed this mellowness in the gray tone of his sketch. He has included, too, a considerable range of ground, bringing in the opposite hills, the town, and the river. As far as the eye can reach, the land is under cultivation. In yonder wide plain there is not one wild acre; and, out beyond the limits of the little town, the farmhouses are numerous and close together.

After leaving Port Jervis we touched at Lackawaxen, to get a sketch of the Delaware and Hudson Canal Aqueduct, and thence continued our journey to Deposit, in which vicinity the scenery becomes grander and wilder.

NANTICOKE DAM.

Lake George, near Caldwell.

day in the year, so it is popularly asserted—while its shores lift themselves into bold highlands. The lake is fairly embowered among high hills—a brilliant mirror set in among cliffs and wooded mountains, the rugged sides of which perpetually reflect their wild features in its clear and placid bosom. "Peacefully rest the waters of Lake George," says the historian Bancroft, "between their rampart of highlands. In the pellucid depth the cliffs and the hills and the trees trace their images; and the beautiful region speaks to the heart, teaching affection for Nature."

Approaching Lake George from the south, the tourist takes the railway at Albany for Caldwell. The spacious hotel on the site of the famous old Fort William Henry stands directly at the head of the lake, with a noble expanse of its waters spread out before it. The scene never seems to lose its charm. Always there is that glorious stretch of lake and shore bursting upon the sojourner's vision; he cannot put foot upon the piazza, he cannot throw open his hotel-window, he cannot come or depart, without there ever spreading before him, in the soft summer air, that perfect landscape, paralleled for beauty only by a similarly idyllic picture at West Point, amid the Highlands of the Hudson.

At Caldwell there is a superb

wooded, and their prevailing hue is a dull, purplish brown. To
ns are better wooded, but the slope is very considerable and the
Between these the river winds in a serpentine form, creating a
: of transcendent loveliness. Here we are actually entering the
y, so renowned for its beauties. The hills are not high, never

Wyoming Valley.

:t, but the banks of the river and the river itself present such
color as kindle the admiration of the most apathetic. The
which is the higher, are more picturesque, and it is impos-
until the river is crossed.
the river just opposite Kingston, of which the bridge takes

SCENES ON LAKE GEORGE.

om the centre of this there is a lovely view. At this point the river
curve, like the flashing of a silver-sided fish, and disappears, showing,
h the trees, broad patches of gleaming white. But this is only a slight
·eal place for a striking view is from Prospect Rock, about two miles
, nearly at the top of the first range of hills on the southern side of
post of observation is on the summit of a jutting crag, and from its
sed boulders one can survey the whole of the Wyoming Valley, which,
vestward to Pittston eastward, lies stretched before the eye of the vis-
picture. It is not broad; for, from Prospect Rock to the topmost
range of opposing hills, the distance, as the crow flies, is not more
d the farthest peak visible not six. But this is a gain rather than a
; that are so wide as to be bounded by the horizon are always sad-
step the landscape leads you beyond the winding river, and beyond
o vast distances, which melt by imperceptible gradations into the gra-

ey, with its narrow boundaries of northern hills, tossing their crests
lowy sea, steeped in clear, distinct hues of a purplish brown, and hav-
urvature plainly in sight, compel the eyes to rest on the green and
ed with countless houses, ever scattered sparsely or gathered thickly
Through the points of brilliant light with which the sun lights up
Susquehanna glides like a gracious lady-mother, making soft sweeps
s there, but ever bordered by fringes of deep, emerald green. The
save where the towns toss up to heaven their towers and spires
hes, and where behind, as if in hiding, black mounds and grimy
llieries. Too far to be vexed with details, too near not to see dis-
rospect Rock views the landscape under just such circumstances
herefore, all who have stood upon these masses of sandstone and
d-shadows sweeping over the broad plain, and have seen the sun
d the stillness of twilight overstretching the happy valley, have
atisfied and rendered at ease. But this was not always a happy
been when this fair stretch of smiling green was smoking with
es, and the green turf was gory with the blood of men defend-
he invader and his savages; when the Susquehanna shuddered
er stream, and the mountains echoed back in horror the shrieks
ture at the Indian's stake. Out of misery came bliss; out of
homes, and captured wives and daughters, came tranquil happi-
rity almost unequalled. The whole valley is one vast deposit
now only in the dawning of its prosperity. What it will be
ne it passeth here to tell.

MPHREMAGOG, SOUTH FROM OWL'S HEAD.

Shelving-Rock Falls.

sketched the scene from the north, showing it just as the declining afternoon sun is sending a flood of radiance through the hollow, forming a rich and glowing contrast of light and shadow. From Sabbath-Day Point the view up the lake is grand, Black Mountain assuming a commanding place in the picture. The next most noticeable point is Anthony's Nose—a bold, high hill. Two miles beyond is Rogers's Slide, another abrupt rocky height, at a point where the lake becomes very narrow. The steamer hugs the precipitous, rocky shore, the narrow passage forming almost a gateway to the main body of the lake for those who enter its waters from the north. This mountain derives its name from an incident that befell, according to tradition, one Rogers, a ranger conspicuous in the French and Indian War. The story runs that, in "the winter of 1758, he was surprised by some Indians and put to flight. Shod with snow-shoes, he eluded pursuit, and, coming to this spot, saved his life by an ingenious device. Descending the mountain until he came to the edge of the precipice, he threw his haversack down upon the ice, unbuckled his snow-shoes, and, without moving them, turned himself about and put them

LAKE MEMPHREMAGOG, NORTH FROM OWL'S HEAD.

Beyond Rogers's Slide the lake is narrow, the shores low and uninteresting, the water shoal, and soon the northern border of the lake is reached. The waters of Lake George flow through a narrow channel at Ticonderoga, about midway between the two lakes, tumbling down a rocky descent in a very picturesque fall.

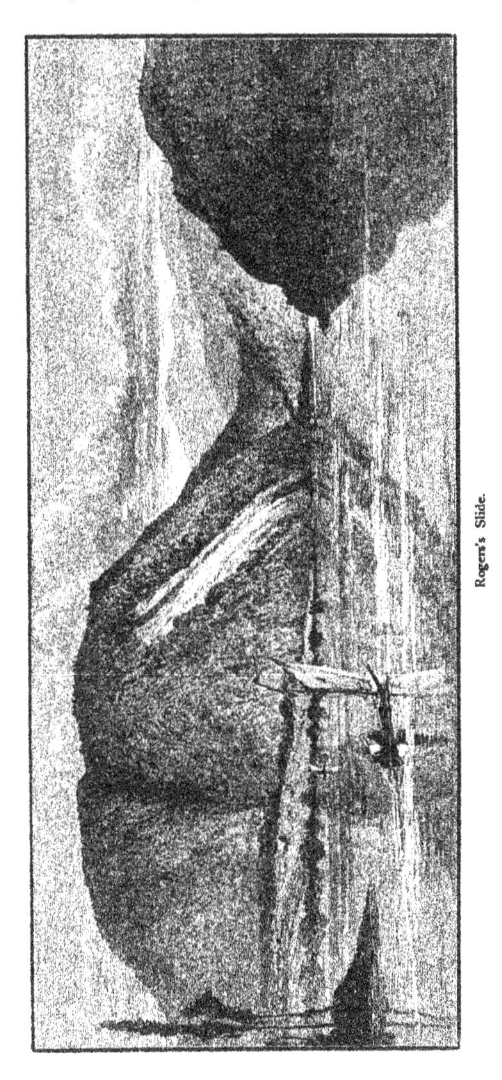

Rogers's Slide.

Lake George has many associations as well as charms. Few places in our country are more connected with historical reminiscences, or so identified with legend and story. Just as Scott has made the Highlands of Scotland teem with the shadows of his imagination, Cooper has peopled the shores of this lake with the creations of his fancy. In all American literature there is no figure so enveloped in poetic mystery, so full of statuesque beauty, as Cooper's Uncas; and, on these shores, the too frequent vulgar nomenclature should give place to some heroic name like that of the brave and beautiful Mohican.

Lake George fills a large place in the colonial history of New York. The lake was first seen by white men in 1646, the discoverer being Father Jagues, who was on his way from Canada to the Mohawk country to perfect a treaty with the Indians. He arrived in a canoe at the outlet of the lake on the eve of the festival of Corpus Christi, and hence named it "Lac du Sacrement."

We hear of the lake being visited by scouting parties, and forming the channel of communication between the Canadian French and the Indian tribes southward; but it was not until the French War